Optimal Reference Shaping for Dynamical Systems

Theory and Applications

Optimal Reference Shaping for Dynamical Systems

Theory and Applications

Tarunraj Singh

State Univesity of New York at Buffalo

Buffalo, New York, U.S.A.

CRC Press

Taylor & Francis Group

Boca Raton London New York

CRC Press is an imprint of the
Taylor & Francis Group, an **informa** business

MATLAB® is a trademark of The MathWorks, Inc. and is used with permission. The MathWorks does not warrant the accuracy of the text or exercises in this book. This book's use or discussion of MATLAB® soft ware or related products does not constitute endorsement or sponsorship by The MathWorks of a particular pedagogical approach or particular use of the MATLAB® software.

CRC Press
Taylor & Francis Group
6000 Broken Sound Parkway NW, Suite 300
Boca Raton, FL 33487-2742

First issued in paperback 2019

© 2010 by Taylor and Francis Group, LLC
CRC Press is an imprint of Taylor & Francis Group, an Informa business

No claim to original U.S. Government works

ISBN-13: 978-1-4398-0562-6 (hbk)
ISBN-13: 978-0-367-38497-5 (pbk)

Library of Congress Cataloging-in-Publication Data

Singh, Tarunraj.
　　Optimal reference shaping for dynamical systems : theory and applications / author, Tarunraj Singh.
　　　　p. cm.
　　"A CRC title."
　　Includes bibliographical references and index.
　　ISBN 978-1-4398-0562-6 (hardcover : alk. paper)
　　1. Feedback control systems. 2. Dynamics. 3. System theory. I. Title.

TJ216.S465 2010
629.8'3--dc22
　　　2009031424

Visit the Taylor & Francis Web site at
http://www.taylorandfrancis.com

and the CRC Press Web site at

To my siblings Prabhjyot and Gururaj who have inspired me more
than they can imagine

Contents

Preface

The area of feedback control is rich with analysis and synthesis tools that cater to continuous and discrete time systems, linear and nonlinear systems, and stochastic and deterministic systems. Feedback control has created technology that has landed men on the moon; created commodity products, such as CD players, where the optical pickup needs to be centered on the data track with an accuracy of a few μ meters; cruise control for cars, and so forth; and in exotic applications such as tele-manipulation in medical surgery. It is hard to identify an engineering discipline that has not been impacted by feedback control. Research into feedback control continues with engineers, economists, biologists, and others studying parochial problems.

Feedforward control refers to techniques that modify or shape the reference input to the system in anticipation of the expected response of the system. If the system being controlled is well understood, the expected output will be close to the actual output. In such scenarios, feedforward control can be selected to meet a desired objective such as rest-to-rest maneuvering or reject periodic disturbances that can be measured. In this book, the terms *feedforward control*, *input shaping*, and *pre-filtering* will be used to refer to the mechanism of selecting an open-loop control profile to satisfy a desired objective. The systems being studied will be limited to those that are stable or marginally stable. Three examples in the following paragraphs are used to illustrate that the benefit of using feedforward control includes the ability of achieving performance that is unable to be matched by feedback control.

Biomimetics is a term used to label research that endeavors to emulate nature. For example, artificial neural networks are an attempt to emulate how the networks of biological neurons learn, process, and control. Other examples include understanding how bats navigate to permit design of autonomous robots for rescue operations in post-earthquake disasters. A recent study to understand the cause of stuttering has led to a conjecture that the process of communicating using sounds includes two modalities, a feedforward command based on a learned model and a feedback mode where the auditory output is compared to the commanded signal and the error is used for correcting speech.* Numerical simulation has revealed that the weakening of the feedforward component of the process mandates an increased reliance on the feedback component, which results in stuttering due to the delays in the feedback loop. The low incidence of stuttering among the deaf is a supporting empirical piece of evidence.

---

*Simulations of Feedback and Feedforward Control in Stuttering, O. Civier and F. H. Guenther, *Proceedings of the 7th Oxford Dysfluency Conference*, St. Catherine's College, Oxford University, June 29–July 2, 2005.

How does the human brain synthesize a motor control profile to move a baseball bat to make contact with a ball moving at 100 miles/hour? Tracking the motion of the ball requires oculomotor control, which includes efferent delays in commanding the motion of the eyes resulting in about 130 milliseconds (ms) of delay in the feedback loop. The visuo-motor loop for moving the baseball bat includes delays to the order of 200 to 250 ms.[†] A baseball traveling at 100 miles/hour takes 412 ms to reach home plate from the pitching rubber, a distance of 60.5 feet. By the time the ball is released by the pitcher and the batter moves the bat, the ball has passed the halfway point, not giving the batter an opportunity to correct for the curve ball. Solely employing a feedback controller will not achieve the desired performance of striking the baseball. Scientists have hypothesized that this complex motion control is possible because of the coexistence of a feedback and feedforward mode in a motor control system. The feedforward component of the control relies on a learned inverse model, while the feedback mode uses a forward model as an estimator to generate estimated states that are compared to reference states and the error is used in the feedback control mode. Abnormal behavior of either of these components can result in inefficient movement execution. Morasso and Sanguineti[‡] conclude that errors in the feedforward component of the control are responsible for inaccuracies in movement execution.

Cells, when exposed to stress can lead to denaturation of proteins. Denaturation refers to the process where proteins lose their three-dimensional (3D) structure which can cause disruptions in the functioning of the cell and in the worst case can cause the death of the cell.[§] The cell can produce defense proteins which restore the denatured protein. These defense proteins are produced in a feedforward manner to rapidly respond to the applied stress. However, uncertainties in the estimate of the stress can lead to an over- or underproduction of the defense protein. Feedback is the mechanism used by the cell to produce the defense proteins in response to the measured damage. Its drawback is the delay in responding to the imposed stress. Clearly, in addition to feedback mechanisms, nature exploits feedforward control, which is based on knowledge of the system model.

A *Google* search of the phrase "feedback control" results in nearly 118 million hits. A search on the phrase "feedforward control" results in about 300,000 hits. This naïve evidence is quite revealing of the disparity in the investment for developing feedforward control compared to feedback control. There are some engineering applications where feedforward control has been used to successfully deal with challenging problems. For instance, in CD and DVD players the center of the spindle motor is seldom coincident with the center of the data spiral on the storage medium. This causes a radial tracking error which is periodic with a frequency which is coin-

---

[†]Internal models for motor control and trajectory planning, M. M. Kawato, *Current Opinion in Neurobiology*, Volume 9, Number 6, December 1, 1999, pp. 718–727.

[‡]Feedforward versus feedback control: the case study of cerebellar ataxia, *http://www.laboratorium.dist.unige.it/ piero/Papers/ISHF01.PDF.*

[§]Optimal choice between feedforward and feedback control in gene expression to cope with unpredictable danger, E. Shudoa, P. Haccoub, Y. Iwasaa, *Journal of Theoretical Biology*, Volume 223, 2003, pp. 149–160.

cident with the spindle speed. The optical pickup must track the center of the spiral within a few μ meters precision to reliably read the data. Feedforward controllers are used to compensate for the periodic disturbance by identifying the amplitude and phase of the periodic radial disturbance, and have been successfully demonstrated on optical storage devices.

This book is a modest attempt to fill the gap in the area of shaping the reference input to dynamical systems whose responses are characterized by lightly damped modes. This is essentially an open-loop control strategy and requires the system being controlled to be stable or marginally stable. However, one should note that the plant being controlled can include a feedback component and the design would consider the closed-loop dynamics of the plant as the system of interest. This book does not exclusively consider open-loop control. It presents robust feedback controller design as an extension of an approach that can be exploited for robust open-loop controller design. The goal of this book is to provide in a tutorial manner various algorithms for attenuating the residual oscillations that are excited by reference inputs to dynamical systems. Some sections of this book are suitable as complementary material to undergraduate courses focusing on feedback control. This book would be suitable for an introductory graduate course in mechanical, aerospace, electrical, civil, and chemical engineering, despite the use of examples which look suitable for the mechanical and aerospace communities only. This is because oscillatory behavior exists, among others, in electrical, chemical, or civil engineering applications.

Outline of the Book

This book is organized as follows: Chapter 1 develops models for applications whose dynamics are dominated by lightly damped poles. Since most of the illustrative examples in the rest of the book are benchmark problems, the objective of the development of dynamic models in Chapter 1 is to provide the reader with a suite of real-world problems which can be used as exercises. Chapter 2 is devoted to the development of time-delay filters (Input Shapers™), which are used to shape the reference input to dynamical systems undergoing rest-to-rest maneuvers, so as to eliminate or minimize the residual energy which would correspond to accurate positioning of the final states. The time-delay filter design technique is illustrated on single- and multi-mode systems and for continuous and discrete-time systems. Chapter 3 reviews basic concepts in *calculus of variations* and can be skipped by the reader who is familiar with this material. Chapter 4 focuses exclusively on four control problems, all of which include saturation constraints for the control input. The four problems are: (1) Time-optimal control, (2) Fuel/time-optimal control, (3) Fuel limited time-optimal control, and (4) Jerk limited time-optimal control. A systematic increase in the level of difficulty starting from a rigid body system and leading to a damped floating oscillator benchmark problem is used to illustrate the change in the structure of the optimal control profile. Chapters 2 and 4 present a technique for desensitizing the open-loop profile or shaped profile to uncertainties in model parameters. The robustness is achieved by forcing the state sensitivities evaluated about the nominal model parameters to zero. Chapter 5 poses the problem of design of a robust time-delay filter as a minimax optimization problem. This technique can exploit knowledge of the distribution of the uncertain parameters. Friction is a constant presence in many mechanical and aerospace applications and is detrimental for precision position control. Chapter 6 focuses on the development of input constrained design of open-loop control profiles which accounts for friction in the design of point-to-point control profiles. Rigid body and systems with lightly damped modes are used to illustrate the control techniques. The final chapter of this book deals with numerical techniques for solving the problem of design of shaped inputs. Gradient-based and convex programming-based approaches are presented and illustrated on simple examples. MATLAB® code is occasionally provided for pedagogical purposes.

MATLAB® is a registered trademark of The MathWorks, Inc. For product information, please contact:

The MathWorks, Inc.
3 Apple Hill Drive
Natick, MA 01760-2098 USA
Tel: 508 647 7000
Fax: 508-647-7001
E-mail: info@mathworks.com
Web: www.mathworks.com

Tarunraj Singh

Acknowledgments

Two people who have been inspirational and to whom I am particulary grateful are John Junkins and Rao Vadali. They have been wonderful mentors and colleagues, and I treasure their friendship. Their tremendous intellect and achievements, coupled with their modesty, make them paragons for anyone who wishes to succeed in academia. Raj Dubey and Farid Golnaraghi introduced me to the occasionally frustrating, mostly rewarding, world of academic research and I am grateful for that opportunity.

The writing of this book has coincided with my teaching of a course on vibration control. I am thankful to my current and former students who were the prime movers of various research projects, the results of which when amalgamated formed the outline of this book. They include Rolf Hartmann, Dominik Fuessel, Robert Call, Joachim Noll, Dirk Tenne, Yong-Lin Kuo, Timothy Hindle, Jae-Jun Kim, Marco Muenchhof, Ulrich Staehlin, Rajaey Kased, Thomas Conord, Umamaheswara Konda, Jayaram Gopalakrishnan, Matt Vossler, Nidal al-Masoud, Shin-Whar Liu, Hasan Alli, Ravi Kumar, Jennifer Haggerty, Christoph Verlohren, and Jeroen van de Wijdeven. I have benefited tremendously from conversing with numerous colleagues who have helped conceive of solutions to challenging problems. In particular, I would like to mention Bill Singhose, Warren Seering, Glenn Heppler, Maarten Steinbuch, Puneet Singla, Peter Scott, John Crassidis, Timothy Chang, William O'Connor, Peter Meckl, Martin Berg, Santosh Devasia, and Arun Banerjee. Any constructive criticism, comments, and corrections are welcome via e-mail at: *tsingh@buffalo.edu*.

Thanks to my academic homes at the University at Buffalo, University of Waterloo, and Texas A & M University, who have provided me with a rich academic environment to pursue my endeavors.

Finally, heartfelt thanks to Bob Stern, executive editor at Taylor & Francis/CRC Press who helped to bring this project to fruition. Special thanks also go to the project editor Linda Leggio for her assistance.

1

Introduction

Imagination is more important than knowledge.

Albert Einstein,1921 Nobel Laureate in Physics

THE primary goal of this book is to present techniques to control self-induced vibrations (oscillations) in linear or quasi-linear dynamical systems. This implies that oscillatory behavior, which is a manifestation of underdamped modes of the system, are excited by the reference input to the system. For example, the oscillatory motion of a portainer (container crane) payload, when the hoist carrying trolley is used to maneuver containers, is an illustration of the self-excited vibrations. This is in contrast to vibrations of systems excited by exogenous inputs, for example, wind excited vibrations that led to the destruction of the Tacoma Narrows Bridge. Precise positioning of the containers in an expeditious manner is mandatory for efficient operation of ports. Since the mass of the containers which have to be loaded and discharged are not uniform, control techniques which are insensitive to variations in container mass are desirous. Techniques which endeavor to desensitize the controller to variations (uncertainties) in system model will be presented and the robustness versus performance trade-off will be discussed. In this book, the term *vibration* will be loosely used to represent any kind of oscillatory motion.

A large number of applications are characterized by self-excited vibrations. These include hard disk drives, high-speed tape drives, high-speed elevators for skyscrapers, containers crane, slosh dynamics of fluids in containers, vehicle platoons in intelligent transportation, wafer scanners for manufacturing integrated circuits using photolithography etc. A growing number of applications have exploited techniques presented in this book or variations of ones presented, with beneficial results. In this chapter, some of the aforementioned applications will be presented in some detail to provide examples which the reader can use to test techniques which will be presented in this book.

1.1 Hard Disk Drives

IBM has consistently pioneered breakthrough technology in magnetic storage devices. Following experimenting with rotating cylindrical drums, the first disk drives whose operating concepts are similar to the ones in use today were conceived of in the 1950s. The first production hard disk drive (IBM 305 RAMAC) was introduced in September 1956, by IBM which stored 5 megabytes on 50 disks of 24 inch diameters. This corresponded to an areal density of 2000 bits/square inch. Areal density is defined as the product of TPI (tracks per inch) and BPI (bits per inch), i.e., the number of bits per inch of a track. Prior to the IBM 305 RAMAC, the read/write head of the hard disk drives were in contact with the surface of this disk to compensate for the low sensitivity of the electronics of that vintage. This resulted in the read/write head and the magnetic coating of the disks, wearing out. The IBM 305 RAMAC was the first disk drive which created an air cushion on which the read/write head floated. This was achieved by supplying pressurized air to create an air bearing between the head and the magnetic media [1]. From that breakthrough technology, the design of heads has progressed to a point where a self-generated air bearing is used to separate the head from the storage medium with a gap distance which is equal to *mean free path*, which is the average distance between collisions of air molecules, with the relative velocity between the head and the magnetic media greater than 40 meters/sec. The air bearing generated by the relative motion of the disk and the head, in conjunction with the suspension permits the head to adapt to imperfections on the disk surface by maintaining a fairly constant *flying height*.

Over the same period, the read head technology has changed from the inductive read/write heads which were first used to sense the magnetization changes to the current thin film inductive write GMR (giant magneto resistive) read heads. IBM introduced the 0663-E12 drive in 1991, which had a form factor of 3.5 inches that illustrated the use of MR (magneto resistive) read heads to greatly increase the areal density. The MR head consists of magnetic material whose resistance changes as a function of the incident magnetic field. This was followed by the introduction of the GMR read head technology, by IBM in December 1997. The industry's first GMR (Peter Grünberg and Albert Fert who discovered GMR won the Nobel Prize in Physics in 2007) based storage device [2] had an areal density of 2.69 Gbits/in^2.

The exponential growth of areal density has permitted the development of drives with smaller form factors, which has also permitted increasing the spindle speed. The increased spindle speeds (15,000 rpm, Savvio 15K Seagate Drive [3]) have resulted in reduced *latency* and the increased areal density has resulted in the requirement of high performance seeking and following controllers. Latency is defined as the time required to position the proper sector under the read/write head and is therefore, wasted time.

As areal density increases, the precision requirement of the seek and follow controllers increases. Reduction of seek times is another driving factor in the development of next generations of hard disk drives. This has to occur in conjunction with the reduction in size of the hard disk drive. Reduction in seek time for a given voice coil motor corresponds to reducing the inertia of the actuator. However, this is associated with a reduction in the natural frequencies of vibrations resulting in the vibratory modes contributing to an increase in the seek time because of residual vibration of the head. Thus, there exists a motivation to develop control algorithms to minimize the seek time with a constraint of eliminating residual vibrations.

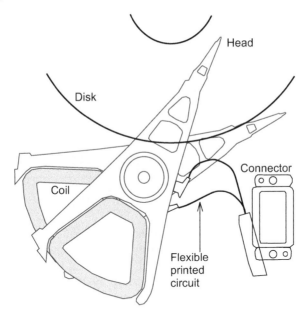

Figure 1.1: Head Disk Assembly (Reprinted with permission [4].)

Figure 1.1 illustrates a head disk assembly (HDA) which consists of the voice coil actuator, arm, and the slider. The arm carrying the read/write head is excited by rapid maneuvering by the voice coil motor. Traditionally, transfer function for the arm includes a rigid body mode and multiple underdamped flexible modes, which are used to design controllers. The rigid body model can be represented as [5]:

$$G_r(s) = \frac{K}{Js^2} \tag{1.1}$$

or

$$G_r(s) = \frac{K}{Js(s+a)} = \frac{1e6}{s(s+1.255e5)} \tag{1.2}$$

which captures the effect of windage on the rigid body motion.

For high precision motion, the vibration of the flexible arm is represented as a series of underdamped second order transfer functions [5]:

$$G_f(s) = \prod_{i=1}^{n} \frac{a_i \omega_i s + b_i \omega_i^2}{s^2 + 2\zeta_i \omega_i s + \omega_i^2} \tag{1.3}$$

The combined transfer function is:

$$G(s) = G_r(s)G_f(s) \tag{1.4}$$

Table 1.1 lists the model parameters which characterize the system dynamics as nonminimum phase with lightly damped modes.

Mode #	a_i	b_i	ω_i	ζ_i
1	0.0000115	-0.00575	$2\pi 70$	0.05
2	0	0.023	$2\pi 2200$	0.005
3	0	0.8185	$2\pi 4000$	0.05
4	0.0273	0.1642	$2\pi 9000$	0.005

Table 1.1: Disk Drive Parameters [5]

1.2 High-Speed Tape Drives

Since its introduction in 1972 by IBM and 3M, magnetic tape storage has been a cost effective means of storing and retrieving data. It continues to be used due to its 10 to 1 cost advantage over disk drives [6]. The current roadmap for tape storage targets a capacity of 16 Terabytes in a cartridge by 2015 [7]. This has prompted research into utilization of thinner media, increasing the read/write velocity from the current 4 m/s to 12–15 m/s and increasing the track density [3]. The paramount problem which needs to be addressed is precise high velocity and low tension control of the tape media.

Panda and Lu [8] present a tutorial on controller for tape drives. They present a simple multi-input multi-output model to represent the dynamics of the tape tension and velocity as a function of the torques used to drive the supply and take-up reels. Mathur and Messner [9] describe the effect of air entrainment between layers of tape on the system dynamics for high-speed low-tension tape drives. The model they proposed is presented here.

Figure 1.2 illustrates schematically a high-speed tape drive.

Force balance of the two spools result in the equations:

$$\dot{J}_1 \omega_1 + J_1 \dot{\omega}_1 + c_1 \omega_1 = K_1 \tau_1 + Tr_1 \tag{1.5}$$
$$\dot{J}_2 \omega_2 + J_2 \dot{\omega}_2 + c_2 \omega_2 = K_2 \tau_2 - Tr_2 \tag{1.6}$$

where J_i is the inertia, ω_i is the angular velocity, and r_i the radius of the ith reel. T is

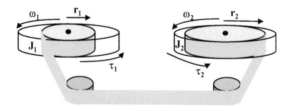

Figure 1.2: Schematic of a Tape Drive

the tension in the tape and c_i and K_i are the damping constant and torque constants respectively. The tape inertia can be represented as:

$$J(r) = J_0 + \int_{r_i}^{r} \rho t_w (2\pi R) R^2 \, dR = J_0 + \frac{\rho t_w \pi}{2} (r^4 - r_i^4) \qquad (1.7)$$

and the rate of change of inertia is:

$$\dot{J}(r) = 2\rho t_w \pi r^3 \dot{r} \qquad (1.8)$$

where ρ and t_w are the tape density and width respectively. The rate of change of the reel radius is given by the equations:

$$\dot{r}_1 = -\frac{\varepsilon \omega_1}{2\pi} \qquad (1.9)$$

$$\dot{r}_2 = \frac{\varepsilon \omega_2}{2\pi} \qquad (1.10)$$

where ε is the tape thickness. Substituting for $\dot{J}_i$ into Equation (1.6), we have:

$$J_1 \dot{\omega}_1 = \varepsilon \rho t_w r_1^3 \omega_1^2 - c\omega_1 + K_1 \tau_1 + Tr_1 \qquad (1.11)$$

$$J_2 \dot{\omega}_2 = -\varepsilon \rho t_w r_2^3 \omega_2^2 - c\omega_2 + K_2 \tau_2 - Tr_2 \qquad (1.12)$$

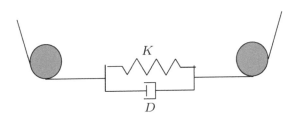

Figure 1.3: Schematic of a Tape Model

To determine the tension T, the equation for the stretching of the tape is calculated as:

$$S(t) = S_0 + \int_{t_0}^{t} r_2(\tau) \omega_2(\tau) \, d\tau - \int_{t_0}^{t} r_1(\tau) \omega_1(\tau) \, d\tau. \qquad (1.13)$$

The tape tension is defined as (Figure 1.3):

$$T(t) = KS + D\dot{S} \qquad (1.14)$$

The tension dynamics can now be represented as:

$$\dot{T} = K(\omega_2 r_2 - \omega_1 r_1) + D(\dot{\omega}_2 r_2 - \dot{\omega}_1 r_1 + \omega_2 \dot{r}_2 - \omega_1 \dot{r}_1) \quad (1.15)$$

Since, the radius of reels changes slowly implying that the inertia changes slowly, one can derive the linearized system dynamics by ignoring the nonlinearities, resulting in the state space model:

$$\begin{bmatrix} \dot{\omega}_1 \\ \dot{\omega}_2 \\ \dot{T} \end{bmatrix} = \begin{bmatrix} -\frac{c}{J_1} & 0 & \frac{r_1}{J_1} \\ 0 & -\frac{c}{J_2} & -\frac{r_2}{J_2} \\ -r_1 K + \frac{Dcr_1}{J_1} & r_2 K - \frac{Dcr_2}{J_2} & -D(\frac{r_1^2}{J_1} + \frac{r_2^2}{J_2}) \end{bmatrix} \begin{bmatrix} \omega_1 \\ \omega_2 \\ T \end{bmatrix} + \begin{bmatrix} \frac{K_1}{J_1} & 0 \\ 0 & \frac{K_2}{J_2} \\ -\frac{r_1 K_1 D}{J_1} & \frac{r_2 K_2 D}{J_2} \end{bmatrix} \begin{bmatrix} \tau_1 \\ \tau_2 \end{bmatrix}$$

$$(1.16)$$

Motor inertia J_0	2.2e-5 kgm^2
Full tape inertia J_f	2.74e-5 kgm^2
Radius of empty reel r_i	0.01112 m
Radius of full reel r_i	0.0304 m
Tape density ρ	1.7e3 kg/m^3
Tape thickness ε	8.9e-6 m
Tape width t_w	12.54e-3 m
Motor torque constant $K_1 = K_2$	1.62e-2 Nm/A
Motor damping constant c	9.56e-5 kg m^2
Tape damping coefficient D	0.9 Ns/m
Tape stiffness K	615 N/m

Table 1.2: Tape Drive Parameters [10]

The model parameters for a tape drive are listed in Table 1.2.

1.3   High-Speed Elevator

With the rapid growth of increasingly taller skyscrapers, there has been a progressive increase in the speed of the elevator. The 1930 Empire State Building elevators are rated at a speed of 300 meters/min (mpm). Since then, rated speed of elevators have grown nearly linearly with time culminating with the 1010 mpm speed of the high-speed elevators in the Taipei 101 in 2004 [11]. With the increasing speed of the elevators have come numerous other issues which require active control. These include atmospheric pressure control to regulate the rate of change of cabin pressure to be the same which has been shown to be physiologically comfortable. Aerodynamic shaping of the cabin and counterweight to minimize noise and finally since the long cable act as springs, vibration control of the cabin. Fortune [12] lists design limits to ensure minimizing discomfort of passengers:

1. vertical acceleration/deceleration ≤ 1.0–1.5 m/s^2

2. jerk rates ≤ 2.5 m/s^3

3. sound ≤ 50 dBa

4. horizontal sway 15–20 mg

5. ear-pressure change ≤ 2000 Pa

The first two requirements have to be taken into account in designing controllers for the motion of the cabin.

Figure 1.4 illustrates schematically a simplified lumped parameter model of a high-speed high-rise elevator. The kinetic energy T and the potential energy V are:

$$T = \frac{1}{2} \left(J\dot{\theta}^2 + M_1 \dot{x}_1^2 + M_2 \dot{x}_2^2 \right) \tag{1.17}$$

$$V = \frac{1}{2} \left(K_1 (x_1 - r\theta)^2 + K_2 (x_2 + r\theta)^2 \right) \tag{1.18}$$

which results in the Euler-Lagrange equations of motion:

$$M_1 \ddot{x}_1 + K_1 (x_1 - r\theta) + c_1 \dot{x}_1 = 0 \tag{1.19}$$

$$M_2 \ddot{x}_2 + K_2 (x_2 + r\theta) + c_2 \dot{x}_2 = 0 \tag{1.20}$$

$$J\ddot{\theta} - K_1 r (x_1 - r\theta) + K_2 r (x_2 + r\theta) + c\dot{\theta} = \tau \tag{1.21}$$

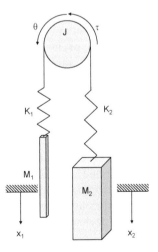

Figure 1.4: Schematic of an Elevator

where M_1, M_2 and J represent the masses of the counterweight, cabin, and inertia of the drive sheave. K_1, K_2 represent the stiffness of the cables suspending the counterweight and the cabin respectively. c_1, c_2 and c represent the damping coefficients to capture the loss of energy due to the velocity of the counterweight, cabin, and sheave. The parameters used for numerical simulation of the elevator model [13] are listed in Table 1.3.

M_1	M_2	J	K_1	r
5300 kg	3400 kg	1622 kgm^2	1.74e7 N/m	1 m
K_2	c_1	c_2	c	
1.74e7 N/m	7.2e4 Ns/m	7.2e4 Ns/m	100 kg m/s	

Table 1.3: Elevator Model Parameters

1.4   Cranes

Coulton [14] dates the use of cranes to the 6th century BC by the Greeks. From the 6th century BC to the present day cranes which include gantry cranes, tower cranes, luffing tower cranes, and boom cranes, have used mechanical advantage generated by pulleys, hydraulic drives, and so forth, to hoist large loads. The lightly damped nature of the cable/payload system results in significant pendulation of the payload. This lightly damped oscillatory motion of the payload can be excited by the control system in the process of reorienting the payload. This motion could also be excited by exogenous sources such as wind, and wave motion for offshore cranes. The non-collocated nature of the actuation and the point to be controlled (payload) results in complex dynamics. Productivity of cranes in ports, construction zones, etc., is contingent on efficient and robust elimination of residual vibration of the payload. Controlling swing motion during the maneuver is also important to ensure collisions are avoided with surrounding objects. Hubbell et al. [15] describes various control problem in cranes which include: sway control, trim/list/skew control, and snag control. Sway control can be accomplished by recording expert crane operators motion and reusing the recorded signal or by using real-time measurement of the system states for feedback control. The trim/list/skew control is achieved by differential motion of the ropes supporting the payload.

The cable in any crane can either be modeled as a distributed parameter system which results in the wave equations describing the dynamic behavior of the cable [16, 17]. This model is rarely used in control design. Lumped parameter models which assume that the cable can be represented as a massless rigid link is often used to model the dynamics of payload/crane system [16, 18]. Figure 1.5 is a schematic of a gantry crane.

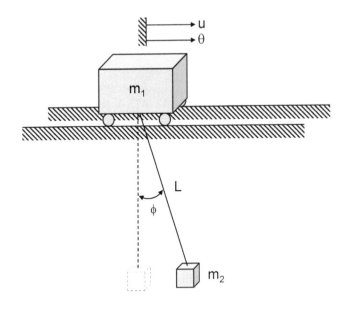

Figure 1.5: Gantry Crane

The Euler-Lagrange equation of motion can be derived using the kinetic and potential energies of the system. The kinetic energy is:

$$T = \frac{1}{2}m_1\dot{\theta}^2 + \frac{1}{2}m_2(\dot{\theta} + L\dot{\phi})^2 \tag{1.22}$$

and the potential energy is:

$$V = m_2gL(1 - \cos(\phi)). \tag{1.23}$$

where g is the acceleration due to gravity. Assuming small angular displacement ϕ of the cable, and no damping, the linearized model of the gantry crane is:

$$(m_1 + m_2)\ddot{\theta} + m_2L\ddot{\phi} = u \tag{1.24}$$

$$m_2\ddot{\theta} + m_2L\ddot{\phi} + m_2g\phi = 0 \tag{1.25}$$

This simple model has been used by various researchers to design and illustrate controllers. Table 1.4 lists typical crane parameters.

m_1	m_2	L	g
1000 kg	8000 kg	1–25 m	9.81 m/s^2

Table 1.4: Gantry Crane Parameters

1.5   Slosh Modeling

Slosh is blamed for the failure of NASA's ATS-V satellite in 1969 [19]. Hubert [20] describes how a small amount of liquid adversely affected the stability of the spin-stabilized satellite, resulting in the loss of the mission. Similarly, rollover of trucks carrying liquid cargo has been attributed to sloshing effect in partially filled tanks [21]. The resulting spill of toxic material can at best cause inconvenience by requiring evacuation and shutting down of transportation lines and at worst result in death and damage to property. Bulk Transporter [22] cites 31 rollover incidents related to transportation of petroleum by one petroleum refiner, in 2001. There is clearly a need to model slosh and develop techniques to alleviate the effect of slosh on the stability of the dynamic system.

Slosh is of concern in packaging industry as well where open containers carrying fluids have to be rapidly moved from filling stations to sealing stations to maximize throughput [23]. Casting industries also need to move large quantities of molten metal which mandates development of motion profiles which minimize slosh [24].

Lamb [25] describes the representation of velocity components of irrotational incompressible fluid in terms of a potential $\phi(x,y,t)$, i.e.,

$$u = -\frac{\partial \phi}{\partial x}, \quad v = -\frac{\partial \phi}{\partial y}. \tag{1.26}$$

The velocity potential has to satisfy the equation:

$$\frac{\partial^2 \phi}{\partial x^2} + \frac{\partial^2 \phi}{\partial y^2} = 0 \tag{1.27}$$

where the motion of the fluid can be restricted to two dimensions [25–27], with

$$\frac{\partial \phi}{\partial p} = 0 \tag{1.28}$$

representing the boundary conditions at the fixed boundary in the pth coordinate.

Figure 1.6 illustrates schematically a simplified pendulum equivalent model for slosh dynamics of a fluid in a rectangular tank.

Assuming a separation of variable representation of ϕ as:

$$\phi(x,y,t) = \psi(x)\eta(y)q(t), \tag{1.29}$$

Equation (1.27) can be rewritten as:

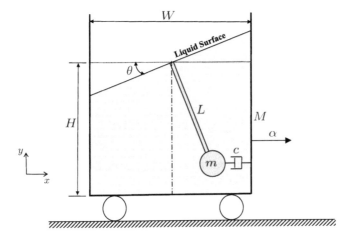

Figure 1.6: Slosh in a Rectangular Tank

$$\frac{d^2\psi(x)}{dx^2}\eta(y)q(t) + \frac{d^2\eta(y)}{dy^2}\psi(x)q(t) = 0 \tag{1.30}$$

$$\Rightarrow \frac{1}{\psi}\frac{d^2\psi(x)}{dx^2} = -\frac{1}{\eta}\frac{d^2\eta(y)}{dy^2} = -\lambda^2. \tag{1.31}$$

Solving Equation (1.31), we have

$$\psi(x) = A\cos(\lambda x) + B\sin(\lambda x) \tag{1.32}$$
$$\eta(y) = C\cosh(\lambda y) + D\sinh(\lambda y). \tag{1.33}$$

For the rectangular tank shown in Figure 1.6, we have:

$$\frac{\partial\phi}{\partial x}\bigg|_{x=0,W} = 0, \Rightarrow \frac{d\psi}{dx}\bigg|_{x=0,W} = 0 \tag{1.34}$$

$$\frac{\partial\phi}{\partial y}\bigg|_{y=0} = 0, \Rightarrow \frac{d\eta}{dy}\bigg|_{y=0} = 0, \tag{1.35}$$

assuming that the reference frame is attached to the lower left-hand corner of the tank. The boundary conditions lead to:

$$B = 0, \lambda W = n\pi, D = 0 \tag{1.36}$$

Normalizing ψ such that its 2 norm over x is 1, we have

$$\psi(x) = \sqrt{\frac{2}{W}} \cos(\frac{n\pi}{W}x) \tag{1.37}$$

$$\eta(y) = \cosh(\frac{n\pi}{W}y) \tag{1.38}$$

$$\phi(x,y,t) = \sum_{n=1}^{\infty} \sqrt{\frac{2}{W}} \cos(\frac{n\pi}{W}x) \cosh(\frac{n\pi}{W}y)q_n(t). \tag{1.39}$$

For a tank without external excitation, and ignoring the square of the velocity in the expression for pressure $P(x,y,t)$ [25], we have

$$\frac{P(x,y,t)}{\rho} = \frac{\partial\phi}{\partial t} - gy. \tag{1.40}$$

Representing the displacement of the fluid above the steady state free surface by δ, we have

$$\frac{P(x,y,t)}{\rho} = \frac{\partial\phi}{\partial t} - g(\delta + H), \tag{1.41}$$

whose time derivative is:

$$\dot{\delta} = -\frac{1}{g\rho}\frac{\partial P}{\partial t} + \frac{1}{g}\frac{\partial^2\phi}{\partial t^2} + \underbrace{\frac{1}{g}\frac{\partial^2\phi}{\partial t \partial\delta}\dot{\delta}}_{\text{nonlinear term}} \tag{1.42}$$

Ignoring the nonlinear term and since

$$\dot{\delta} = -\frac{\partial\phi}{\partial y}, \tag{1.43}$$

we have

$$\frac{1}{g}\frac{\partial^2\phi}{\partial t^2} + \frac{\partial\phi}{\partial y} = \frac{1}{g\rho}\frac{\partial P}{\partial t} \tag{1.44}$$

Approximating the surface of the fluid to lie at $y = H$, and substituting for $\phi(x,y,t)$ into Equation (1.44), we have

$$\frac{1}{g}\sum_{n=1}^{\infty}\sqrt{\frac{2}{W}}\cos(\frac{n\pi}{W}x)\cosh(\frac{n\pi}{W}H)\ddot{q}_n(t) + \sum_{n=1}^{\infty}\frac{n\pi}{W}\sqrt{\frac{2}{W}}\cos(\frac{n\pi}{W}x)\sinh(\frac{n\pi}{W}H)q_n(t) = 0 \tag{1.45}$$

since the pressure at the fluid surface is not a function of time. The inner product in Hilbert space given by integrating the product of Equation (1.45) with the mode shape $\psi(x) = \sqrt{\frac{2}{W}}\cos(\frac{k\pi}{W}x)$ leads to

$$\frac{1}{g}\cosh(\frac{k\pi}{W}H)\ddot{q}_k(t) + \frac{k\pi}{W}\sinh(\frac{k\pi}{W}H)q_k(t) = 0 \tag{1.46}$$

$$\ddot{q}_k(t) + \underbrace{\frac{kg\pi}{W}\tanh(\frac{k\pi}{W}H)}_{\omega_k^2}q_k(t) = 0. \tag{1.47}$$

Equation (1.47) illustrates that the free surface dynamics can be represented by a series of second order underdamped systems. Assuming that the first mode captures the dominant part of the slosh dynamics, the equivalent pendulum can be represented as shown in Figure 1.6. The length of the pendulum is given by the equation:

$$L = \frac{W}{\pi}\coth(\frac{\pi}{W}H). \tag{1.48}$$

The kinetic energy T, potential energy V, and Rayleigh dissipation energy $\mathscr{F}$ are given as:

$$T = \frac{1}{2}m\left(\dot{x} + L\cos(\theta)\dot{\theta}\right)^2 + \frac{1}{2}M\dot{x}^2 \tag{1.49}$$

$$V = mgL\left(1 - \cos(\theta)\right) \tag{1.50}$$

$$\mathscr{F} = \frac{1}{2}c\left(L\cos(\theta)\dot{\theta}\right)^2 + \frac{1}{2}c_x\left(\dot{x}\right)^2 \tag{1.51}$$

where m and M represent the masses of the fluid and the container, and c_x and c are damping constants associated with the motion of the container and the fluid respectively. The Euler-Lagrange equations

$$\frac{d}{dt}\frac{\partial T}{\partial\dot{q}} - \frac{\partial(T-V)}{\partial q} + \frac{\partial\mathscr{F}}{\partial\dot{q}} = Q \tag{1.52}$$

where q and Q are the generalized displacement and generalized force respectively, lead to the equations of motion for the slosh problem:

$$(m+M)\ddot{x} + mL\ddot{\theta} + c_x\dot{x} = u \tag{1.53}$$

$$m\left(\ddot{x} + L\ddot{\theta}\right)L + cL^2\dot{\theta} + mgL\theta = 0 \tag{1.54}$$

by ignoring the nonlinear terms and assuming small angular motion for the pendulum.

Table 1.5 lists model parameters which can be used to test various control algorithms that will be presented in the following chapters.

m	M	W	H	L	c	c_x
100 kg	20 kg	1 m	0.5 m	0.3471 m	5 Ns/m²	10 Ns/m

Table 1.5: Slosh Dynamics Parameters

1.6   Vehicle Platooning

Intelligent vehicle highway systems (IVHS) is a catch-all concept to address various issues related to relieving congestions in metropolitan highways [28]. Motivation of increasing throughput, reducing fuel consumed, reducing lost productivity time for drivers, are compelling auto manufacturers to develop technology to alleviate the aforementioned issues. IVHS has the potential of improving safety of the highway systems as well. The cost of augmenting the highway system to cater to increasing traffic is prohibitive in view of the financial and environmental impact. With the increased integration of sensors in vehicles and with reduced cost of microprocessors, development of self-aware vehicles is viable.

Bae and Gerdes [29] present a simple model to represent the dynamics of a platoon of tractor trailers. The goal of their research was to prevent actuator saturation of the vehicles and ensure stability of the coupled tractor platoon. This includes *string stability* which implies that disturbances in the upstream inter-vehicle distance should be attenuated as it travels downstream. The series of tractors can be represented by lumped masses coupled by virtual springs and dashpots, resulting in a system with underdamped response. Modifying the reference command to the lead vehicle was shown to minimize the oscillatory behavior of the inter-tractor distance.

Figure 1.7 illustrates a series of tractor trailers moving in tandem. The protocol for the platoons includes providing a reference command to the lead vehicle and the follower vehicles only use knowledge of position and velocity of the preceding vehicle to make decisions about motion control.

Each vehicle is modeled as a first order system to represent the actuator delay. Thus the transfer function relating the vehicle acceleration to the control input $U(s)$ is:

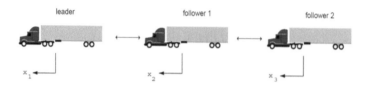

Figure 1.7: Tandem Motion of Tractors (Reprinted with Permission [29].)

$$\frac{s^2 X_i(s)}{U_i(s)} = \frac{b_i}{s + a_i} \tag{1.55}$$

where X_i is the displacement of the ith vehicle. The lead vehicle uses the control law:

$$U_1 = A_r + K_1(V_r - \dot{X}_1) \tag{1.56}$$

where A_r and V_r are the reference acceleration and velocity. Figure 1.8 is the block diagram representation of the closed-loop system for the lead truck. The resulting

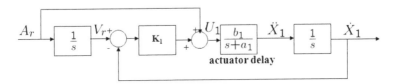

Figure 1.8: Lead Truck Dynamics

transfer function is:

$$\frac{U_1(s)}{A_r(s)} = \frac{(s + K_1)(s + a_1)}{s(s + a_1) + K_1 b_1} \tag{1.57}$$

Similarly, the dynamics of the ith truck can be derived. Figure 1.9 is the block diagram representation of the closed-loop dynamics of the ith follower truck. The

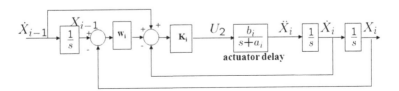

Figure 1.9: I^{th} Truck Dynamics

reference input to the ith truck is the velocity of the preceding truck. The controllers for the follower trucks are:

$$U_i = K_i((\dot{X}_{i-1} - \dot{X}_i) + w_i(X_{i-1} - X_i)) \tag{1.58}$$

The closed-loop dynamics can be shown to be:

$$\frac{U_i(s)}{\dot{X}_{i-1}(s)} = \frac{s(s + w_i)(s + a_i)K_i}{s(s(s + a_i) + K_i b_i) + w_i K_i b_i} \tag{1.59}$$

Bae and Gerdes [29] modeled a three tractor trailer platoon using parameters presented in Table 1.6.

Vehicle #	a	b	k	K_i	w_i
1	31.4	11	NA	5	1.2
2	31.4	7	4	5	1.2
3	31.4	10	4	5	1.2

Table 1.6: Platoon Dynamics Parameters [29]

1.7   Summary

This chapter details numerous examples which are characterized by underdamped oscillatory motion when the dynamic system is tracking a reference input. A large spectrum of applications are presented to illustrate that the techniques presented in subsequent chapters are not limited to control of vibrations of structures. Rather, they can be used to attenuate oscillatory motion excited by the reference input, which changes the setpoint to the control system. Dynamic models with numerical values are presented to permit the reader to design and evaluate controllers presented in this book on examples of their interest.

References

[1] E. D. Daniel, C. D. Mee, and M. K. Clark. *Magnetic Recording: The First 100 Years*. IEEE Press, New York, 1999.

[2] E. Grochowski and R. F. Hoyt. Future trends in hard disk drives. *IEEE Transactions on Magnetics*, 32(3):1850–1854, May 1999.

[3] http://www.seagate.com/www/en-us/products/servers/cheetah/.

[4] T. Semba and M. T. White. An identification method of seek-induced vibration modes in hard disk drives. In *16th IFAC World Congress*, Prague, Czech Republic, 2005.

[5] G. F. Franklin, J. D. Powell, and M. Workman. *Digital Control of Dynamic Systems*. Addison Wesley, 1997.

[6] http://www-03.ibm.com/systems/storage/tape/why/index.html.

[7] http://www.insic.org/tape.htm.

[8] S. P. Panda and Y. Lu. Tutorial on control systems design in tape drives. In *ACC*, Denver, CO, 2003.

[9] P. D. Mathur and W. C. Messner. Controller development for a prototype high-speed low-tension tape transport. *IEEE Trans. on Cont. Systems Tech.*, 6(4):534–542, July 1998.

[10] Y. Lu. Advanced control for tape transport. Dissertation, CMU, 2002.

[11] H. Mizuguchi, T. Nakagawa, and Y. Fujita. Breaking the 1000 mpm barrier. *Elevator World*, pages 71–76, September 2005.

[12] J. W. Fortune. Mega high-rise elevators. *Elevator World*, 43(7):63, December 1995.

[13] R. Roberts. Control of high-rise/high-speed elevators. In *ACC*, Philadelphia, PA, June 1998.

[14] J. J. Coulton. Lifting in early Greek architecture. *The Journal of Hellenic Studies*, 95:1–19, 1974.

[15] J. T. Hubbell, B. Koch, and D. McCormick. Modern crane control enhancements. *Ports '92*, pages 757–767, 1992.

[16] H. Alli and T. Singh. Passive control of overhead cranes. *Journal of Vibration and Control*, 5:443–459, 1999.

[17] C. D. Rahn, F. Zhang, S. Joshi, and D. M. Dawson. Asymptotically stabilizing angle feedback for a flexible cable gantry crane. *ASME J. of Dyn. Systems, Measurements, and Cont.*, 121(3):563–566, 1999.

[18] E. M. Abdel-Rahman, A. H. Nayfeh, and Z. N. Masoud. Dyanmics and control cranes: A review. *Journal of Vibration and Control*, 9:863–908, 2003.

[19] J. P. B. Vreeburg. Spacecraft maneuvers and slosh control. *Control Systems Magazine, IEEE*, 25(3):12–16, June 2005.

[20] C. Hubert. Behavior of spinning space vehicles with onboard liquids. In *2003 Flight Mechanics Symp.*, pages 1–14, NASA Goddard, MD, June 2003.

[21] T. Acarman and U. Ozguner. Rollover prevention for heavy trucks using frequency shaped sliging model control. In *2003 IEEE Conference on Control Applications*, pages 7–12, Istanbul, Turkey, June 2003.

[22] $http://bulktransporter.com/mag/transportation\_cargo\_tank\_rollovers/$.

[23] M. Grundelius and B. Bernhardsson. Control of liquid slosh in an industrial packaging machine. In *IEEE International Conference on Control Applications*, pages 1654–1659, Hawaii, HI, August 1999.

[24] Y. Noda, K. Yano, and K. Terashima. Tracking to moving object and sloshing suppression control using time varying filter gain in liquid container transfer. In *SICE Annual Conference*, pages 2283–2288, Fukui University, Japan, August 2003.

[25] H. Lamb. *Hydrodynamics*. Cambridge Mathematical Library, Cambridge, UK, 1993.

[26] R. Venugopal and D. S. Bernstein. State space modeling and active control of slosh. In *IEEE International Conference on Control Applications*, pages 1072–1077, Dearborn, MI, 1996.

[27] R. A. Ibrahim. *Liquid Sloshing Dynamics: Theory and Applications*. Cambridge University Press, New York, 2005.

[28] A. Martin, H. Marini, and S. Tosunoglu. Intelligent vehicle/highway system: A survey part 1. In *Florida Conference on Recent Advances in Robotics*, Gainesville, FL, April 29–30 1999.

[29] H. S. Bae and J. C. Gerdes. Command modification using input shaping for automated highway systems with heavy trucks. California PATH Research Report UCB-ITS-PRR-2004-48, Stanford University, 2004.

2

Time-Delay Filter/Input Shaping

I find that the harder I work, the more luck I seem to have.

Thomas Jefferson, 3rd U.S. President; 1801–1809; diplomat, politician, and scholar

PRECISE position control and rapid rest-to-rest motion is the desired objective in a variety of applications. The desire for reducing the maneuver time requires reducing the inertia of the structure which usually reduces the stiffness. This results in the system response being dominated by the low resonance frequencies. The maneuver time can also be reduced by increasing the available control force which mandates larger actuators resulting in increased size and cost. If judiciously selected feedback and/or feedforward controllers can be used to address the issues raised by the lightly damped modes, one can achieve the benefit of rapid maneuvering with lightweight structures.

For illustrative purposes, consider a hard disk drive: time dedicated to transitioning from one cylinder to another has to be reduced to decrease access time. This mandates driving the voice coil motor to its limits, which consequently results in exciting the vibratory modes. Since, the position error of the read/write head has to be less than 10% of the inter-track distance which translates to 25 nm for a drive with 100,000 tracks per inch, damping of any vibration is necessary. Clearly, there is a need to develop strategies, that minimize the residual vibration for such point-to-point maneuvers.

The requirement of precise position control implies that the vibration of the structure should be zero or near zero at the end of the maneuver. Some applications that are characterized by the low-frequency, low-damping dynamics include wafer scanners [1], read/write arms of hard disk drives [2], gantry cranes [3], spacecraft with flexible appendages [4], and so forth. The natural impulse of control engineers is to design feedback control which has a rich history of successful application and proven ability in common place devices such as refrigerators, to esoteric applications such as control of hydrocephalus by pressure relief valves using sensors which monitor intracranial pressure. Feedback controllers can broadly be classified as regulators where the objective is to maintain a set point, or tracking control where the set point is a function of time. For tracking control, one has the option of manipulating the reference profile so as to satisfy tracking objectives, desensitizing the system response to modeling uncertainties, or satisfying desired transient state constraints.

The shaping of the reference profile requires no additional hardware, which makes its inclusion into legacy systems quite easy.

The problem of rapid maneuvering of a flexible structure between two quiescent attitudes with the objective of minimizing structural vibration has been addressed by many researchers. Swigert [5] parameterized the control profile using a summation of ramp and harmonic (basis) functions and solved for coefficients of the basis function to satisfy boundary conditions and minimize the energy consumed. Scientists interested in designing minimum time control profiles have exploited the knowledge that the optimal solution for most systems requires the control to be at its limits (maximum or minimum) and switch between the limits a finite number of times. The switching between the limits which is referred to as bang-bang control has been represented by hyperbolic tangent functions or splines which permits the use of gradient based optimization algorithms to solve for the switch times. The motivation for using the hyperbolic tangent or the spline functions to approximate the discontinuous switch is to minimize "ringing" of the higher modes of the structure which can degrade the performance determined by the terminal state errors. A fairly comprehensive treatment of this family of problems has been presented by Junkins and Turner [6], Singh and Singhose [7], and Scrivener and Thompson [8].

Thus, there are two main issues which influence the performance of any control technique for the point-to-point maneuvering of structures whose transients are dominated by oscillations. These are:

- the natural frequency and damping ratio and the associated uncertainties of the underdamped modes and

- the higher frequency modes which are generally unmodeled

Model based controllers exploit the knowledge of the system model to design feedback or open-loop control strategies. The effect of the errors in model parameters on the stability and performance of the control system has to be analyzed prior to its deployment. Similarly, the effect of the unmodeled dynamics which can result in instability of some feedback control systems, has to be carefully studied. To illustrate this, consider a system whose transfer function is:

$$\frac{Y(s)}{U(s)} = \frac{100}{(s^2+1)(s^2+100)} \tag{2.1}$$

which is characterized by two undamped modes of frequencies $\omega = 1$, and 10 rad/sec. Assume that the higher frequency corresponds to an unmodeled mode. The step responses of the system considering only the low frequency mode and both the modes are shown by the dashed and solid line respectively, in the first subplot of Figure 2.3. It is clear that the high frequency mode's contribution to the system response is significantly smaller than that of the low frequency mode. Further, the system response is marginally stable since there is no mechanism to dissipate the energy in the system. A proportional-derivative (PD) controller can be used to stabilize the system represented by the low frequency dynamics as shown in Figure 2.1. The effect of

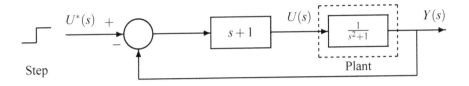

Figure 2.1: PD Control of Reduced Model

the stabilizing PD controller on the behavior of the system represented by both the modes is shown in the second subplot of Figure 2.3.

The response to an unit step input is shown by the dashed line in Figure 2.3 which asymptotically approaches the final value of 0.5. Simulating the same PD controller with the unmodeled mode, as shown in Figure 2.2 results in an unstable response, illustrated by the solid line in Figure 2.3.

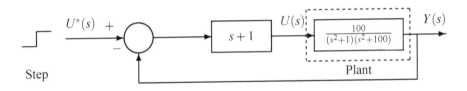

Figure 2.2: PD Control of Complete Model

This simple example illustrates the fact that the effect of unmodeled dynamics on the open-loop control system is small. However, the desired objective of forcing the output to a final value is unachievable. On the other hand, the effect of the unmodeled dynamics on the closed-loop system is manifested in the form of instability. Since the control designer can anticipate the response of the open-loop system to a given input, the question of shaping the input to satisfy the final value constraints is an obvious problem to be studied.

Posicast Control

A very interesting technique was proposed by Tallman and Smith [9] which involved splitting the input excitation into several segments such that the sum of all

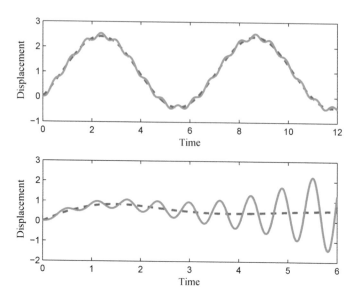

Figure 2.3: Effect of Unmodeled Dynamics

transient terms equals zero after the last excitation. This technique was referred to as the *Posicast* technique. Figure 2.4 illustrates schematically the response of a *Posicast* controller.

The transfer function of the system illustrated in Figure 2.4 is one with undamped poles. As shown in the upper part of the figure, the response of the system to a step input of magnitude 0.5 is harmonic, with a peak magnitude of unity. Consider the same step input applied to the system delayed by a time T as shown in the bottom half of Figure 2.4. As anticipated, the response is harmonic and its initiation is delayed by T. If the delay T is selected to be one-half of the period of oscillation of the system output ($\frac{\pi}{\omega}$), summing the output of the systems to the two step inputs results in a nonoscillatory response as shown by the middle graph. Thus, replacing the unit step input with a staircase input which is generated by summing the two steps of magnitude 0.5, with one delayed by T seconds, results in a finite time point-to-point transition. This staircase input can be represented as:

$$u(t) = 0.5\mathcal{H}(t) + 0.5\mathcal{H}\left(t - \frac{\pi}{\omega}\right) \tag{2.2}$$

where $\mathcal{H}(.)$ is the Heaviside unit step function.

The successful application of the *Posicast* controller is contingent on the precise knowledge of the system's natural frequency. Any errors in the estimated natural frequency which corresponds to an incorrect selection of T results in poor performance of the controller. This fact prevented the use of the *Posicast* controller for numerous applications which are characterized by underdamped system response.

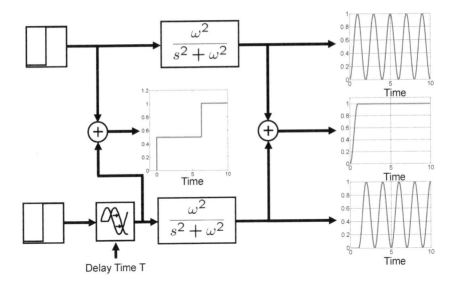

Figure 2.4: *Posicast* Control

In this chapter, we first design a time-delay controller for a single mode system and illustrate that the resulting solution is the same as the *Posicast* controller. To minimize the loss of performance due to errors in estimated model parameters, it will be shown that multiple instances of this controller in cascade lead to controllers that are less sensitive to parameter errors. We then design a time-delay controller, where the time-delay is user selected, this is a generalization of the time-delay controller proposed by Singh and Vadali [10]. This is followed by studying some properties of the controller regarding the sign of the gains of the time-delay elements and periodicity of the controller. We refer to the magnitude of the delayed step input, as the time-delay gain. A general procedure for the design of user selected time-delay controllers for multi-mode systems is presented. The controller presented in Section 2.3.2 is based on minimizing the time-delay with the gains of the time-delay elements constrained to lie between 0 and 1. An indexing mechanism with a PD controller is used to illustrate the time-delay controller.

Effects of high frequency unmodeled modes have been identified as a cause of degradation of performance of controlled systems. Rolling off the frequency response function as a function of frequency leads to simple techniques to alleviate the problem caused by the high frequency unmodeled modes. These techniques are expounded and tested on an illustrative example in Section 2.4. Finally, the design of time-delay filters for discrete-time systems is presented and illustrated on the benchmark flexible transmission problem.

An important point to note is that the techniques presented in this chapter deal with shaping the reference input to the plant. The plant could represent an open-loop or a closed-loop system. The only assumption is that the plant is stable or marginally stable and its response is dominated by underdamped or undamped oscillations.

2.1   Time-Delay Filters

2.1.1   Proportional Plus Delay (PPD) Control

As illustrated earlier, shaping the step input to a stair case profile results in a rest-to-rest maneuver of an undamped system in finite time. To analyze the finite time rest-to-rest response of the system, consider the Laplace transform of the control $u(t)$:

$$U(s) = \frac{0.5}{s} + \frac{0.5}{s}\exp\left(-s\frac{\pi}{\omega}\right) = \frac{1}{s}\left(0.5 + 0.5\exp\left(-s\frac{\pi}{\omega}\right)\right) \tag{2.3}$$

which can be represented as the output of a time-delay filter with a transfer function:

$$\frac{U(s)}{R(s)} = G(s) = 0.5 + 0.5\exp\left(-s\frac{\pi}{\omega}\right) \tag{2.4}$$

subject to a unit step input $(R(s))$. The zeros of the transfer function $G(s)$ are located at:

$$z = \pm(2k+1)\omega j, \quad k = 0, 1, 2, ... \tag{2.5}$$

where $j = \sqrt{-1}$. It can be seen that one pair of zeros lie at $\pm\omega j$ which cancel the poles of the open-loop system which lie at $\pm\omega j$. This structure of a time-delay filter will be used to generalize the *Posicast* controller.

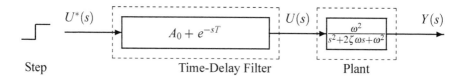

Figure 2.5: Time-Delay Filter Structure

Figure 2.5 represents a time-delay filter which shapes a step input to a second-order underdamped system. We need to determine A_0 and T so that the poles of the system are canceled by the zeros of the time-delay filter. The zeros of the filter are given by the following equation

$$A_0 + \exp(-sT) = 0 \tag{2.6}$$

where we have normalized the relative amplitudes of the proportional and time-delayed signal. A_0 is the amplitude of the proportional signal and T is the delay time of the time-delayed signal. Representing the Laplace variable s as

$$s = \sigma + j\omega \tag{2.7}$$

and substituting Equation (2.7) into Equation (2.6) and equating the real and imaginary parts to zero, we have

$$A_0 + \exp(-\sigma T)\cos(\omega T) = 0 \tag{2.8}$$

and

$$\exp(-\sigma T)\sin(\omega T) = 0 \tag{2.9}$$

Solving Equation (2.9) we have

$$\omega = (2n+1)\frac{\pi}{T}, 2n\frac{\pi}{T} \tag{2.10}$$

which when substituted into Equation (2.8), illustrates that only $\omega = (2n+1)\frac{\pi}{T}$ produces a positive value of A_0, which leads to

$$\sigma = -\frac{1}{T}\ln(A_0) \tag{2.11}$$

The zeros of the time-delay filter are

$$s = \begin{cases} \frac{-\ln(A_0)+(2n+1)\pi j}{T} & \text{if } A_0 > 0 \\ \frac{-\ln(-A_0)+2n\pi j}{T} & \text{if } A_0 < 0 \end{cases} \quad n = -\infty, \ldots, \infty \tag{2.12}$$

In this development, we assume A_0 is positive and use the zeros corresponding to that assumption.

To cancel the system poles at $s = -\zeta_i\omega_i + j\omega_i\sqrt{1-\zeta_i^2}$, we have from (2.10) (setting n = 0)

$$\omega = \omega_i\sqrt{1-\zeta_i^2} = \frac{\pi}{T} \tag{2.13}$$

$$\Rightarrow T = \frac{\pi}{\omega_i\sqrt{1-\zeta_i^2}} \tag{2.14}$$

and

$$\sigma = -\zeta_i\omega_i \tag{2.15}$$

Substituting Equation (2.14) into Equation (2.11), we have

$$-\ln(A_0) = -\zeta_i\omega_i\frac{\pi}{\omega_i\sqrt{1-\zeta_i^2}} \tag{2.16}$$

which leads to

$$A_0 = \exp(-\frac{\zeta_i \pi}{\sqrt{1 - \zeta_i^2}}) \tag{2.17}$$

This corresponds exactly to the solution of the *Posicast* controller. The control profile can also be written as

$$u(s) = (s^2 + 2\zeta_i \omega_i s + \omega_i^2)(s^2 + 2\zeta_i \omega_i s + 9\omega_i^2 - 8\zeta_i^2 \omega_i^2)...$$
$$(s^2 + 2\zeta_i \omega_i s + n^2 \omega_i^2 - (n^2 - 1)\zeta_i^2 \omega_i^2) \; n = 1, 3, 4, ... \tag{2.18}$$

The pole-zero locations of the controlled system are shown in Figure 2.6. Thus, the single time-delay filter can also be used to cancel poles of the system that are odd multiples of the two primary poles.

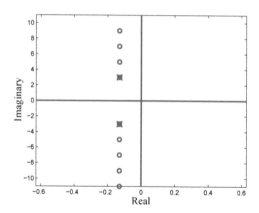

Figure 2.6: Pole-Zero Locations of the Controlled System in the S-plane

The final value of the single time-delay filtered system driven by a unit step input is given by

$$\lim_{s \to 0} \frac{1}{s} s \left(\frac{(A_0 + \exp(-sT))\omega_i^2}{s^2 + 2\zeta_i \omega_i s + \omega_i^2} \right) = A_0 + 1 \tag{2.19}$$

To force the final value of the input after passing through the time-delay filter to be the same as that entering the filter, we normalize the amplitudes of the direct and time delayed signal, so that the time-delay filter's transfer function is $(A_0 + \exp(-sT))/(A_0 + 1)$.

2.1.2 Proportional Plus Multiple Delay (PPMD) Control

The single time-delay filter, by virtue of canceling the poles corresponding to the oscillatory behavior of the system, provides us with a technique to produce non-oscillatory response. The cancelation of the poles of the system is contingent on the availability of accurate data regarding them. To improve the robustness of the time-delay filter to errors in estimated poles, a two-time-delay filter is proposed as illustrated in Figure 2.7. The zeros of the filter are given by the equation:

$$A_0 + A_1 \exp(-sT) + \exp(-2sT) = 0 \tag{2.20}$$

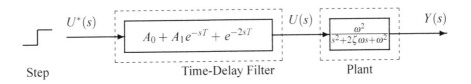

Figure 2.7: Two-Time-Delay Controlled System

The tacit assumption made is that the second delay is twice the first delay. Letting $s = \sigma + j\omega$ and equating the real and imaginary parts to zero, we have

$$A_0 + A_1 \exp(-\sigma T)\cos(\omega T) + \exp(-2\sigma T)\cos(2\omega T) = 0 \tag{2.21}$$
$$A_1 \exp(-\sigma T)\sin(\omega T) + \exp(-2\sigma T)\sin(2\omega T) = 0 \tag{2.22}$$

respectively. Equation (2.22) can be rewritten as

$$\sin(\omega T)(A_1 \exp(-\sigma T) + 2\exp(-2\sigma T)\cos(\omega T)) = 0 \tag{2.23}$$

the solution of which is

$$\omega = (2n+1)\frac{\pi}{T}, 2n\frac{\pi}{T} \tag{2.24}$$

or

$$\omega = \frac{1}{T}\cos^{-1}\left(-\frac{A_1}{2}\exp(\sigma T)\right) \tag{2.25}$$

Substituting $\omega = (2n+1)\pi/T$ into Equation (2.21), we have

$$A_0 - A_1 \exp(-\sigma T) + \exp(-2\sigma T) = 0 \tag{2.26}$$

To design a time-delay filter that is robust to errors in estimated frequencies, we equate the derivative of Equation (2.26) with respect to σ to zero.

$$A_1 \exp(-\sigma T) - 2\exp(-2\sigma T) = 0 \tag{2.27}$$

We arrive at the same equation if we differentiate Equation (2.23) with respect to ω and equate it to zero. Solving Equation (2.27), we have

$$A_1 = 2\exp(-\sigma T) \tag{2.28}$$

Substituting Equation (2.28) into Equation (2.26), we have

$$A_0 = \exp(-2\sigma T) \tag{2.29}$$

To cancel the system poles, we equate

$$\sigma = -\zeta_i \omega_i \tag{2.30}$$

and

$$\omega = \omega_i \sqrt{1 - \zeta_i^2} = \frac{\pi}{T} \tag{2.31}$$

Thus we have

$$A_0 = \exp\left(\frac{2\pi\zeta_i}{\sqrt{1 - \zeta_i^2}}\right) \tag{2.32}$$

$$A_1 = 2\exp\left(\frac{\pi\zeta_i}{\sqrt{1 - \zeta_i^2}}\right) \tag{2.33}$$

which are also the impulse amplitudes obtained by the solution of the three-impulse shaped-input technique [11].

Substituting Equation (2.25) into Equation (2.21), we have

$$A_0 + A_1 \exp(-\sigma T)\left(-\frac{A_1}{2}\exp(\sigma T)\right) + \exp(-2\sigma T)\left(2(-\frac{A_1}{2}\exp(\sigma T))^2 - 1\right) = 0 \tag{2.34}$$

which simplifies to

$$A_0 - \exp(-2\sigma T) = 0 \tag{2.35}$$

or

$$\sigma = -\frac{1}{2T}\ln(A_0) \tag{2.36}$$

Substituting T and A_0 from Equations (2.31) and (2.32), respectively, we have

$$\sigma = -\zeta_i \omega_i \tag{2.37}$$

and from Equation (2.25), we have

$$\omega = \omega_i \sqrt{1 - \zeta_i^2} \tag{2.38}$$

which indicates that the use of either Equation (2.24) or Equation (2.25) leads to the same solution. Thus, the effect of the second delay is to provide a second set of zeros, with all the zeros being coincident with the first infinite set, as indicated in Figure 2.6. The controller should be normalized in the same fashion as the single time-delayed controller. This controller can also be represented as

$$u(s) = (s^2 + 2\zeta_i\omega_i s + \omega_i^2)^2 (s^2 + 2\zeta_i\omega_i s + 9\omega_i^2 - 8\zeta_i^2\omega_i^2)^2 ...$$
$$(s^2 + 2\zeta_i\omega_i s + n^2\omega_i^2 - (n^2-1)\zeta_i^2\omega_i^2)^2 \; n = 1,3,4,... \tag{2.39}$$

Since the zeros are repeated, the controller transfer function should also be equal to

$$\frac{U(s)}{R(s)} = \left(\exp\left(-\frac{\zeta_i \pi}{\sqrt{1-\zeta_i^2}} \right) + \exp(-sT) \right)^2 \tag{2.40}$$

or

$$\frac{U(s)}{R(s)} = \left(\exp\left(2\frac{\zeta_i \pi}{\sqrt{1-\zeta_i^2}} \right) + 2\exp\left(\frac{\zeta_i \pi}{\sqrt{1-\zeta_i^2}} \right) \exp(-sT) + \exp(-2sT) \right) \tag{2.41}$$

which is the same solution represented by Equations (2.20), (2.32), and (2.33).

The robustness of the time-delay filter to uncertainties in estimated damping and natural frequency can be analyzed by studying the Taylor series expansion of the transfer function of the time-delay filter. The Taylor series around the point s_{nom} is

$$G(s) = \sum_{k=0}^{\infty} \frac{1}{k!} \frac{\partial^k G(s)}{\partial s^k} \bigg|_{s=s_{nom}} (s - s_{nom})^k \tag{2.42}$$

$$= G(s)|_{s=s_{nom}} + \frac{\partial G(s)}{\partial s} \bigg|_{s=s_{nom}} (s - s_{nom}) + \tag{2.43}$$

$$\frac{1}{2} \frac{\partial^2 G(s)}{\partial s^2} \bigg|_{s=s_{nom}} (s - s_{nom})^2 + \mathcal{HOT}(s) \tag{2.44}$$

where $\mathcal{HOT}$ denotes the higher order terms. Thus, it can be seen that by locating n pairs of zeros of the time-delay filter at the nominal value of the poles of the system (s_{nom}), the first n terms of the Taylor series are forced to zero which implies that the magnitude of $G(s)$ in the vicinity of s_{nom} is reduced.

Example 2.1: To illustrate the PPD and PPMD controllers, consider the position

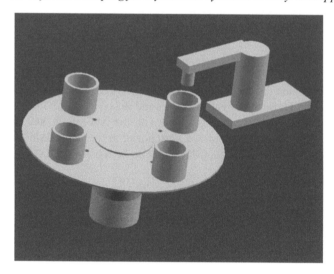

Figure 2.8: Indexing Mechanism

control of a double integrator which can represent the model of a motor (Figure 2.8). The system model is

$$J\ddot{\theta} = u, \tag{2.45}$$

where J corresponds to the inertia of the motor. A proportional plus derivative controller

$$u = -K_1(\theta - r) - K_2\dot{\theta} \tag{2.46}$$

where r is the reference input, is guaranteed to be stable provided the gains K_1 and K_2 are greater than zero. The closed-loop transfer function is

$$\frac{\Theta(s)}{R(s)} = \frac{K_1}{Js^2 + K_2s + K_1} \tag{2.47}$$

and the closed-loop poles are located at

$$s = \frac{-K_2 \pm \sqrt{K_2^2 - 4JK_1}}{2J}. \tag{2.48}$$

If $4JK_1 > K_2^2$, the resulting closed-loop poles result in an underdamped system response. The natural frequency and damping ratios of the closed-loop response are

$$\omega_n = \sqrt{\frac{K_1}{J}}, \quad \zeta = \frac{K_2}{2\sqrt{JK_1}} \tag{2.49}$$

which are used to determine the transfer function of the time-delay filters.

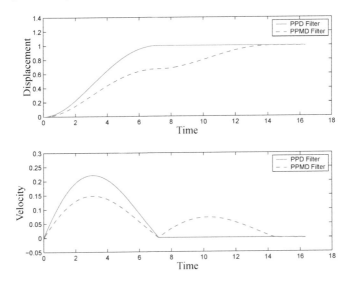

Figure 2.9: System Response to the PPD and PPMD Time-Delay Filters

Figure 2.9 illustrates the evolution of the angular displacement and the angular velocities of the system where it is assumed that the inertia of the system is exactly known. The solid line and the dashed line correspond to the PPD and the PPMD time-delay filters respectively. It is clear that both the filters generate zero residual energy at the end of the maneuver. However, the PPMD filter takes twice as long to complete the maneuver as the PPD filter.

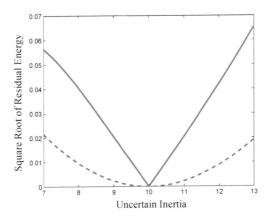

Figure 2.10: Time-Delay Filter Sensitivity

To study the variation in the performance of the PPD and PPMD time-delay filters in the presence of errors in estimated inertia of the system, the square root of the residual energy of the systems driven by filters designed for the nominal systems are determined and plotted in Figure 2.10. The solid line which corresponds to the PPD filter illustrates that the performance of the PPD filter deteriorates rapidly as the error in the estimated inertia increases. The results of the PPMD filter illustrated by the dashed line confirm the insensitivity of the PPMD filter in the vicinity of the nominal model of the system.

2.2   Proportional Plus User Selected Multiple Delay Control

In the time-delay filter designed in Section 2.1, the time-delays appears non-linearly in the equations, which increases the difficulty of arriving at closed-form solutions for anything except a one or two mode system. The time-delay of the multiple time-delay controller is a function of the natural frequencies of the system. To eliminate this constraint, we propose, in this section, a class of time-delay filters where the time-delay is selected by the user, thus giving greater latitude to the designer. The unknown gains of the time-delayed signals, which appear linearly, can be easily solved for. Consider an underdamped second-order plant and a two-time-delay filter (Figure 2.12) where T is selected by the user. We choose the second time-delay to be twice that of the first although this is not necessary. We need to determine A_0, A_1, and A_2 to cancel the poles of the system. We require that the zeros of the time-delay filter be coincident with the system poles. The equations

$$A_0 + A_1 \exp(-\sigma T)\cos(\omega T) + A_2 \exp(-2\sigma T)\cos(2\omega T) = 0 \tag{2.50}$$

$$A_1 \exp(-\sigma T)\sin(\omega T) + A_2 \exp(-2\sigma T)\sin(2\omega T) = 0 \tag{2.51}$$

where $\sigma = -\zeta_i \omega_i$ and $\omega = \omega_i \sqrt{1 - \zeta_i^2}$ are used to solve for A_0, A_1, and A_2. The requirement that the final value of the prefiltered signal be the same at the reference signal leads to the constraint:

$$A_0 + A_1 + A_2 = 1. \tag{2.52}$$

Solving Equations (2.50), (2.51), and (2.52), we have

$$A_0 = \frac{\exp(-2\sigma T)}{\exp(-2\sigma T) - 2\exp(-\sigma T)\cos(\omega T) + 1} \tag{2.53}$$

$$A_1 = \frac{-2\exp(-\sigma T)\cos(\omega T)}{\exp(-2\sigma T) - 2\exp(-\sigma T)\cos(\omega T) + 1} \tag{2.54}$$

$$A_2 = \frac{1}{\exp(-2\sigma T) - 2\exp(-\sigma T)\cos(\omega T) + 1} \tag{2.55}$$

Figure 2.11 illustrates the variation of the log of the absolute value of the transfer

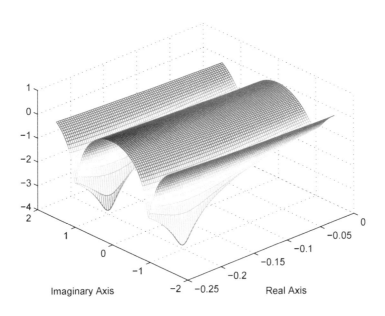

Figure 2.11: Log Magnitude of Time-Delay Filter Transfer Function

function of the time-delay filter over the s-domain. It is clear that the function goes to zero at the estimated location of the poles of the system. However, there is a rapid increase in the magnitude as one moves away from the poles of the system. Thus, any error in estimated frequency or damping, will lead to an oscillatory response to a step input. To build robustness into the time-delay filter, we require that, in addition to the cancelation of the poles of the system, the variation of Equations (2.50) and (2.51) with respect to frequency be zero. This leads to a total of five equations with three unknowns. We therefore introduce two more delays in the controller whose transfer function is now

$$A_0 + A_1 \exp(-sT) + A_2 \exp(-2sT) + A_3 \exp(-3sT) + A_4 \exp(-4sT) \quad (2.56)$$

The gains of the time-delay filter can be determined such that the following five equations are satisfied:

$$A_0 + A_1 \exp(-\sigma T)\cos(\omega T) + A_2 \exp(-2\sigma T)\cos(2\omega T) +$$
$$A_3 \exp(-3\sigma T)\cos(3\omega T) + A_4 \exp(-4\sigma T)\cos(4\omega T) = 0 \quad (2.57)$$
$$A_0 + A_1 \exp(-\sigma T)\sin(\omega T) + A_2 \exp(-2\sigma T)\sin(2\omega T) +$$

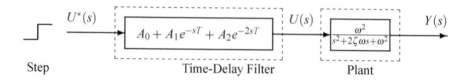

Step   Time-Delay Filter   Plant

Figure 2.12: User Selected Time-Delay Controlled System

$$A_3 \exp(-3\sigma T)\sin(3\omega T) + A_4 \exp(-4\sigma T)\sin(4\omega T) = 0 \qquad (2.58)$$

$$\frac{d}{d\omega}(A_0 + A_1 \exp(-\sigma T)\cos(\omega T) + A_2 \exp(-2\sigma T)\cos(2\omega T) +$$

$$A_3 \exp(-3\sigma T)\cos(3\omega T) + A_4 \exp(-4\sigma T)\cos(4\omega T)) = 0 \qquad (2.59)$$

$$\frac{d}{d\omega}(A_0 + A_1 \exp(-\sigma T)\sin(\omega T) + A_2 \exp(-2\sigma T)\sin(2\omega T) +$$

$$A_3 \exp(-3\sigma T)\sin(3\omega T) + A_4 \exp(-4\sigma T)\sin(4\omega T)) = 0 \qquad (2.60)$$

$$A_0 + A_1 + A_2 + A_3 + A_4 = 1. \qquad (2.61)$$

It can be shown that two, two-time-delay filters in series are equivalent to the resulting filter. The transfer function of the resulting time-delay filter is

$$\frac{U(s)}{R(s)} = \left(\frac{\exp(-2\sigma T) - 2\exp(-\sigma T)\cos(\omega T)\exp(-sT) + \exp(-2sT)}{\exp(-2\sigma T) - 2\exp(-\sigma T)\cos(\omega T) + 1} \right)^2 \qquad (2.62)$$

It is to be noted that we will arrive at the same solution if we force the variations of Equation (2.56) with respect to σ, to zero. Thus, the robustness is concurrently increased with respect to errors in estimated frequency and damping.

2.2.1   Signs of the Time-Delay Gains

It might sometimes be required to limit the magnitude of the time-delay gains to be positive, while satisfying Equation (2.52). This is desirable from a practical standpoint as we would not require the actuator to track large steps, which could occur if the time-delay gains were unconstrained. Hence, we require from Equation (2.54), that the time-delay T satisfy

$$\frac{\pi}{2} < \omega T < \frac{3\pi}{2} \qquad (2.63)$$

which is equivalent to

$$\frac{T_s}{4} < T < \frac{3T_s}{4} \qquad (2.64)$$

where T_s, is the system period. To illustrate that Equations (2.53), (2.54) and (2.55) represent a generalization of the time-delay control [10], we study the case when $T = T_s/4$. This leads to

$$A_0 = \frac{\exp\left(\frac{\zeta\pi}{\sqrt{1-\zeta^2}}\right)}{\exp\left(\frac{\zeta\pi}{\sqrt{1-\zeta^2}}\right)+1} \tag{2.65}$$

$$A_1 = 0 \tag{2.66}$$

$$A_2 = \frac{1}{\exp\left(\frac{\zeta\pi}{\sqrt{1-\zeta^2}}\right)+1} \tag{2.67}$$

and the filter degenerates to the two-impulse shaped-input controller. Further, when $T = T_s/2$, we have

$$A_0 = \frac{\exp\left(\frac{2\zeta\pi}{\sqrt{1-\zeta^2}}\right)}{\exp\left(2\frac{\zeta\pi}{\sqrt{1-\zeta^2}}\right)+2\exp\left(\frac{\zeta\pi}{\sqrt{1-\zeta^2}}\right)+1} \tag{2.68}$$

$$A_1 = \frac{2\exp\left(\frac{\zeta\pi}{\sqrt{1-\zeta^2}}\right)}{\exp\left(2\frac{\zeta\pi}{\sqrt{1-\zeta^2}}\right)+2\exp\left(\frac{\zeta\pi}{\sqrt{1-\zeta^2}}\right)+1} \tag{2.69}$$

$$A_2 = \frac{1}{\exp\left(2\frac{\zeta\pi}{\sqrt{1-\zeta^2}}\right)+2\exp\left(\frac{\zeta\pi}{\sqrt{1-\zeta^2}}\right)+1} \tag{2.70}$$

which is equivalent to the three impulse shaped-input controller. It can also be shown that when $T = 3T_s/4$ the resulting filter is the same as when $T = T_s/4$ except that the system response delay is increased by one period.

Figure 2.13 illustrates the variations of the gains of the time-delay filter as a function of the user selected time-delay. It can be seen from the figure that there exist periodic intervals over which all the gains are positive. The figure is generated for a system mode with a frequency of 1 and a damping ratio ζ of 0.1.

2.2.2 Periodicity

The time-delay filter leads to an infinite number of zeros, one pair of which cancels a pair of poles of the system. To determine the zeros of the time-delay filter as a function of the user selected time T, we write Equations (2.50) and (2.51) with the appropriate values of A_0, A_1, and A_2 to arrive at

$$\exp(-2\sigma_s T) - 2\exp(-\sigma_s T)\cos(\omega_s\sqrt{1-\zeta_i^2}T)\exp(-\sigma T)\cos(\omega T)$$
$$+\exp(-2\sigma T)\cos(2\omega T) = 0 \tag{2.71}$$

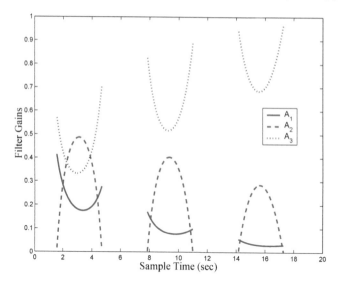

Figure 2.13: Time-Delay Filter Gains

and

$$-2\exp(-\sigma_s T)\cos(\omega_s\sqrt{1-\zeta_i^2}T)\exp(-\sigma T)\sin(\omega T)+\exp(-2\sigma T)\sin(2\omega T)=0 \tag{2.72}$$

From Equation (2.71), which is quadratic in $\exp(-\sigma T)$ we have

$$\exp(-\sigma T)=\exp(-\sigma_s T)\frac{\cos(\omega_s\sqrt{1-\zeta_i^2}T)\cos(\omega T)\pm\Phi}{\cos(2\omega T)} \tag{2.73}$$

and from Equation (2.72), we have

$$\begin{aligned}-2\cos(\omega_s\sqrt{1-\zeta_s^2}T)+\\\frac{\cos(\omega_s\sqrt{1-\zeta_i^2}T)\cos(\omega T)\pm\Phi}{\cos(2\omega T)}\sin(2\omega T)=0\end{aligned} \tag{2.74}$$

where

$$\Phi=\left[\cos^2(\omega_s\sqrt{1-\zeta_i^2}T)\cos^2(\omega T)-\cos(2\omega T)\right]^{0.5}$$

Solving Equations (2.73) and (2.74), we have

$$\sigma=\sigma_s \tag{2.75}$$

and

$$\omega = \pm\omega_s\sqrt{1-\zeta_s^2} + \frac{2n\pi}{T} \qquad (2.76)$$

Thus, the smaller the value of T, the larger are the intervals between zeros of the time-delay controller. In the special case when $T = \pi/\omega_s\sqrt{1-\zeta_s^2}$ the imaginary parts of the zeros of the filter are odd multiples of the imaginary parts of the system poles.

The time-delay filter can be mapped into the z-domain, where one can show that the same objective is met, i.e., cancelation of the poles of the system by the time-delay filter zeros. However, in the z-domain, we would solve a polynomial equation to arrive at the zeros of the filter, which are finite in number, unlike in the s-domain where we solve a transcendental equation to results in an infinite set of zeros of the filter.

2.3 Time-Delay Control of Multi-Mode Systems

The technique for the design of time-delay filters for single-mode systems can be extended to multiple-mode systems by cascading the time-delay filters designed for each mode. This will not lead to the smallest number of delay times to cancel the dynamics of the two underdamped modes. Figure 2.14 illustrates the cascaded time-delay filter for a multi-mode system.

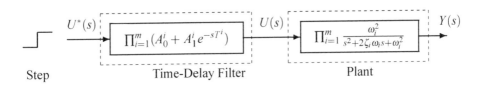

Step   Time-Delay Filter   Plant

Figure 2.14: Multi Mode Time-Delay Controlled System

where

$$A_0^i = \frac{\exp\left(\frac{\zeta_i \pi}{\sqrt{1-\zeta_i^2}}\right)}{\exp\left(\frac{\zeta_i \pi}{\sqrt{1-\zeta_i^2}}\right)+1} \qquad (2.77)$$

$$A_1^i = \cfrac{1}{\exp\left(\cfrac{\zeta_i \pi}{\sqrt{1-\zeta_i^2}}\right) + 1} \tag{2.78}$$

$$T^i = \frac{\pi}{\omega_i \sqrt{1 - \zeta_i^2}}. \tag{2.79}$$

2.3.1   Concurrent Time-Delay Filter Design for Multi-Mode Systems

Time-delay filters for multi-mode system can be designed by selecting the parameters of the filter to concurrently satisfy the pole cancelation constraints. Analytical solutions to multi-mode systems can rarely be derived. This prompts posing an optimization problem to numerically design the time-delay filter. Consider a system with m underdamped modes (poles). The time-delay filter can be parameterized as:

$$\sum_{i=0}^{N} A_i e^{-sT_i} \tag{2.80}$$

where $T_0 = 0$. The number of delays N is a function of the system to be controlled. The optimization problem can be stated as:

$$\min_{A_i, T_i} \ J = T_N$$

subject to

$$\sum_{i=0}^{N} A_i \exp(\zeta_j \omega_j T_i) \cos(\omega_j \sqrt{1 - \zeta_j^2} T_i) = 0 \ \forall j = 1, 2, \ldots, m$$

$$\sum_{i=0}^{N} A_i \exp(\zeta_j \omega_j T_i) \sin(\omega_j \sqrt{1 - \zeta_j^2} T_i) = 0 \ \forall j = 1, 2, \ldots, m$$

$$\sum_{i=0}^{N} A_i = 1$$

$$T_i > T_{i-1} > 0 \ \forall i = 1, 2, \ldots, N$$

$$0 < A_i < 1 \ \forall i = 1, 2, \ldots, N$$

This formulation results in an impractical solution if A_i's are not constrained. It can lead to large values for A_i which can result in a reference profile which demands large inputs if the constraints are $|A_i| < C$, for $C > 1$. By enforcing the constraint $0 < A_i < 1$, the resulting staircase reference profile demands a smaller peak control input as compared to the unfiltered reference input.

Example 2.2: The closed-loop dynamics of a gantry crane characterized by the model:

$$9150\ddot{\theta} + 80000\ddot{\phi} = u \tag{2.82}$$

$$80000\ddot{\theta} + 800000\ddot{\phi} + 784800\phi = 0 \tag{2.83}$$

subject to the collocated PD controller:

$$u = -600(\theta - 1) - 100\dot{\theta} \tag{2.84}$$

is characterized by closed-loop poles located at:

$$s = -0.0049 \pm 0.2488i \text{ and } -0.0386 \pm 2.8745i.$$

Time-delay filters to cancel the two modes separately are:

$$G_1(s) = 0.5105 + 0.4895\exp(-1.0929s) \tag{2.85}$$
$$G_2(s) = 0.5154 + 0.4846\exp(-12.6263s) \tag{2.86}$$

which when convolved lead to the time-delay filter

$$G(s) = G_1(s)G_2(s) = 0.2631 + 0.2523\exp(-1.0929s)$$
$$+ 0.2474\exp(-12.63s) + 0.2372\exp(-13.72s). \tag{2.87}$$

The solid line in Figure 2.15 illustrates the staircase reference profile which induces the nonoscillatory motion of the pendulum illustrated in Figure 2.16 by the solid line.

Designing a time-delay filter parameterized using two time-delays and solving for the parameters of the filter to concurrently cancel the two pairs of underdamped poles of the closed-loop system results in the filter:

$$G(s) = 0.5111 + 0.3240\exp(-12.0961s) + 0.1649\exp(-13.6687s) \tag{2.88}$$

Dashed lines in Figures 2.15 and 2.16 illustrate the shaped reference profile and the corresponding pendular displacement. The reduction in maneuver time for this example is minimal. The concurrent design however, uses fewer time-delays in shaping the reference profile.

2.3.2 User Selected Time-Delay

The design of the time-delay filters in Section 2.2 was simplified as the unknowns appeared linearly in the constraint equations. This concept can easily be extended to the design of time-delay filters for systems with more than one mode of vibration. The easiest design would be to connect the time-delay filters for each of the modes in series. This, however, will not lead to the minimum number of delays in the

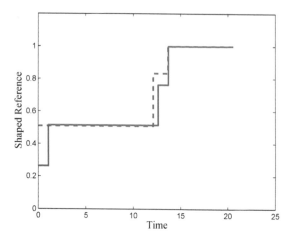

Figure 2.15: Shaped-Reference Profile

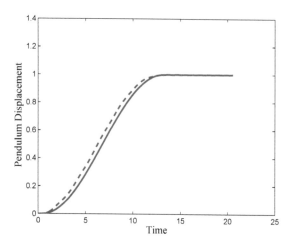

Figure 2.16: Motion and Control Profile

controller. In addition, the system response will not be the shortest possible. We can reformulate the constraints so that the selection of the gains of the time-delay filter satisfy the constraints simultaneously. To control one mode, we required a two-time-delay filter, and we need to add two more time-delays for the control of each additional mode. The transfer function of the time-delay filter for a m-mode system is

$$\sum_{i=0}^{2m} A_i \exp(-isT) \tag{2.89}$$

The constraints A_i have to satisfy the constraints

$$\sum_{i=0}^{2m} A_i \exp(-i\sigma_j T)\cos(i\omega_j T) = 0 \tag{2.90}$$

$$\sum_{i=0}^{2m} A_i \exp(-i\sigma_j T)\sin(i\omega_j T) = 0 \tag{2.91}$$

for j = 1 to m, where σ_j is the damping constant and ω_j is the damped natural frequency of the jth mode. We also require

$$\sum_{i=0}^{2m} A_i = 1 \tag{2.92}$$

to ensure that the zero frequency of the time-delay filter is unity. We now have 2m + 1 equations in 2m + 1 unknowns, which can be solved easily as the unknowns appear linearly in the equations. The matrix equation to be solved is

$$\begin{bmatrix} 1 & e^{-\sigma_1 T}\cos(\omega_1 T) & e^{-2\sigma_1 T}\cos(2\omega_1 T) & \cdots & e^{-2m\sigma_1 T}\cos(2m\omega_1 T) \\ 0 & e^{-\sigma_1 T}\sin(\omega_1 T) & e^{-2\sigma_1 T}\sin(2\omega_1 T) & \cdots & e^{-2m\sigma_1 T}\sin(2m\omega_1 T) \\ \cdots & \cdots & & \cdots & \cdots \\ 1 & e^{-\sigma_m T}\cos(\omega_m T) & e^{-2\sigma_m T}\cos(2\omega_m T) & \cdots & e^{-2m\sigma_m T}\cos(2m\omega_m T) \\ 0 & e^{-\sigma_m T}\sin(\omega_m T) & e^{-2\sigma_m T}\sin(2\omega_m T) & \cdots & e^{-2m\sigma_m T}\sin(2m\omega_m T) \\ 1 & 1 & 1 & \cdots & 1 \end{bmatrix}$$

$$\begin{Bmatrix} A_0 \\ A_1 \\ \vdots \\ A_{2m-2} \\ A_{2m-1} \\ A_{2m} \end{Bmatrix} = \begin{Bmatrix} 0 \\ 0 \\ \vdots \\ 0 \\ 0 \\ 1 \end{Bmatrix}. \tag{2.93}$$

For certain values of T, the above matrix might become singular. In such cases, the solution can be obtained by using the pseudo-inverse technique. The pseudo-inverse solution will be exact for a row deficiency, but for a column rank deficiency, a least square approximation solution results, which does not satisfy the constraints. In this case, we need to select a different T so that column rank deficiency is avoided.

2.3.3 Minimum Time-Delay

In Section 2.3.2, a set of linear equations had to be solved to arrive at the time-delay filter. We choose to modify the proposed technique such that all the time-delay gains are positive. When the gains are unconstrained, the resulting control input could *ring* the unmodeled dynamics of the system. This is undesirable for flexible structures. With all the time-delay gains being positive, we are assured that they will lie in the range 0–1. We reformulate the design process so as to minimize the time-delay T, i.e.,

$$\min_{A_i,T} J = T^2 \tag{2.94a}$$

subject to $\qquad\qquad\qquad\qquad\qquad\qquad\qquad\qquad\qquad\qquad$ (2.94b)

$$\sum_{i=0}^{2m} A_i \exp(-i\sigma_j T) \cos(i\omega_j T) = 0, \quad \forall j = 1,2,..,m \tag{2.94c}$$

$$\sum_{i=0}^{2m} A_i \exp(-i\sigma_j T) \sin(i\omega_j T) = 0, \quad \forall j = 1,2,..,m \tag{2.94d}$$

$$\sum_{i=0}^{2m} A_i = 1 \tag{2.94e}$$

$$A_i \geq 0 \quad \forall i \tag{2.94f}$$

Solution of the optimization problem leads to the desired time-delay filter. The use of multiple time-delay filters in cascade adds robustness to errors in system parameters.

2.4   Jerk Limited Input Shapers

Jerk is the time rate of change of acceleration, i.e., the derivative of the control input and is a measure of the impact level. Impact in mechanical systems is manifested in the form of noise and reduces life due to fatigue. For various flexible structures, higher level of jerk correlates to exciting the higher modes of the system which is commonly referred to as *ringing*. The input shaping/time-delay filter design requires information of the natural frequency and damping ratio of specific modes to be controlled. If it is necessary to roll off the energy over the high frequency spectrum, which includes unmodeled modes, additional constraints need to be included in the filter design. It is interesting to note that human ballistic motion which refer to movement initiated by robust muscular contractions followed by relaxed muscles while the momentum carries the limbs to their final destination, has been shown to be jerk limited.

This section will address the problem of jerk limited input shapers for prefiltering command inputs to vibratory systems without rigid body modes. The papers by Muenchhof and Singh [12, 13] addressed the problem of design of control profiles for systems with rigid body modes. This section will start by addressing the design of time-delay filters where the delay time and the gains of the delayed signals are parameters to be solved for. This will be followed by the presentation of a general concept to design input shapers by including additional dynamics to the time-delay filter such as harmonic oscillators and first order dynamics to permit smooth ramping up and ramping down of control profiles.

2.4.1 Undamped Systems

This section deals with the design of jerk limited time-delay filter (Input Shaper) which is schematically represented in Figure 2.17. The development which follows is for a single mode system, but can be easily extended for multiple mode systems [14].

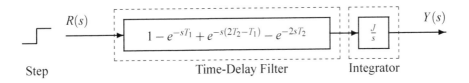

Step   Time-Delay Filter   Integrator

Figure 2.17: Jerk Limited Time-Delay Controlled System

The transfer function of the filter shown in Figure 2.17 without the integrator element is

$$G(s) = (1 - \exp(-sT_1) + \exp(-s(2T_2 - T_1)) - \exp(-2sT_2)). \qquad (2.95)$$

The output of the transfer function G(s) subject to a unit step input is shown in Figure 2.18 and its time integral is represented as

Figure 2.18: Parameterized Control Profile

$$y(t) = J(t - (t-T_1)\mathcal{H}(t-T_1) + (t - (2T_2 - T_1))\mathcal{H}(t - (2T_2 - T_1)) - (t - 2T_2)\mathcal{H}(t - 2T_2))),$$

$$\text{(2.96)}$$

where J is the permissible jerk and $\mathcal{H}()$ is the Heaviside step function. $y(t)$ should equal 1 at steady state for a DC (zero frequency) gain of unity which results in the constraint equation

$$y(2T_2) = J(2T_2 - (2T_2 - T_1) + (2T_2 - (2T_2 - T_1))) = 1, \qquad \text{(2.97)}$$

or

$$T_1 = \frac{1}{2J}. \qquad \text{(2.98)}$$

which implies that the first switch T_1 is only a function of the permitted jerk. To cancel the undamped poles of the system, we require a pair of zeros of the time-delay filter to cancel the poles of the system. This results in the constraint equations

$$1 - \cos(\omega T_1) + \cos(\omega(2T_2 - T_1)) - \cos(2\omega T_2) = 0 \qquad \text{(2.99)}$$

and

$$-\sin(\omega T_1) + \sin(\omega(2T_2 - T_1)) - \sin(2\omega T_2) = 0 \qquad \text{(2.100)}$$

These two constraint equations are satisfied if

$$\sin(\omega T_2) = \sin(\omega(T_2 - T_1)). \qquad \text{(2.101)}$$

Substituting Equation (2.98) into Equation (2.101), and simplifying we have

$$\tan(\omega T_2) = -\cot\left(\frac{\omega}{4J}\right) \qquad \text{(2.102)}$$

which results in the closed-form solutions

$$T_2 = \frac{(2n+1)\pi}{2\omega} + \frac{1}{4J}. \qquad \text{(2.103)}$$

For specific values of ω and J, T_1 can equal T_2, which corresponds to the first and the second switch collapsing. From Equations (2.98) and (2.101), this corresponds to

$$\sin(\omega T_2) = 0, \Rightarrow \cos\left(\frac{\omega}{4J}\right) = 0 \qquad \text{(2.104)}$$

or

$$\frac{\omega}{4J} = (2m+1)\frac{\pi}{2}, \; m=1,2,3... \qquad \text{(2.105)}$$

So, for a given J or ω, we can solve for ω or J respectively for which T_1 and T_2 are equal, which corresponds to a simple ramp input to the system.

Figures 2.19 and 2.20 illustrate the variation of the switch times and the final time of the time-delay filter as a function of varying jerk and frequency, respectively. It is clear from Figure 2.20, that the first and second switches coincide, which corresponds to the solid line intersecting the dashed line. Figure 2.20 is generated for J = 3, for which we have from Equation (2.105), ω = 6π, 18π, 30π, for which the switches collapse, which corroborates the results in Figure 2.20.

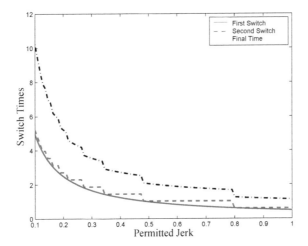

Figure 2.19: Switch Time Variation vs. Jerk for $\omega = 15$

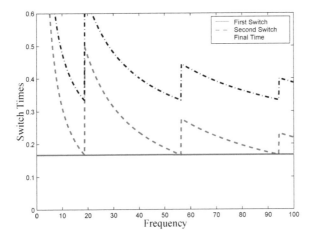

Figure 2.20: Switch Time Variation vs. Frequency

2.4.2 Damped Systems

Unlike the jerk limited time-delay filters for undamped systems, the jerk limited time-delay filter for damped systems cannot be solved in closed form. The problem can be solved numerically by posing the design as an optimization problem.

The jerk limited time-delay filter is parameterized as

$$G(s) = \frac{J}{s}(1 - \exp(-sT_1) + \exp(-sT_2) - \exp(-sT_3)). \qquad (2.106)$$

To satisfy the requirement that the final value of the jerk limited time-delay filter be unity when it is driven by an unit step input results in the constraint equation

$$y(T_3) = J(T_3 + T_1 - T_2) = 1. \tag{2.107}$$

To cancel the damped poles of the system at $s = \sigma \pm j\omega$, we require a pair of zeros of the time-delay filter to be collocated with the plant's damped poles. This results in the constraint equations

$$1 - e^{-\sigma T_1} \cos(\omega T_1) + e^{-\sigma T_2} \cos(\omega T_2) - e^{-\sigma T_3} \cos(\omega T_3) = 0 \tag{2.108}$$

and

$$-e^{-\sigma T_1} \sin(\omega T_1) + e^{-\sigma T_2} \sin(\omega T_2) - e^{-\sigma T_3} \sin(\omega T_3) = 0. \tag{2.109}$$

The optimization problem can be stated as minimization of T_3 subject to the three equality constraints given by Equations (2.107), (2.108), and (2.109).

2.5   Robust Jerk Limited Time-Delay Filter

Most systems have errors in estimated damping and natural frequencies which can result in significant residual errors when a rest-to-rest maneuver is performed. It is therefore imperative to design filters which can handle uncertainties in estimated model parameters. There are multiple approaches to achieve robustness. The simplest includes reducing the sensitivity of the residual energy of the modes to uncertain parameters, based on the nominal values of uncertain parameters. If bounds and distributions of the uncertain parameters are available to the designer, the minimax approach proposed by Singh [15] can be used to arrive at filters which minimize the maximum magnitude of the residual energy in the domain of interest. This will be presented in detail in Chapter 5. In this development, robustness is achieved by placing multiple zeros of the time-delay filter at the location of the uncertain poles of the plant.

The added requirement of robustness results in a filter with increased number of parameters to be determined. The approach for the design of robust jerk limited time-delay filters is developed for damped systems with the knowledge that the undamped systems are a subset of the damped system. The robust jerk limited time-delay filter is parameterized as

$$G(s) = \frac{J}{s}(1 - \exp(-sT_1) + \exp(-sT_2) - \exp(-sT_3) + \exp(-sT_4) - \exp(-sT_5)). \tag{2.110}$$

To satisfy the requirement that the final value of the jerk limited time-delay filter be unity when it is driven by an unit step input results in the constraint equation

$$y(T_5) = J(T_5 - T_4 + T_3 - T_2 + T_1) = 1. \tag{2.111}$$

To cancel the damped poles of the system at $s = \sigma \pm j\omega$, we require a pair of zeros of the time-delay filter to cancel the damped poles of the system. This results in the constraint equations

$$1 - e^{-\sigma T_1} \cos(\omega T_1) + e^{-\sigma T_2} \cos(\omega T_2) - e^{-\sigma T_3} \cos(\omega T_3) +$$

$$e^{-\sigma T_4} \cos(\omega T_4) - e^{-\sigma T_5} \cos(\omega T_5) = 0 \qquad (2.112)$$

and

$$-e^{-\sigma T_1} \sin(\omega T_1) + e^{-\sigma T_2} \sin(\omega T_2) - e^{-\sigma T_3} \sin(\omega T_3) +$$

$$e^{-\sigma T_4} \sin(\omega T_4) - e^{-\sigma T_5} \sin(\omega T_5) = 0. \qquad (2.113)$$

Robustness to uncertainties in natural frequencies or damping ratios is achieved by placing a second pair of zeros of the time-delay filter at the estimated location of the oscillatory poles of the system, which results in the equations

$$-T_1 e^{-\sigma T_1} \sin(\omega T_1) + T_2 e^{-\sigma T_2} \sin(\omega T_2) - T_3 e^{-\sigma T_3} \sin(\omega T_3) +$$

$$T_4 e^{-\sigma T_4} \sin(\omega T_4) - T_5 e^{-\sigma T_5} \sin(\omega T_5) = 0 \qquad (2.114)$$

and

$$-T_1 e^{-\sigma T_1} \cos(\omega T_1) + T_2 e^{-\sigma T_2} \cos(\omega T_2) - T_3 e^{-\sigma T_3} \cos(\omega T_3) +$$

$$T_4 e^{-\sigma T_4} \cos(\omega T_4) - T_5 e^{-\sigma T_5} \cos(\omega T_5) = 0. \qquad (2.115)$$

The optimization problem can now be stated as the minimization of T_5 subject to the constraint given by Equations (2.111) through (2.115).

To illustrate the reduced sensitivity of the residual energy to variations in the frequency, the response of the system was studied for various values of model frequencies with a filter designed for a frequency of 15 rad/sec and a permitted jerk of 4. Figure 2.21 illustrates the improved performance of the robust jerk limited time-delay filter.

2.6 Jerk Limited Time-Delay Filters for Multi-Mode Systems

The proposed approach can be used for the control of systems with multiple underdamped modes. A generic formulation is developed below. The number of parameters to be optimized for can be reduced for undamped systems by exploiting the symmetric characteristics of the time-delay filter. The transfer function of the time-delay filter is

$$G(s) = \frac{J}{s} \sum_{i=0}^{N} (-1)^i \exp(-sT_i) \qquad (2.116)$$

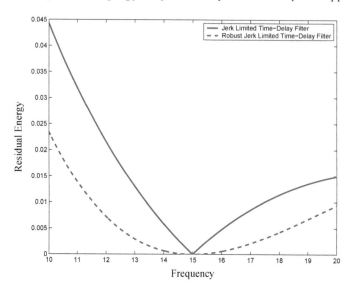

Figure 2.21: Sensitivity Curve

where $T_0 = 0$ and N is an odd number. The unknowns (T_i) have to satisfy the constraint equation

$$\sum_{i=1}^{N}(-1)^{i+1}T_i = \frac{1}{J} \tag{2.117}$$

which satisfies the requirement that the final value of the output of the filter when it is subject to an unit step input is unity. To cancel the undamped or underdamped poles at

$$s_k = \sigma_k \pm j\omega_k \text{ for } k = 1, 2, 3, \dots \tag{2.118}$$

the following constraints have to be satisfied

$$\sum_{i=1}^{N}(-1)^i \exp(-\sigma_k T_i)\cos(\omega_k T_i) = 0 \text{ for } k = 1, 2, 3, \dots \tag{2.119}$$

and

$$\sum_{i=1}^{N}(-1)^i \exp(-\sigma_k T_i)\sin(\omega_k T_i) = 0 \text{ for } k = 1, 2, 3, \dots \tag{2.120}$$

The optimal solution is one which satisfies all the constraints and minimizes T_N.

To desensitize the filter to errors in estimated damping or frequency, the following constraint equations:

$$\sum_{i=1}^{N}(-1)^i T_i \exp(-\sigma_k T_i)\sin(\omega_k T_i) = 0 \text{ for } k = 1, 2, 3, \dots \tag{2.121}$$

and

$$\sum_{i=1}^{N}(-1)^i T_i \exp(-\sigma_k T_i)\cos(\omega_k T_i) = 0 \text{ for } k = 1, 2, 3, ... \quad (2.122)$$

which are the derivatives of Equations (2.119) and (2.120) with respect to σ_k or ω_k, are added to the optimization problem. It can be seen that desensitizing the filter with respect to damping simultaneously desensitizes the filter to the frequency as well.

The design of jerk limited time-delay filters for user specified time-delays follows the process proposed described in Section 2.3.2. It is clear that additional number of delays are required since the delay times are no longer variables in the optimization process.

To illustrate the design of multi-mode jerk limited input shapers, consider the system

$$\frac{y(s)}{u(s)} = \frac{225}{s^4 + 34s^2 + 225} \quad (2.123)$$

which is characterized by two modes with frequencies 3 and 5. For a jerk constraint of 3, the jerk limited input shaper is designed. The dashed line and the solid line in Figure 2.22 illustrates the response of the system to a step input and the shaped input respectively. It is clear that the residual vibration of the two modes is eliminated after shaping the input.

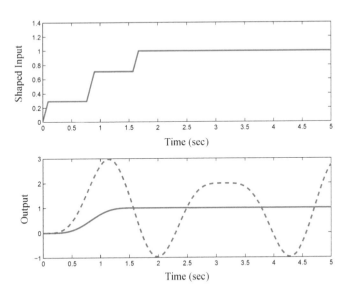

Figure 2.22: Shaped Input and Comparison of System Response

2.7   Filtered Input Shapers

The technique presented in this section, where an integrator is concatenated to a time-delay filter to satisfy the constraint of jerk limited filter design can be extended by cascading other transfer functions such as that of first-order systems, harmonic systems, and so forth.

2.7.1   First-Order Filtered Input Shaper

Instead of using an integrator in conjunction with a time-delay filter to account for the limit on the permitted jerk, one can concatenate a first-order filter to a time-delay filter to generate a smooth input which can then be used to drive a time-delay filter designed to cancel the underdamped poles of the system of interest. Figure 2.23 illustrates the proposed filter structure where T is a user selected time-delay which in the case of a discrete time implementation, can be an integral multiple of the sampling interval.

$$r(s) \longrightarrow \boxed{-\frac{e^{aT}}{1-e^{aT}} + \frac{1}{1-e^{aT}}e^{-sT}} \longrightarrow \boxed{\frac{a}{s+a}} \longrightarrow \boxed{A_0 + A_1 e^{-sT_1}} \longrightarrow u(s)$$

Figure 2.23: First-Order Filtered Time-Delay Filter

2.7.2   Sinusoid Filtered Input Shaper

Filtering with a transfer function of a scaled sinusoid results in an input which emulates a step input but with zero initial and final slopes. The scaling of the sinusoid transfer function is to satisfy the requirement that the DC gain of the transfer function is unity. The sinusoid filtered time-delay filter is illustrated in Figure 2.24, which can be rewritten as shown in Figure 2.25.

$$r(s) \longrightarrow \boxed{A_0 + A_1 e^{-sT_1} + A_2 e^{-sT_2} + A_3 e^{-sT_3}} \longrightarrow \boxed{\frac{\omega^2}{s^2+\omega^2}} \longrightarrow u(s)$$

Figure 2.24: Sinusoid Filtered Time-Delay Filter

Here the first time-delay filter cancels the oscillatory response of the scaled harmonic oscillator. This truncated harmonic response is then input to the second time-delay filter which is designed to cancel the oscillatory mode of the system. Figure 2.26 illustrates the control profile. The benefit of this approach can be gauged

$r(s) \longrightarrow \boxed{0.5 + 0.5e^{-s\pi/\omega}} \longrightarrow \boxed{\dfrac{\omega^2}{s^2 + \omega^2}} \longrightarrow \boxed{A_0 + A_1 e^{-sT_1}} \longrightarrow u(s)$

Figure 2.25: Sinusoid Filtered Time-Delay Filter

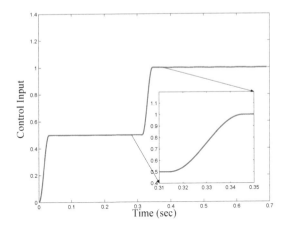

Figure 2.26: Sinusoid Filtered, Input Shaped Control Input

from the frequency response plots of the sinusoid filtered time-delay filter. Figure 2.27 illustrates the frequency response plots of the time-delay filter, jerk limited time-delay filter and a sinusoid filtered time-delay filter. The sinusoid filtered time-delay filter has been designed such that the maximum jerk of the control profile is equal to the maximum permitted jerk. It can easily be seen that the magnitude plots of the sinusoid filtered time-delay filters rolls off much more rapidly compared to the time-delay filter and the jerk limited time-delay filter. Thus, this input will not significantly excite the unmodeled dynamics.

2.7.3   Jerk Limits

Consider a part of the sinusoid filtered time-delay filter illustrated in Figure 2.28. The output p of the time-delay filter subject to a unit step input is

$$p(t) = \sin^2(\omega/2t) + \sin^2(\omega/2(t - \pi/\omega))\mathscr{H}\left(t - \frac{\pi}{\omega}\right) \qquad (2.124)$$

and the rate of change of p which is the jerk is

$$\dot{p}(t) = \frac{\omega}{2}\sin(\omega t) + \frac{\omega}{2}\sin(\omega(t - \pi/\omega))\mathscr{H}\left(t - \frac{\pi}{\omega}\right) \qquad (2.125)$$

which implies that the maximum magnitude of the jerk is $\frac{\omega}{2}$ and occurs at time $t = \frac{\pi}{2\omega}$. This is the upper bound for the jerk. It can be seen that the jerk is zero at the start and the end of the maneuver which results in a very practical control profile. When the signal p is passed through the second time-delay filter, based on the damping present in the oscillatory pole to be canceled, the jerk can lie in the limit

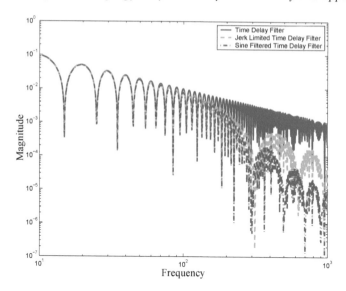

Figure 2.27: Frequency Response

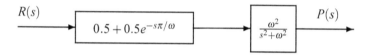

Figure 2.28: Time-Delay Filter

$$\frac{\omega}{4} \leq \text{Maximum Jerk} \leq \frac{\omega}{2}. \tag{2.126}$$

If the pole to be canceled is undamped the maximum jerk is $\frac{\omega}{4}$ since A_0 and A_1 are equal to 0.5. When the poles to be canceled contain damping, A_0 is greater than 0.5 and A_1 is less than 0.5, resulting in the maximum jerk lying in the range specified by Equation (2.126).

This constraint is valid when the time-delay filter is designed to cancel the unwanted underdamped pole. However, if the underdamped pole has to be controlled using a robust time-delay filter, the limits on the jerk changes, since the robust time-delay filter uses smaller gains.

Example 2.3:

To illustrate the benefit of using the filtered input shaper, consider a two-mass system shown in Figure 2.29, which includes the dominant mass m which is driven by a force transmitted through the spring of stiffness k. A secondary mass m_s connected by a stiff spring is driven by the motion of mass m. The Euler-Lagrange equations of motion are:

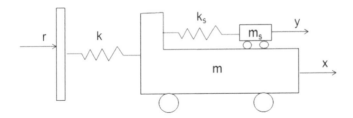

Figure 2.29: Parasitic Vibratory System

$$(m + m_s)\ddot{x} + m_s(\ddot{x} + \ddot{y}) + kx = kr \qquad (2.127)$$
$$m_s(\ddot{x} + \ddot{y}) + k_s y = 0 \qquad (2.128)$$

where x is the displacement of the mass m in an inertial frame, y is the differential displacement of mass m_s with respect to mass m and r is the reference displacement imposed on the system. Figure 2.30 illustrates the displacement of the two masses. The parasitic response of mass m_s is of high frequency and can be outside the bandwidth of the actuator. It is desirable to minimize the excitation of the parasitic oscillations by shaping the reference input to the system.

Assuming $m = 1$, $m_s = 0.1$, $k = 1$, $k_s = 50$, the natural frequencies of the system are $\omega_i = 0.95$ and 23.45 rad/sec. Designing a finite jerk Input Shaper, a first-order filtered Input Shaper and a sinusoid filtered Input Shaper with a jerk limit of 4, we can compare the performance of these jerk limited input shapers to the standard PPD time-delay filter (ZV Input Shaper). Figure 2.31 illustrates the shaped reference profiles derived based on the four filtering approaches listed above. Figure 2.32 presents the corresponding displacement of the main mass m and the parasitic mass m_s. The solid line, the dotted line, the dashed lines, and the dash-dot lines correspond to the input shaper, jerk limited input shaper, the first-order filtered input shaper, and the sinusoid filtered input shaper, respectively. It is clear that the parasitic oscillations of the sinusoid filtered input shaper are significantly smaller than the rest of the shapers. The cost trade-off with reference to maneuver time is small as can be seen in the evolution of the position of mass m.

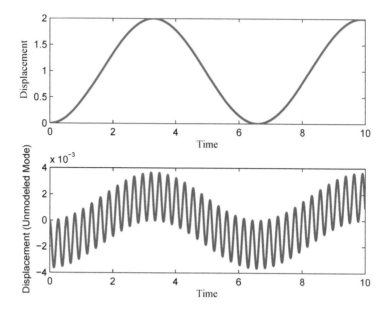

Figure 2.30: Step Response

2.8   Discrete-Time Time-Delay Filters

The pole-zero cancelation concept for the design of pre-filters can be extended to discrete-time systems where the pre-filter is parameterized as a FIR (finite impulse response) filter. Consider a general FIR filter of the form

$$H(z) = \sum_{i=0}^{k} c_i z^{-i} \tag{2.129}$$

which has k poles at the origin of the z-plane which implies stability. This filter is also referred to as a *moving average* filter. The transfer function $H(z)$ has k zeros whose locations are a function of c_i. Selecting the c_is so as to cancel the poles of the system to be controlled results in the counterpart of the time-delay filter in the discrete-time domain.

Figure 2.33 illustrates the implementation of a time-delay filter in the discrete-time domain for a nth order system. The n poles of the system which are the roots of the polynomial $\sum_{i=0}^{n} a_i z^{-i}$ are represented as $z_j = \sigma_j \pm \omega_j \sqrt{-1}$. Note, some poles of the system might be real. The optimization problem for the determination of the parameters (c_i) of the FIR filter can be posed as:

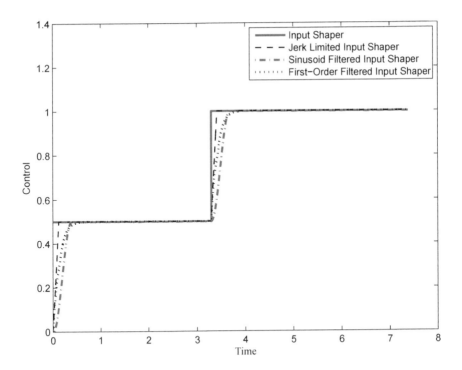

Figure 2.31: Shaped Reference Profiles

$$\min_{c_i} \quad J = \sum_{i=0}^{k} (i+1)^{\Lambda} |c_i|$$

subject to

$$\mathcal{R}\left(\sum_{i=0}^{k} c_i(\sigma_j + \omega_j\sqrt{-1})^{-i}\right) = 0 \; \forall j = 1,2,\ldots,n$$

$$\mathcal{I}\left(\sum_{i=0}^{k} c_i(\sigma_j + \omega_j\sqrt{-1})^{-i}\right) = 0 \; \forall j = 1,2,\ldots,n$$

$$\sum_{i=0}^{k} c_i = 1$$

$$0 < c_i < 1 \; \forall i = 1,2,\ldots,k$$

where $\mathcal{R}$ and $\mathcal{I}$ correspond to the real and imaginary parts of the argument respectively. The cost function is the weighted ℓ_1 norm of the coefficients. Minimizing the

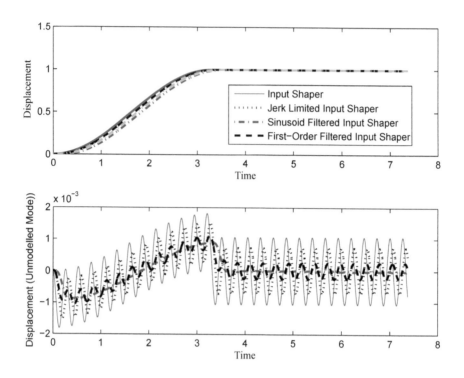

Figure 2.32: System Response

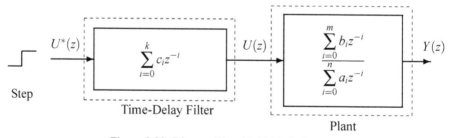

Figure 2.33: Discrete-Time Multi Mode System

ℓ_0 norm results in a sparse solution, i.e., the number of non-zero elements is minimized. Since minimizing the ℓ_0 norm is difficult, we pose the problem as minimizing the weighted ℓ_1 norm which can be posed as a linear programming problem and tends to a sparse solution. The weighting function $(.)^\Lambda$ where $\Lambda > 1$, is to increase the penalty nonlinearly with time so as to require the maneuver to be completed in minimal time.

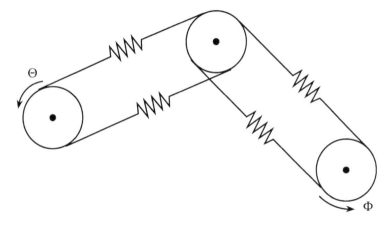

Figure 2.34: Flexible Transmission

Example 2.4:

Landau et al. [16] proposed a flexible transmission system as a benchmark problem for the design and evaluation of digital controllers. The benchmark problem consists of three pulleys connected by flexible belts and is modeled as a system with two very lightly damped modes. The input to the system is the position of the first pulley and the output is the rotation of the third pulley shown in Figure 2.34. The discrete time model for the nominal model is:

$$\frac{\Phi(z)}{\Theta(z)} = G(z) = \frac{z^{-2}(0.10276z^{-1}+0.18123z^{-2})}{1-1.99185z^{-1}+2.20265z^{-2}-1.84083z^{-3}+0.89413z^{-4}}$$

$$(2.131)$$

with a sampling frequency of 20 Hz. In Figure 2.36, the dashed line illustrates the undamped step response of the system. Solving the ℓ_1 norm minimization problem subject to the constraint that the FIR filter cancel the underdamped poles of the system located at:

$$z = 0.0853 \pm 0.9552i \text{ and } 0.9106 \pm 0.3782i \qquad (2.132)$$

with a $\Lambda = 3$, results in the FIR filter

$$H(z) = 0.4715 + 0.0052z^{-2} + 0.0680z^{-6} + 0.2571z^{-7} + 0.1982z^{-10}. \qquad (2.133)$$

Figure 2.35 illustrates the shaped reference input and the solid line in Figure 2.36 is the corresponding evolution of the output, which illustrates the nonoscillatory characteristic of the system response.

As in the design of time-delay filters for continuous-time systems, robustness to errors in estimated location of the poles of the system can be achieved by locating multiple zeros of the time-delay filter at the estimated location of the poles of the

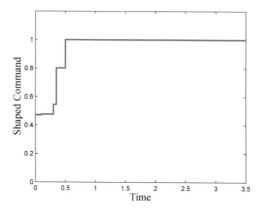

Figure 2.35: Shaped Reference Profiles

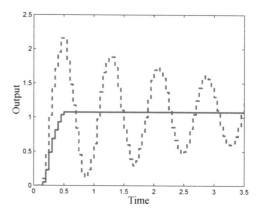

Figure 2.36: System Response

system. This constraint can be stated as:

$$\frac{dH(z)}{dz} = \sum_{i=0}^{k} (-i)c_i z^{-(i+1)} = 0 \qquad (2.134)$$

for all zs corresponding to the uncertain poles.

Example 2.5:   Landau et al. [16] present models of the flexible transmission system under no-load and full-load conditions, which are represented as:

$$\frac{\Phi(z)}{\Theta(z)}\bigg|_{NL} = G(z) = \frac{z^{-2}(0.2826z^{-1} + 0.5066z^{-2})}{1 - 1.4183z^{-1} + 1.5893z^{-2} - 1.3160z^{-3} + 0.8864z^{-4}} \qquad (2.135)$$

and

$$\frac{\Phi(z)}{\Theta(z)}\bigg|_{FL} = G(z) = \frac{z^{-2}(0.0640z^{-1} + 0.1040z^{-2})}{1 - 2.0967z^{-1} + 2.3196z^{-2} - 1.9335z^{-3} + 0.8712z^{-4}} \quad (2.136)$$

The robust time-delay filter is determined by solving the problem:

$$\min_{c_i} \; J = \sum_{i=0}^{20} (i+1)^3 |c_i|$$

subject to

$$\sum_{i=0}^{20} c_i (0.0853 \pm 0.9552\sqrt{-1})^{-i} = 0$$

$$\sum_{i=0}^{20} c_i (0.9106 \pm 0.3782\sqrt{-1})^{-i} = 0$$

$$\sum_{i=0}^{20} -ic_i (0.0853 \pm 0.9552\sqrt{-1})^{-i-1} = 0$$

$$\sum_{i=0}^{20} -ic_i (0.9106 \pm 0.3782\sqrt{-1})^{-i-1} = 0$$

$$\sum_{i=0}^{20} c_i = 1$$

$$0 < c_i < 1 \; \forall i = 1, 2, \dots, k$$

which includes 9 equality constraints and 21 unknown parameters to be solved for.

The robust time-delay filter is designed for the nominal model of the system and the performance of the nonrobust and robust controllers are tested on the no-load and full-load cases. Figures 2.37 and 2.38 illustrate the response of the no-load and full-load cases, respectively. The solid and dashed lines corresponds to the nonrobust and robust cases, respectively. Clearly, the robust-filter outperforms the nonrobust filter. The solid line in Figure 2.39 is the robust shaped reference profile and illustrates that the maneuver time increases compared to the nonrobust control profile shown by the dashed line.

2.9 Summary

A simple technique that is used to design pre-filters to attenuate residual vibration of underdamped systems by canceling the corresponding poles of the system has been presented. The single time-delay pre-filter is identical to the *zero vibration Input Shaper* and the *Posicast* controller. Cascading two single time-delay filters in

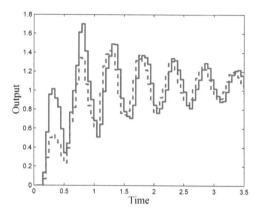

Figure 2.37: System Response (No Load)

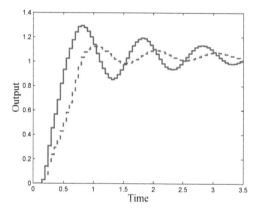

Figure 2.38: System Response (Full Load)

series results in the *zero vibration derivative Input Shaper* which provide robustness to uncertainties in damping ratio and natural frequency simultaneously. The technique to design time-delay filters is extended to pre-filters where the delay times are specified by the user. This is compatible with implementing it using a discrete time control system. The time-delay filtering techniques can also be used to minimize residual vibration of multi-mode systems.

There are numerous applications where the goal of the control system is to perform point-to-point maneuvers. The reference input to such systems is a step input. Filtering the step input through a time-delay filter results in a staircase form of a reference input which minimizes the magnitude of residual energy compared to the application of the step input to the nominal system. The staircase input still might excite higher modes and unmodeled dynamics and furthermore, actuators cannot track step

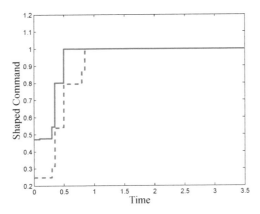

Figure 2.39: Shaped Reference Profiles

inputs. The section on jerk limited time-delay filters provides a technique to account for the jerk limits on the actuator and also provide a mechanism to generate a reference profile which reduces the excitation of the higher frequency unmodeled modes as shown by the frequency response plots of the jerk limited time-delay filtered signals. The jerk limited time-delay filter is generated by including an integrator into the pre-filter. This motivated the design of filtered Input Shapers where first- and higher-order filters are integrated with the standard time-delay filter and the resulting reference profiles are smooth, which further reduces the excitation of unmodeled dynamics.

Although the development and illustration of the time-delay filtering technique has been illustrated for point-to-point control, the pre-filtering of any reference input results in nonoscillatory tracking of the reference input. For instance, requiring the glass plate of a flat-bed scanner to follow a trapezoidal velocity profile can be achieved by filtering the reference velocity profile through a time-delay filter. Similarly, inserting a time-delay filter between a drive pendant of a gantry crane and the actuators can reduce the pendulation of the crane to arbitrary inputs.

Exercises

2.1 Design a time-delay filter for a system with a transfer function

$$\frac{Y(s)}{R(s)} = \frac{5}{s+5}$$

such that the system output $y(t)$ reaches the final value in finite time when the reference input is a unit step function.

2.2   Derive a closed form expression for the parameters of a time-delay filter for a system with a transfer function.

$$\frac{Y(s)}{R(s)} = \frac{a\omega^2}{(s+a)(s^2 + 2\zeta\omega s + \omega^2)}$$

The form of the time-delay filter is:

$$G(s) = A_0 + A_1 e^{-sT_1} + e^{-2sT_1} \qquad (2.138)$$

2.3   Derive a closed-form expression for the parameters of a time-delay filter to cancel two undamped pair of poles. The form of the time-delay filter is:

$$G(s) = A_0 + A_1 e^{-sT_1} + e^{-sT_2}$$

and the system model is:

$$\frac{Y(s)}{R(s)} = \frac{36}{(s^2 + 4)(s^2 + 9)}$$

2.4   Figure 2.40 schematically illustrates the Furuta pendulum which can be used to represent a tower crane. Assume a PD control for the actuator which drives the horizontal beam and design an input shaper to bring the pendulum to rest at the end of the maneuver.

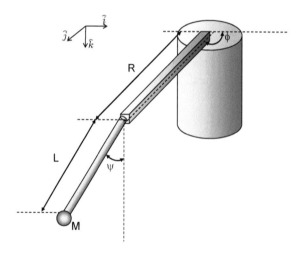

Figure 2.40: Furuta Pendulum

The kinetic energy T and potential energy V for the pendulum is:

$$T = \frac{1}{2}J_L\dot{\psi}^2 + \frac{1}{2}J_L(\dot{\phi}\sin(\psi))^2 + \frac{1}{2}J_R\dot{\phi}^2 + \frac{1}{2}Mr_p^2$$
$$V = mgL(1 - \cos(\psi))$$

where r_p is:

$$r_p = (R\cos(\phi) + L\sin(\psi)\sin(\phi))\hat{i} + (R\sin(\phi) - L\sin(\psi)\cos(\phi))\hat{j} + L\cos(\psi)\hat{k}$$

where $\hat{i}$, $\hat{j}$, and $\hat{k}$ are unit vectors in the x, y, and z directions, respectively. Assume M = 0.01 kg, R = 0.2 m, L = 0.5 m, J_L = 0.05 kg m^2, and J_R = 0.001 kg m^2.

2.5 Design a time-delay filter for eliminating residual vibrations of the system shown in Figure 2.41 for a rest-to-rest maneuver. The initial displacement of the mass and inertia are zero and the final rotation of the inertia is required to be 1 radian.

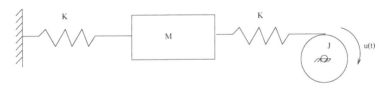

Figure 2.41: Two-Degree of Freedom System

The equations of motion of the system are:

$$M\ddot{x} + Kx + K(x - R\theta) = 0$$
$$J\ddot{\theta} - KR(x - R\theta) = u$$

where M is the mass of the block, J is the inertia of the pulley whose radius is R. $u(t)$ is the torque input to the pulley. Assume $M = 1$, $J = 1$, $R = 1$, and $K = 1$.

2.6 The optical pick-up of a compact disc (CD) player consists of a sledge for large displacements and a radial servo which is limited to move over a few hundred tracks. The optical pickup is a two-degree of freedom system with one degree catering to the motion of an objective lens for focusing and a second degree for fine radial positioning. Vidal et al. [17] model the pick-up as a two mass-spring system. Figure 2.42 illustrates a schematic of the pickup system where mass m_1 corresponds to the objective and m_2 is the mass of the objective suspension links. The model relating the displacement of the objective to voltage input to the pickup actuator is:

$$\frac{X(s)}{U(s)} = \frac{7.054e - 006s^2 + 0.003088s + 68.5}{3.148e - 007s^4 + 0.0001402s^3 + 3.217s^2 + 16.83s + 1.145e005}$$

Assume 16,000 tracks per inch (TPI) of track density, the intertrack distance is 1.5875e-7 meters. To move the objective by 100 tracks requires a step input reference signal of magnitude 0.0265 volts. Design a time-delay filter to complete the seek motion of 100 tracks with no residual vibrations.

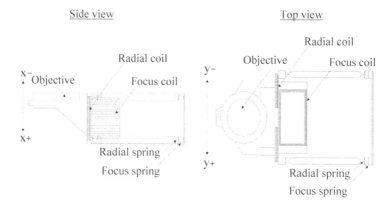

Figure 2.42: CD Pickup (Reprinted with permission [18].)

2.7   To illustrate the design of time-delay filters for a multi-mode system, consider the control of a flexible belt-drive system (Figure 2.43). The system model is

$$J_1\ddot{\theta} + c\dot{\theta} + 2kR^2(\theta - \phi) = \tau$$
$$J_2\ddot{\phi} + c\dot{\phi} - 2kR^2(\theta - \phi) = 0$$

where J_1 and J_2 corresponds to the inertia of the motor and load, respectively. The belt stiffness is k and the radius of the pulleys are R. A proportional plus derivative controller

$$\tau = -K_1(\theta - r) - K_2\dot{\theta}$$

where r is the reference input, is guaranteed to be stable for all gains K_1 and K_2 greater than zero. Assuming the model and controller parameters to be:

$$J_1 = 1, \ J_2 = 2, \ c = 0.1, \ R = 1, \ K = 2, \ K_1 = 2, \ K_2 = 1,$$

the closed-loop transfer function is:

$$\frac{\Phi(s)}{R(s)} = \frac{4}{s^4 + 1.15s^3 + 8.055s^2 + 2.5s + 4}$$

and the closed-loop poles are located at:

$$s = -0.1391 + 0.7290i, \quad \text{and} \quad -0.4359 \pm 2.6593i.$$

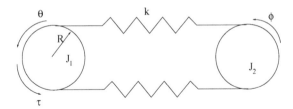

Figure 2.43: Flexible Transmission

Design a time-delay filter to move the inertia from an initial position of rest and bring it to rest with a final position of one radian. Illustrate the improvement in the robustness to uncertainties in belt stiffness k, by using a robust time-delay filter.

References

[1] D. de Roover, F. B. Sperling, and O. H. Bosgra. Point-to-point control of a mimo servomechanism. In *ACC*, Philadelphia, PA, 1998.

[2] Y. Mizoshita, S. Hasegawa, and K. Takaishi. Vibration minimized access control for disk drives. *IEEE Transactions on Magnetics.*, 32(3):1793–1798, 1996.

[3] D. B. Rathbun, M. C. Berg, and K. W. Buffinton. Pulse width control for precise positioning of structurally flexible systems subject to stiction and Coulomb friction. *ASME J. of Dyn. Systems, Measurements, and Cont.*, 126(1):131–138, 2004.

[4] M. J. Doherty and R. H. Tolson. Input shaping to reduce solar array stuctural vibrations. Joint Institute for Advancement of Flight Sciences report NASA/CR-1998-208698, NASA Langley Research Center, 1998.

[5] C. J. Swigert. Shaped torque techniques. *J. of Guid., Cont. and Dyn.*, 3(5):460–467, 1979.

[6] J. L. Junkins and J. D. Turner. *Optimal Spacecraft Rotational Maneuvers*. Amsterdam: Elsevier, 1986.

[7] T. Singh and W. Singhose. Tutorial on input shaping/time delay control of maneuvering flexible structures. In *ACC*, Anchorage, Alaska, May 2002.

[8] S. L. Scrivener and R. C. Thompson. Shaped input for multimode system. *J. of Guid., Cont. and Dyn.*, 17(2):225–233, 1994.

[9] G. H. Tallman and O.J.M. Smith. Analog study of dead-beat posicast control. 45:14–21, September 1958.

[10] T. Singh and S. R. Vadali. Robust time delay control. *ASME J. of Dyn. Systems, Measurements, and Cont.*, 115(2):303–306, 1993.

[11] N. Singer and W. P. Seering. Pre-shaping command inputs to reduce system vibration. *ASME J. of Dyn. Systems, Measurements, and Cont.*, 112(1):76–82, 1990.

[12] M. Muenchhof and T. Singh. Jerk limited time optimal control of flexible structures. *ASME J. of Dyn. Systems, Measurements, and Cont.*, 125(1):139–142, 2003.

[13] M. Muenchhof and T. Singh. Desensitized jerk limited-time optimal control of multi-input systems. *J. of Guid., Cont. and Dyn.*, 25(3):474–481, 2002.

[14] T. Singh. Jerk limited input shapers. *ASME Journal of Dynamic Systems, Measurement and Control*, 126(1):215–219, March 2004.

[15] T. Singh. Minimax design of robust controllers for flexible systems. *J. of Guid., Cont. and Dyn.*, 25(5):868–875, 2002.

[16] I. D. Landau, D. Rey, A. Karimi, A. Voda, and A. Franco. A flexible transmission system as a benchmark for robust digital control. *European Journal of Control*, 1:77–96, 1995.

[17] W. Vidal, J. Stoustrup, P. Andersen, T. S. Pedersen, and H. F. Mikkelsen. Parametric uncertainty with perturbations restricted to be real on 12 CD mechanisms. In *American Control Conference*, Denver, CO, 2003.

[18] E. V. Sanchez. *Robust and Fault Tolerant Control of CD-Players*. PhD thesis, Aalborg University, Aalborg, Denmark, 2003.

3

Optimal Control

Success is going from failure to failure without a loss of enthusiasm.

Winston Churchill (1874–1965); Nobel Laureate, writer, painter, soldier, journalist, and politician

CLASSICAL control refers to techniques that include Bode diagrams, Nyquist plots, Nichols charts, and so forth, used to design controllers based on measures such as gain and phase margins and bandwidth. Following the development of the root-locus technique by Evans after World War II, time-domain approaches which use measures such as percent overshoot, rise, settling, and peak time were used to design controllers. The post–World War II period led to numerous breakthroughs in controller design, which have been labeled *modern control*. Contributions by Bellman, LaSalle, Pontryagin, Kalman, Bryson, and others laid the foundation for what is currently referred to as *optimal control*. Optimal control has been extensively used in aerospace, electrical, automotive and other applications. Recently, problems of precision control of systems characterized by low frequency lightly damped modes have been formulated as optimal control problems.

Chapter 2 presented a simple technique to shape the reference input to a stable or marginally stable system, which minimizes the excitation of the underdamped modes about the desired reference trajectory. The technique was illustrated on rest-to-rest maneuvers only. The time-delay filter was designed so as to cancel the underdamped poles of the system which resulted in rest-to-rest motion with no residual vibrations. To address the problem of uncertainties in estimated model parameters, the time-delay filter was required to place multiple sets of zeros at the nominal location of the uncertain underdamped poles, which resulted in improved performance in the proximity of the nominal model. The time-delay filter design was not posed as an optimal control problem, but can be designed by posing an appropriate optimal control problem. This chapter will introduce the basic theory of optimal control and well known controllers such as linear quadratic regulator (LQR), and frequency-shaped LQR will be used to design optimal controllers for benchmark spring-mass systems.

3.1   Calculus of Variations

Optimal control theory is founded on the fundamentals of *calculus of variations* whose roots can be traced to the Brachistochrone problem posed by Bernouilli in 1696 CE. The pioneering work of Euler (1741) which provides a geometrical approach to interpret the variational problem provided the seed for the development of what is now referred to as *calculus of variations*. In 1755, 19-year-old Lagrange wrote Euler a letter where he regenerated Euler's results using an elegant method of *variations*, which Euler embraced and called *calculus of variations*.

Calculus of variations was designed to extend results from the differential calculus, notably necessary and sufficient conditions for extreme values of ordinary functions over finite-dimensional vector spaces, to extreme values of *functions of functions*, or functionals, over infinite-dimensional function spaces. We begin by deriving the Euler-Lagrange equations for a fixed time, fixed boundary conditions problem. This is followed by the development of the constraints for the optimization of a functional subject to differential equation constraints. Finally, the Hamiltonian formulation, which results in the canonical representation of the necessary conditions of optimality is derived.

Consider the problem where it is desired to solve for $x(t)$ which minimizes the cost function

$$J(x(t),t) = \int_{t_0}^{t_f} \phi(x(t),\dot{x}(t),t)\, dt. \tag{3.1}$$

$J(x(t),t)$ is a functional, that is a mapping from a function space into a scalar. The function space (in this case of continuously differentiable functions) is infinite-dimensional and the functional $J(x(t),t)$ maps it to a real scalar space. It is assumed that ϕ and $x(t)$ are continuously differentiable with respect to their arguments up to order 2, i.e., they are C^2. t_0 and t_f represent the initial and final times, respectively, and $x(t_0)$ and $x(t_f)$ are assumed to be fixed. $x(t)$ represents the optimal and unknown path which needs to be determined.

The *increment* of the functional J is defined as

$$\Delta J \triangleq J(x(t) + \delta x(t),t) - J(x(t),t) \tag{3.2}$$

where $\delta x(t)$ is called the *variation* of the function $x(t)$ as shown in Figure 3.1. Since $x(t_0)$ and $x(t_f)$ are prescribed, $\delta x(t_0)$ and $\delta x(t_f)$ are zero. The variation is assumed to be smooth and of class C^2. The *increment* can therefore be represented as

$$\Delta J = \int_{t_0}^{t_f} \phi(x(t) + \delta x(t),\dot{x}(t) + \delta\dot{x}(t),t)\, dt - \int_{t_0}^{t_f} \phi(x(t),\dot{x}(t),t)\, dt. \tag{3.3}$$

Just as in ordinary calculus, where the first differential is defined as the linear linear part of the increment in a function due to infinitesimal changes in its argument, the

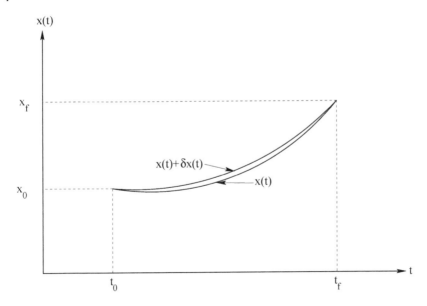

Figure 3.1: Optimal and Varied Path

first variation likewise is defined as the linear part of the increment ΔJ and is represented as δJ. Expanding Equation (3.3) using the Taylor series in $\delta x(t)$, $\delta \dot{x}(t)$, we have

$$\delta J = \int_{t_0}^{t_f} \left[\frac{\partial \phi}{\partial x}^T \delta x(t) + \frac{\partial \phi}{\partial \dot{x}}^T \delta \dot{x}(t) \right] dt \tag{3.4}$$

where

$$\delta \dot{x}(t) = \frac{d}{dt} \delta x(t), \tag{3.5}$$

and $\frac{\partial \phi}{\partial x}$ and $\frac{\partial \phi}{\partial \dot{x}}$ represent the gradient of ϕ with respect to $x(t)$ and $\dot{x}(t)$. Since $\delta \dot{x}(t)$ is dependent on $\delta x(t)$, we integrate the expression which is a function of $\delta \dot{x}(t)$ by parts, to represent the *first variation* in terms of the arbitrary variation $\delta x(t)$. The *first variation* can be written as

$$\delta J = \frac{\partial \phi}{\partial \dot{x}}^T \delta x(t) \Big|_{t_0}^{t_f} + \int_{t_0}^{t_f} \left[\frac{\partial \phi}{\partial x}^T \delta x(t) - \frac{d}{dt} \left(\frac{\partial \phi}{\partial \dot{x}}^T \right) \delta x(t) \right] dt \tag{3.6}$$

Since $x(t_0)$ and $x(t_f)$ are specified, their variations $\delta x(t_0)$ and $\delta x(t_f)$ are zero.

The *Fundamental theorem of the calculus of variations* which is a necessary condition for optimality requires that the *first variation* of J vanish for an optimum, for all permissible arbitrary variations $\delta x(t)$. Thus, Equation (3.6) leads to the Euler-Lagrange equation

$$\frac{\partial \phi}{\partial x} - \frac{d}{dt}\left(\frac{\partial \phi}{\partial \dot{x}}\right) = 0 \tag{3.7}$$

subject to the boundary conditions

$$x(t_0) = x_0 \text{ and } x(t_f) = x_f \tag{3.8}$$

3.1.1   Beltrami Identity

Consider the total derivative of $\phi(x(t),\dot{x}(t),t)$ with respect to time

$$\frac{d\phi}{dt} = \frac{\partial \phi}{\partial t} + \frac{\partial \phi}{\partial x}^T \frac{dx}{dt} + \frac{\partial \phi}{\partial \dot{x}}^T \frac{d\dot{x}}{dt} \tag{3.9}$$

which can be rewritten as

$$\frac{\partial \phi}{\partial x}^T \frac{dx}{dt} = \frac{d\phi}{dt} - \frac{\partial \phi}{\partial t} - \frac{\partial \phi}{\partial \dot{x}}^T \frac{d\dot{x}}{dt} \tag{3.10}$$

Multiply the Euler-Lagrange equation (3.7) with $\frac{dx}{dt}$ to arrive at the equation

$$\frac{dx^T}{dt}\frac{\partial \phi}{\partial x} - \frac{dx^T}{dt}\frac{d}{dt}\left(\frac{\partial \phi}{\partial \dot{x}}\right) = 0 \tag{3.11}$$

Substituting Equation (3.10) into Equation (3.11), we have

$$\frac{d\phi}{dt} - \frac{\partial \phi}{\partial t} - \frac{\partial \phi}{\partial \dot{x}}^T \frac{d\dot{x}}{dt} - \frac{dx^T}{dt}\frac{d}{dt}\left(\frac{\partial \phi}{\partial \dot{x}}\right) = 0 \tag{3.12}$$

which can be rewritten as

$$-\frac{\partial \phi}{\partial t} + \frac{d}{dt}\left(\phi - \frac{dx^T}{dt}\frac{\partial \phi}{\partial \dot{x}}\right) = 0 \tag{3.13}$$

If ϕ is not explicitly a function of time, i.e., $\frac{\partial \phi}{\partial t} = 0$, Equation (3.13) reduces to

$$\phi - \frac{dx^T}{dt}\frac{\partial \phi}{\partial \dot{x}} = K. \tag{3.14}$$

where K is a constant of integration and is referred to as the *Beltrami identity*, which can be exploited to solve a class of problems where the integrand of the cost function is not explicitly a function of the independent variable such as the *brachistochrone problem*.

Example 3.1:

The optimization problem which endeavors to find a curve which connects two points in a gravity potential so as to minimize the time taken to move from one point to the other under the influence of gravity is referred to as the brachistochrone problem [1]. Figure 3.2 illustrates a generic shape connecting the origin to P.

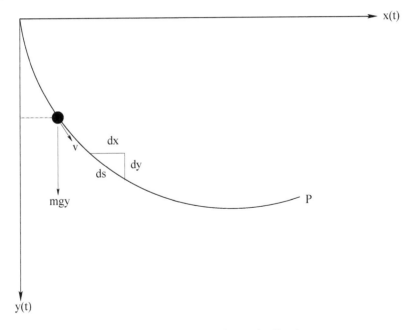

Figure 3.2: Particle Motion under Gravity

For a frictionless bead of mass m sliding on the curve under the influence of gravity, conservation of energy can be exploited to relate the velocity of the bead to the distance traveled in the direction of gravity y. Equating the kinetic energy to the potential energy leads to the equation:

$$\frac{1}{2}mv^2 = mgy \tag{3.15}$$

$$\Rightarrow v = \sqrt{2gy} \tag{3.16}$$

The time taken to move from the origin to the point P is given by the equation

$$T = \int_0^P \frac{ds}{v} \tag{3.17}$$

Substituting the relationship

$$ds^2 = dx^2 + dy^2 \tag{3.18}$$

into Equation (3.17), we have

$$T = \int_0^P \frac{\sqrt{dx^2 + dy^2}}{\sqrt{2gy}} = \int_0^P \frac{\sqrt{1 + \dot{y}^2}}{\sqrt{2gy}}\, dx \tag{3.19}$$

where $\dot{y} = \frac{dy}{dx}$. Since the integrant is not an explicit function of x, the Beltrami identity results in the equation

$$\frac{\sqrt{1+\dot{y}^2}}{\sqrt{2gy}} - \dot{y}\frac{\partial}{\partial\dot{y}}\left(\frac{\sqrt{1+\dot{y}^2}}{\sqrt{2gy}}\right) = \frac{1}{K} \tag{3.20}$$

$$\Rightarrow \frac{1}{\sqrt{2gy}\sqrt{1+\dot{y}^2}} = \frac{1}{K} \tag{3.21}$$

$$\Rightarrow 2gy(1+\dot{y}^2) = K^2. \tag{3.22}$$

Using a parametric solution of the form

$$\dot{y} = \frac{dy}{dx} = cot(\sigma), \tag{3.23}$$

and substituting into Equation (3.22), we have

$$y = \frac{\kappa^2}{1+\dot{y}^2} = \frac{\kappa^2}{1+cot(\sigma)^2} = \kappa^2\sin(\sigma)^2 = \frac{1}{2}\kappa^2(1-\cos(2\sigma)). \tag{3.24}$$

where $\kappa^2 = \frac{K^2}{2g}$. Next, from the relationship

$$dx = \frac{dy}{\dot{y}} = \frac{2\kappa^2\sin(\sigma)\cos(\sigma)}{cot(\sigma)}d\sigma = 2\kappa^2\sin(\sigma)^2\,d\sigma = \kappa^2(1-\cos(2\sigma))\,d\sigma \tag{3.25}$$

we have

$$x = \kappa^2(\sigma - \frac{1}{2}\sin(2\sigma)) + C \tag{3.26}$$

Since $y(0) = 0$, we have from Equation (3.24), $\sigma = 0$, which when substituted into Equation (3.26) results in the integration constant $C = 0$.

The range of the parametric variable σ and the constant K are determined as a function of the desired end point of the bead. The optimal solution for three end points are given in Table 3.1 and the optimal shapes of the curves are presented in Figure 3.3.

End Point Coordinates	κ	σ range
X = 1, Y = −1	1.0704	0–1.2060
X = 2, Y = −1	1.0171	0–1.7542
X = 3, Y = −1	1.1133	0–2.0258

Table 3.1: Brachistochrone Solutions

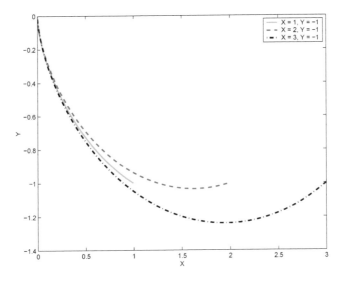

Figure 3.3: Brachistochrone Solutions

Figure 3.3 illustrates the optimal shapes for three different end points. It can be seen that for some end points, the mass has to move up toward the end of the maneuver to complete it in minimum time.

The optimal shape which is a *cycloid* has another interesting property. Starting from rest at any point on the curve, the time required to reach the lowest point is the same. Hence, this curve is also called a *tautochrone* and was first presented by Huygens in 1657 CE [2].

3.1.2 Differential Equation Constraints

In this section, we develop the variational formulation for the optimization of a functional with differential equation constraint. Consider the state-space model of a nonlinear system

$$\dot{x}(t) = f(x(t), u(t), t) \tag{3.27}$$

where $x(t)$ corresponds to the states ($x(t) \in \mathcal{R}^n$) of the system and $u(t)$ the control input ($u(t) \in \mathcal{R}^m$). It is the goal of the optimization algorithm to determine $u(t)$, which minimizes the cost function

$$J = \underbrace{\theta(x_f, t_f)}_{\text{Mayer problem}} + \underbrace{\int_{t_0}^{t_f} \phi(x(t), u(t), t) \, dt}_{\text{Lagrange problem}} \tag{3.28}$$

given the boundary conditions

$$x(0) = x_0, \text{ and } t_0 \text{ specified.} \tag{3.29}$$

Additional constraints at the terminal time of the form

$$\alpha(x_f, t_f) = 0 \tag{3.30}$$

which correspond to a q dimensional vector of algebraic constraints, need to be included into the optimal control problem formulation. Cost function which include a terminal cost and an integral cost are called the *Bolza problems* [3]. As shown in Equation (3.28), cost functions which are functions of the terminal time and states are referred to as *Mayer problems* [3] and those which penalize an integral of the evolving states, control and time are referred to as *Lagrange problems* [3].

Introducing two Lagrange vectors $\lambda(t)$ and v (not a function of time) of dimensions n and q respectively, the augmented cost function can be represented as

$$
\begin{aligned}
J_a = & \theta(x_f, t_f) + v^T \alpha(x_f, t_f) \\
& + \int_{t_0}^{t_f} \left\{ \phi(x(t), u(t), t) + \lambda^T(t) [f(x(t), u(t), t) - \dot{x}(t)] \right\} dt
\end{aligned} \tag{3.31}
$$

The first variation of the augmented cost J_a is

$$
\begin{aligned}
\delta J_a = & \frac{\partial \theta}{\partial x_f} \delta x_f + \frac{\partial \theta}{\partial t_f} \delta t_f + v^T \frac{\partial \alpha}{\partial x_f} \delta x_f + v^T \frac{\partial \alpha}{\partial t_f} \delta t_f + \\
& \int_{t_f}^{t_f + \delta t_f} \left\{ \phi + \lambda^T(t) [f - \dot{x}(t)] \right\} dt + \int_{t_0}^{t_f} \left[\left\{ \frac{\partial \phi}{\partial x} + \lambda^T \frac{\partial f}{\partial x} \right\} \delta x + \right. \\
& \left. \left\{ \frac{\partial \phi}{\partial u} + \lambda^T(t) \frac{\partial f}{\partial u} \right\} \delta u - \lambda^T(t) \delta \dot{x}(t) + (f - \dot{x})^T \delta \lambda \right] dt
\end{aligned} \tag{3.32}
$$

If both t_f and $x(t_f)$ are free, a relationship between them still exists as illustrated in Figure 3.4 for a scalar case. δx_f corresponds to the difference between the ordinates at the end of the optimal and varied curves. A first-order approximation of the relationship can be represented as

$$\delta x(t_f) = \delta x_f - \dot{x}(t_f) \delta t_f \tag{3.33}$$

Further, the integral corresponding to the variation in the final time, can be approximated as

$$\int_{t_f}^{t_f + \delta t_f} \left\{ \phi + \lambda^T(t) [f - \dot{x}(t)] \right\} dt \approx \left\{ \phi + \lambda^T(t) [f - \dot{x}(t)] \right\} \big|_{t = t_f} \delta t_f \tag{3.34}$$

Substituting Equations (3.33) and (3.34) into Equation (3.32), and after integrating the term $\lambda^T(t) \delta \dot{x}(t)$ by parts, Equation (3.32) can be rewritten as

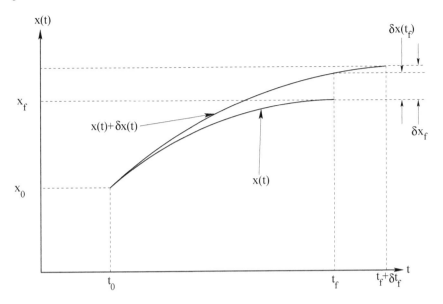

Figure 3.4: Optimal and Varied Path

$$\delta J_a = \frac{\partial \theta}{\partial x_f} \delta x_f + \frac{\partial \theta}{\partial t_f} \delta t_f + v^T \frac{\partial \alpha}{\partial x_f} \delta x_f + v^T \frac{\partial \alpha}{\partial t_f} \delta t_f +$$

$$\{\phi + \lambda^T(t)[f - \dot{x}(t)]\}\big|_{t=t_f} \delta t_f - \lambda^T(t) \underbrace{\delta x(t_f)}_{\delta x(t_f) = \delta x_f - \dot{x}(t_f)\delta t_f} +$$

$$\int_{t_0}^{t_f} \left[\left\{ \frac{\partial \phi}{\partial x} + \lambda^T \frac{\partial f}{\partial x} + \dot{\lambda}^T \right\} \delta x + \left\{ \frac{\partial \phi}{\partial u} + \lambda^T(t) \frac{\partial f}{\partial u} \right\} \delta u + (f - \dot{x})^T \delta \lambda \right] dt$$

$$(3.35)$$

Collecting terms, Equation (3.35) can be written as

$$\delta J_a = \left\{ \frac{\partial \theta}{\partial x_f} + v^T \frac{\partial \alpha}{\partial x_f} - \lambda^T \right\} \delta x_f + \left\{ \frac{\partial \theta}{\partial t_f} + v^T \frac{\partial \alpha}{\partial t_f} + \phi(t_f) + \lambda^T(t_f)f \right\} \delta t_f +$$

$$\cancel{\lambda^T \dot{x}(t)\delta t_f} - \cancel{\lambda^T \dot{x}(t)\delta t_f} +$$

$$\int_{t_0}^{t_f} \left[\left\{ \frac{\partial \phi}{\partial x} + \lambda^T \frac{\partial f}{\partial x} + \dot{\lambda}^T \right\} \delta x + \left\{ \frac{\partial \phi}{\partial u} + \lambda^T(t) \frac{\partial f}{\partial u} \right\} \delta u + (f - \dot{x})^T \delta \lambda \right] dt$$

$$(3.36)$$

From the *Fundamental theorem of the calculus of variations*, we arrive at the necessary conditions for optimality

$$-\dot{\lambda}^T = \frac{\partial \phi}{\partial x} + \lambda^T \frac{\partial f}{\partial x} \tag{3.37}$$

$$\dot{x}(t) = f(x(t), u(t), t) \tag{3.38}$$

$$0 = \frac{\partial \phi}{\partial u} + \lambda^T(t) \frac{\partial f}{\partial u} \tag{3.39}$$

$$0 = \left\{ \frac{\partial \theta}{\partial x_f} + v^T \frac{\partial \alpha}{\partial x_f} - \lambda^T \right\}\bigg|_{t_f} \delta x_f \tag{3.40}$$

$$0 = \left\{ \frac{\partial \theta}{\partial t_f} + v^T \frac{\partial \alpha}{\partial t_f} + \phi + \lambda^T(t) f \right\}\bigg|_{t_f} \delta t_f \tag{3.41}$$

where Equations (3.37), (3.38), and (3.39) are the Euler-Lagrange equations and Equations (3.40) and (3.41) are known as the transversality conditions.

Example 3.2:

Determine the solution of the problem:

$$\min_{u(t)} J = \int_0^\pi \underbrace{(u^2)}_{\phi} dt \tag{3.42}$$

subject to the differential equation constraint:

$$\begin{Bmatrix} \dot{x}_1 \\ \dot{x}_2 \end{Bmatrix} = \underbrace{\begin{bmatrix} 0 & 1 \\ -\omega^2 & 0 \end{bmatrix} \begin{Bmatrix} x_1 \\ x_2 \end{Bmatrix} + \begin{Bmatrix} 0 \\ 1 \end{Bmatrix} u}_{f}, \tag{3.43}$$

and the boundary conditions:

$$\begin{Bmatrix} x_1 \\ x_2 \end{Bmatrix}(0) = \begin{Bmatrix} 0 \\ 0 \end{Bmatrix} \text{ and } \begin{Bmatrix} x_1 \\ x_2 \end{Bmatrix}(\pi) = \begin{Bmatrix} 1 \\ 0 \end{Bmatrix}. \tag{3.44}$$

Since the final time and final states are prescribed, the transversality conditions (Equations 3.40 and 3.41) are satisfied. The necessary conditions for optimality, Equations (3.37) through (3.39), lead to:

$$-\begin{Bmatrix} \dot{\lambda}_1 \\ \dot{\lambda}_2 \end{Bmatrix} = \underbrace{\begin{bmatrix} 0 & -\omega^2 \\ 1 & 0 \end{bmatrix}}_{A^T} \begin{Bmatrix} \lambda_1 \\ \lambda_2 \end{Bmatrix} \tag{3.45}$$

$$\begin{Bmatrix} \dot{x}_1 \\ \dot{x}_2 \end{Bmatrix} = \underbrace{\begin{bmatrix} 0 & 1 \\ -\omega^2 & 0 \end{bmatrix}}_{A} \begin{Bmatrix} x_1 \\ x_2 \end{Bmatrix} + \begin{bmatrix} 0 \\ 1 \end{bmatrix} u \tag{3.46}$$

$$0 = 2u + \begin{bmatrix} 0 & 1 \end{bmatrix} \begin{Bmatrix} \lambda_1 \\ \lambda_2 \end{Bmatrix} \tag{3.47}$$

The solution to Equation (3.45) is:

$$\left\{\begin{matrix} \lambda_1 \\ \lambda_2 \end{matrix}\right\}(t) = e^{-A^T t}\left\{\begin{matrix} \lambda_1 \\ \lambda_2 \end{matrix}\right\}(0) = \begin{bmatrix} \cos(\omega t) & \omega\sin(\omega t) \\ -\frac{\sin(\omega t)}{\omega} & \cos(\omega t) \end{bmatrix}\left\{\begin{matrix} \lambda_1 \\ \lambda_2 \end{matrix}\right\}(0) \tag{3.48}$$

Substituting Equations (3.47) and (3.48) into Equation (3.46) leads to:

$$\begin{bmatrix} \cos(\omega\pi) & -\frac{\sin(\omega\pi)}{\omega} \\ \omega\sin(\omega\pi) & \cos(\omega\pi) \end{bmatrix}\left\{\begin{matrix} x_1 \\ x_2 \end{matrix}\right\}(\tau) = $$
$$\begin{bmatrix} \frac{\cos(\omega\tau)\sin(\omega\tau)-\omega\tau}{4\omega^3} & -\frac{(\cos(\omega\tau))^2}{4\omega^2} \\ -\frac{(\cos(\omega\tau))^2}{4\omega^2} & -\frac{\cos(\omega\tau)\sin(\omega\tau)+\omega\tau}{4\omega} \end{bmatrix}\left\{\begin{matrix} \lambda_1 \\ \lambda_2 \end{matrix}\right\}(0) \tag{3.49}$$

Substituting the boundary conditions (Equation (3.44)), we can solve for $\lambda_1(0)$ and $\lambda_2(0)$

$$\left\{\begin{matrix} \lambda_1 \\ \lambda_2 \end{matrix}\right\}(0) = \left\{\begin{matrix} \frac{4\omega^3(\sin(\pi\omega)+\pi\omega\cos(\pi\omega))}{\sin(\pi\omega)^2-\pi^2\omega^2} \\ \frac{4\omega^3\pi\sin(\pi\omega)}{\sin(\pi\omega)^2-\pi^2\omega^2} \end{matrix}\right\} \tag{3.50}$$

which results in the optimal control

$$u(t) = -\frac{\lambda_2}{2} = \frac{\sin(\omega t)}{2\omega}\frac{4\omega^3(\sin(\pi\omega)+\pi\omega\cos(\pi\omega))}{\sin(\pi\omega)^2-\pi^2\omega^2} - \frac{\cos(\omega t)}{2}\frac{4\omega^3\pi\sin(\pi\omega)}{\sin(\pi\omega)^2-\pi^2\omega^2}. \tag{3.51}$$

3.2 Hamiltonian Formulation

Optimal control design is the process of determination of the control which optimizes the cost function while satisfying the dynamical equations of motion and all state and control constraints. The optimal control problem is stated as the minimization of the cost

$$J = \theta(x(t_f),t_f) + \int_0^{t_f} \phi(x,u,t)\, dt \tag{3.52}$$

subject to the constraints

$$\dot{x} = f(x,u,t) \tag{3.53}$$

and the initial and final state constraints. Without loss of generality, we have assumed that the initial time is zero and the final time is t_f. Augmenting the cost function with the state equation, we have

$$J_a = \theta(x(t_f), t_f) + \int_0^{t_f} \phi(x, u, t) + \lambda^T (f - \dot{x}) \, dt. \tag{3.54}$$

Define the Hamiltonian as

$$\mathcal{H} = \phi + \lambda^T f, \tag{3.55}$$

and using calculus of variations, the Euler-Lagrange equations are derived as:

$$-\dot{\lambda} = \frac{\partial \mathcal{H}}{\partial x} \tag{3.56}$$

$$\dot{x} = \frac{\partial \mathcal{H}}{\partial \lambda} \tag{3.57}$$

$$\frac{\partial \mathcal{H}}{\partial u} = 0 \tag{3.58}$$

with the associated Transversality conditions

$$\lambda^T(0)\delta x(0) + \left(\frac{\partial \theta}{\partial x} - \lambda \right)^T \Bigg|_{t_f} \delta x(t_f) + \left(\mathcal{H} + \frac{\partial \theta}{\partial t_f} \right)^T \Bigg|_{t_f} \delta t_f = 0 \tag{3.59}$$

If θ is not an explicit function of the final time, and the problem corresponds to one where the final time is free, the Transversality condition requires $\mathcal{H} = 0$ at the final time, i.e.,

$$\mathcal{H}(t_f) = 0. \tag{3.60}$$

Consider the total time derivative of the Hamiltonian for a system whose state equations do not explicitly depend on time

$$\frac{d\mathcal{H}}{dt} = \overset{0}{\cancel{\frac{\partial \mathcal{H}}{\partial t}}} + \frac{\partial \mathcal{H}}{\partial x}^T \frac{dx}{dt} + \frac{\partial \mathcal{H}}{\partial u}^T \frac{du}{dt} + \frac{\partial \mathcal{H}}{\partial \lambda}^T \frac{d\lambda}{dt} \tag{3.61}$$

which can be rewritten as using Equations (3.56) and (3.57) as

$$\frac{d\mathcal{H}}{dt} = \cancel{\frac{\partial \mathcal{H}}{\partial x}}^T \cancel{\frac{\partial \mathcal{H}}{\partial \lambda}} + \frac{\partial \mathcal{H}}{\partial u}^T \frac{du}{dt} - \cancel{\frac{\partial \mathcal{H}}{\partial \lambda}}^T \cancel{\frac{\partial \mathcal{H}}{\partial x}} \tag{3.62}$$

Since the optimality conditions requires $\frac{\partial \mathcal{H}}{\partial u} = 0$, Equation (3.62) can be reduced to

$$\frac{d\mathcal{H}}{dt} = \overset{0}{\cancel{\frac{\partial \mathcal{H}}{\partial u}}} \frac{du}{dt} = 0 \Rightarrow \mathcal{H}(t) = \text{Constant} \tag{3.63}$$

The constraint that $\mathcal{H}(t_f) = 0$, requires the Hamiltonian be zero for all time which is a useful condition for numerical simulation.

The control law which satisfies Equations (3.56) through (3.59) is called the optimal controller. Equation (3.58) is referred to as the optimality condition which assumes that the control input is unbounded. If the control input is restricted to an admissible set, the Pontryagin's minimum principle (PMP) is sufficient to determine the optimal control law. The PMP states that the optimal control at any instant of time satisfies the constraint

$$\mathcal{H}(x^*, u^*, \Lambda^*, t)) \leq \mathcal{H}(x^*, u, \Lambda^*, t)) \ \forall \text{ admissible } u \tag{3.64}$$

where $(.)^*$ represent the optimal quantities. Occasionally, the principle is presented as Pontryagin's maximum principle since the original development by Pontryagin focused on maximizing a cost function.

Example 3.3:

Design a rest-to-rest optimal controller to minimize the cost function

$$J = \int_0^\pi \left(\dot{u}(t)^2 \right) dt \tag{3.65}$$

for the spring-mass system whose equation of motion is

$$\ddot{y} + \omega^2 y = u, \tag{3.66}$$

which in state space form is

$$\dot{x} = \begin{bmatrix} 0 & 1 \\ -\omega^2 & 0 \end{bmatrix} x + \begin{bmatrix} 0 \\ 1 \end{bmatrix} u \tag{3.67}$$

where x is the state vector $\{y \ \dot{y}\}^T$. The boundary conditions of the rest-to-rest maneuver are

$$x(0) = \dot{x}(0) = 0, \text{ and } x(\pi) = 1, \dot{x}(\pi) = 0. \tag{3.68}$$

Define a new state u, which satisfies the equation:

$$\dot{u} = v \tag{3.69}$$

where v is new control variable to be determined. The cost and state-space model are now given as:

$$\phi = v^2 \tag{3.70}$$

$$f(x(t), u(t), v(t)) = \begin{bmatrix} 0 & 1 & 0 \\ -\omega^2 & 0 & 1 \\ 0 & 0 & 0 \end{bmatrix} \begin{Bmatrix} x \\ u \end{Bmatrix} + \begin{bmatrix} 0 \\ 0 \\ 1 \end{bmatrix} v, \tag{3.71}$$

The Hamiltonian of the system is

$$\mathcal{H} = v^2 + \lambda^T \left(\begin{bmatrix} 0 & 1 & 0 \\ -\omega^2 & 0 & 1 \\ 0 & 0 & 0 \end{bmatrix} \begin{Bmatrix} x \\ u \end{Bmatrix} + \begin{bmatrix} 0 \\ 0 \\ 1 \end{bmatrix} v \right). \tag{3.72}$$

Since the final time t_f, and final states $x(t_f)$ are fixed, δt_f and δx_f are zero. However, the new state u is not specified at the initial and final times. The necessary conditions for optimality are

$$\begin{Bmatrix} \dot{x} \\ \dot{u} \end{Bmatrix} = \frac{\partial \mathcal{H}}{\partial \lambda} = \begin{bmatrix} 0 & 1 & 0 \\ -\omega^2 & 0 & 1 \\ 0 & 0 & 0 \end{bmatrix} \begin{Bmatrix} x \\ u \end{Bmatrix} + \begin{bmatrix} 0 \\ 0 \\ 1 \end{bmatrix} v \tag{3.73}$$

$$-\dot{\lambda} = \frac{\partial \mathcal{H}}{\partial y} = \begin{bmatrix} 0 & -\omega^2 & 0 \\ 1 & 0 & 0 \\ 0 & 1 & 0 \end{bmatrix} \lambda \tag{3.74}$$

$$0 = \frac{\partial \mathcal{H}}{\partial v} = 2v + \begin{bmatrix} 0 & 0 & 1 \end{bmatrix} \lambda \tag{3.75}$$

with the boundary conditions

$$\lambda_3(0) = \lambda_3(\pi) = 0. \tag{3.76}$$

where $y = \{x\ u\}$.

The costates can be represented from Equation (3.74) as

$$\lambda(t) = \begin{bmatrix} \cos(\omega t) & \omega \sin(\omega t) & 0 \\ -\frac{\sin(\omega t)}{\omega} & \cos(\omega t) & 0 \\ -\frac{\cos(\omega t)+1}{\omega^2} & -\frac{\sin(\omega t)}{\omega} & 1 \end{bmatrix} \lambda(0) \tag{3.77}$$

and the optimal control from Equation (3.75) is

$$v = -\frac{1}{2} \begin{bmatrix} 0 & 0 & 1 \end{bmatrix} \lambda(t) = -\frac{1}{2} \begin{bmatrix} -\frac{\cos(\omega t)+1}{\omega^2} & -\frac{\sin(\omega t)}{\omega} & 1 \end{bmatrix} \lambda(0). \tag{3.78}$$

The system states at any the final time are given as:

$$\begin{Bmatrix} x_1 \\ x_2 \\ u \end{Bmatrix}(\pi) = \begin{bmatrix} \cos(\omega\pi) & \frac{\sin(\omega\pi)}{\omega} & \frac{-\cos(\omega\pi)+1}{\omega^2} \\ -\omega\sin(\omega\pi) & \cos(\omega\pi) & \frac{\sin(\omega\pi)}{\omega} \\ 0 & 0 & 1 \end{bmatrix} \begin{Bmatrix} 0 \\ 0 \\ u(0) \end{Bmatrix} +$$

$$\begin{bmatrix} -\frac{\cos(\pi\omega)\omega\pi-3\sin(\omega\pi)+2\pi\omega}{4\omega^5} & -\frac{2\omega\cos(\pi\omega)+\sin(\omega\pi)\omega^2\pi-2\omega}{4\omega^5} & 0 \\ -\frac{2\cos(\pi\omega)-\sin(\omega\pi)\omega\pi+2}{4\omega^4} & -\frac{-\omega\sin(\pi\omega)+\cos(\omega\pi)\omega^2\pi}{4\omega^4} & 0 \\ -\frac{\sin(\pi\omega)+\omega\pi}{2\omega^3} & -\frac{\omega\cos(\pi\omega)-\omega}{2\omega^3} & 0 \end{bmatrix} \begin{Bmatrix} \lambda_1(0) \\ \lambda_2(0) \\ 0 \end{Bmatrix} \tag{3.79}$$

The costates at the final time are given by the equation:

$$\left\{\begin{matrix} \lambda_1 \\ \lambda_2 \\ \lambda_3 \end{matrix}\right\} (\pi) = \begin{bmatrix} \cos(\omega\pi) & \omega\sin(\omega\pi) & 0 \\ -\frac{\sin(\omega\pi)}{\omega} & \cos(\omega\pi) & 0 \\ \frac{-\cos(\omega\pi)+1}{\omega^2} & -\frac{\sin(\omega\pi)}{\omega} & 1 \end{bmatrix} \left\{\begin{matrix} \lambda_1(0) \\ \lambda_2(0) \\ 0 \end{matrix}\right\} \qquad (3.80)$$

Since $\lambda_3(\pi)$ should equal 0, we arrive at a constraint equation:

$$\begin{bmatrix} \frac{-\cos(\omega\pi)+1}{\omega^2} & -\frac{\sin(\omega\pi)}{\omega} \end{bmatrix} \left\{\begin{matrix} \lambda_1(0) \\ \lambda_2(0) \end{matrix}\right\} = 0 \qquad (3.81)$$

Equations (3.79) and (3.81) can be solved for $\lambda_1(0)$, $\lambda_2(0)$ and $u(0)$, resulting in:

$$\left\{\begin{matrix} \lambda_1(0) \\ \lambda_2(0) \\ u(0) \end{matrix}\right\} = \left\{\begin{matrix} 0 \\ 0 \\ 0.5 \end{matrix}\right\} \qquad (3.82)$$

which implies that $v(t) = 0$ for all time t, and the control $u(t) = 0.5$ for all time. Figure 3.5 illustrates the evolution of the system states x_1 (solid line) and x_2 (dashed line), and the control input for $\omega = 1$.

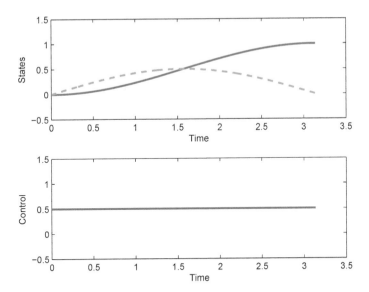

Figure 3.5: System Response and Optimal Control

Example 3.4:

Design a rest-to-rest optimal controller to minimize the cost function

$$J = \int_0^\pi \left(u(t)^2 \right) dt \tag{3.83}$$

for the spring-mass system whose equation of motion is

$$\ddot{y} + \omega^2 y = u, \tag{3.84}$$

which in state-space form is

$$\dot{x} = \begin{bmatrix} 0 & 1 \\ -\omega^2 & 0 \end{bmatrix} x + \begin{bmatrix} 0 \\ 1 \end{bmatrix} u \tag{3.85}$$

where x is the state vector. The boundary conditions of the rest-to-rest maneuver are

$$x(0) = \dot{x}(0) = 0, \text{ and } x(\pi) = 1, \dot{x}(\pi) = 0. \tag{3.86}$$

Since,

$$\phi = u^2 \tag{3.87}$$

$$f(x(t), u(t)) = \begin{bmatrix} 0 & 1 \\ -\omega^2 & 0 \end{bmatrix} x + \begin{bmatrix} 0 \\ 1 \end{bmatrix} u, \tag{3.88}$$

The Hamiltonian of the system is

$$\mathcal{H} = u^2 + \lambda^T \left(\begin{bmatrix} 0 & 1 \\ -\omega^2 & 0 \end{bmatrix} x + \begin{bmatrix} 0 \\ 1 \end{bmatrix} u \right). \tag{3.89}$$

Since the final time and final states are fixed, δt_f and δx_f are zero. The necessary conditions for optimality are

$$\dot{x} = \frac{\partial \mathcal{H}}{\partial \lambda} = \begin{bmatrix} 0 & 1 \\ -\omega^2 & 0 \end{bmatrix} x + \begin{bmatrix} 0 \\ 1 \end{bmatrix} u \tag{3.90}$$

$$-\dot{\lambda} = \frac{\partial \mathcal{H}}{\partial x} = \begin{bmatrix} 0 & -\omega^2 \\ 1 & 0 \end{bmatrix} \lambda \tag{3.91}$$

$$0 = \frac{\partial \mathcal{H}}{\partial u} = 2u + \begin{bmatrix} 0 & 1 \end{bmatrix} \lambda \tag{3.92}$$

The costates can be represented from Equation (3.91) as

$$\lambda(t) = \begin{bmatrix} \cos(\omega t) & \omega \sin(\omega t) \\ -\frac{\sin(\omega t)}{\omega} & \cos(\omega t) \end{bmatrix} \lambda(0) \tag{3.93}$$

and the optimal control from Equation (3.92) is

$$u = -\frac{1}{2}\begin{bmatrix} 0 & 1 \end{bmatrix}\lambda = \frac{1}{2}\left[\frac{\sin(\omega t)}{\omega} - \cos(\omega t)\right]\lambda(0) \tag{3.94}$$

Substituting Equation (3.94) into Equation (3.90), we have

$$x(t) = \frac{\lambda_1(0)}{4}\left(\frac{\sin(\omega t)}{\omega^3} - \frac{t\cos(\omega t)}{\omega^2}\right) - \frac{\lambda_2(0)}{4}\frac{t\sin(\omega t)}{\omega} \tag{3.95}$$

where

$$\lambda(0) = \left\{\begin{array}{c} \frac{4\omega^3(\sin(\pi\omega) + \pi\omega\cos(\pi\omega))}{\sin(\pi\omega)^2 - \pi^2\omega^2} \\ \frac{4\omega^3\pi\sin(\pi\omega)}{\sin(\pi\omega)^2 - \pi^2\omega^2} \end{array}\right\} \tag{3.96}$$

are solved from the boundary conditions. For $\omega = 1$, the system response and the control input is illustrated in Figure 3.6.

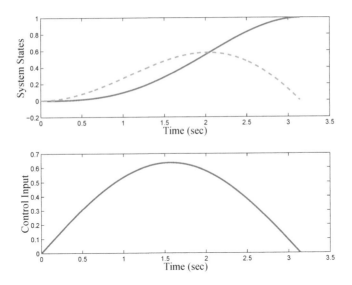

Figure 3.6: System Response and Optimal Control

3.2.1 Linear Quadratic Regulator (LQR)

Chapter 2 dealt with technique to modify the reference input to stable or marginally stable linear systems with the goal of reaching the desired states in finite time. In the presence of model uncertainties, the objective was to minimize the residual vibra-tion of the system response. The shaping or filtering of the reference input is an open-loop solution.

Wiener [4] developed a filter to minimize the effect of noise on a signal. He posed an optimization problem where the goal was to minimize the expected value of the square of the difference of the stationary noisy signal and the truth. Subsequently, Kalman and Koepcke [5] used the same quadratic cost function to develop state feedback controllers for linear systems. The resulting solution (LQR) is one of the most popular and widely studied technique for feedback controller design. This section will illustrate the formulation of the LQR problem.

Consider the cost function

$$J = \frac{1}{2}x(t_f)^T Sx(t_f) + \frac{1}{2}\int_0^{t_f} \left(x^T Qx + u^T Ru\right) dt \tag{3.97}$$

where the matrices S and Q are constant positive definite or positive semi-definite while the matrix R is positive definite, subject to the constraints

$$\dot{x} = Ax + Bu \tag{3.98}$$

$$x(0) = x_0. \tag{3.99}$$

The Hamiltonian of the system is

$$\mathcal{H} = \frac{1}{2}(x^T Qx + u^T Ru) + \lambda^T (Ax + Bu). \tag{3.100}$$

The Euler-Lagrange equations are

$$\dot{x} = \frac{\partial \mathcal{H}}{\partial \lambda} = Ax + Bu \tag{3.101}$$

$$\dot{\lambda} = -\frac{\partial \mathcal{H}}{\partial x} = -A^T \lambda - Qx \tag{3.102}$$

$$\frac{\partial \mathcal{H}}{\partial u} = Ru + B^T \lambda = 0 \tag{3.103}$$

where λ is referred to as the costate.

The control can be solved from Equation (3.103) resulting in the equation

$$u = -R^{-1}B^T \lambda. \tag{3.104}$$

Substituting Equation (3.104) into the state and costate equations (Equations 3.101 and 3.102), the resulting coupled equations are

$$\dot{x} = Ax - BR^{-1}B^T \lambda \tag{3.105}$$

$$\dot{\lambda} = -A^T \lambda - Qx. \tag{3.106}$$

The transversality conditions result in the constraint

$$Sx(t_f) = \lambda(t_f). \tag{3.107}$$

Assuming the structure of Equation (3.107) to relate the states and costates for all time, we have

$$\lambda(t) = Px(t). \tag{3.108}$$

Differentiating Equation (3.108) with respect to time gives

$$\dot{\lambda} = \dot{P}x + P\dot{x}. \tag{3.109}$$

Substituting the state and costate equations into Equation (3.109), we have

$$-A^T Px - Qx = \dot{P}x + P(Ax - BR^{-1}B^T Px) \tag{3.110}$$

which can be rewritten as

$$-\dot{P} = A^T P + PA + Q - PBR^{-1}B^T P \tag{3.111}$$

which is referred to as the matrix differential Riccati equation which is a nonlinear differential equation in P subject to the boundary conditions given by

$$P(t_f) = S. \tag{3.112}$$

Integrating Equation (3.111) backward in time with initial conditions S results in an optimal controller

$$u = -R^{-1}B^T Px = -Kx \tag{3.113}$$

where the feedback gain $K \equiv R^{-1}B^T P$ is a time varying feedback gain which requires the time variation of P to be stored prior to its implementation.

For the infinite time horizon problem, assuming that the closed-loop system is stable, the terminal cost does not contribute to J and the matrix differential Riccati equation can be reduced to the algebraic Riccati equation (ARE) since asymptotically $\dot{P}$ is zero. The time invariant feedback gain of the infinite time horizon LQR problem is determined from the solution of the ARE,

$$A^T P_{ss} + P_{ss}A - P_{ss}BR^{-1}B^T P_{ss} + Q = 0. \tag{3.114}$$

Define

$$P = P_{ss} + V \tag{3.115}$$

and substituting Equation (3.115) into Equation (3.111) results in

$$-\dot{V} = A^T(V + P_{ss}) + (V + P_{ss})A + Q - (V + P_{ss})BR^{-1}B^T(V + P_{ss}) \tag{3.116}$$

which can be rewritten as

$$-\dot{V} = A^T V + VA - VBR^{-1}B^T V - VBR^{-1}B^T P_{ss} - P_{ss}BR^{-1}B^T V$$
$$+ \underbrace{A^T P_{ss} + P_{ss}A + Q - P_{ss}BR^{-1}B^T P_{ss}}_{=0} \qquad (3.117)$$

which results in the equation

$$-\dot{V} = (A - BR^{-1}B^T P_{ss})^T V + V(A - BR^{-1}B^T P_{ss}) - VBR^{-1}B^T V \qquad (3.118)$$

subject to the boundary conditions

$$V(t_f) = S - P_{ss}. \qquad (3.119)$$

Assuming that V^{-1} exists, consider the identity

$$V^{-1}V = I. \qquad (3.120)$$

The time derivative of Equation (3.120) results in

$$\dot{V}^{-1}V + V^{-1}\dot{V} = 0, \Rightarrow \dot{V}^{-1} = -V^{-1}\dot{V}V^{-1}. \qquad (3.121)$$

Defining

$$\Phi = V^{-1} \qquad (3.122)$$

and substituting Equations (3.121) and (3.122) into Equation (3.118) results in the equation

$$\dot{\Phi} = \Phi(A - BR^{-1}B^T P_{ss})^T + (A - BR^{-1}B^T P_{ss})\Phi - BR^{-1}B^T \qquad (3.123)$$

subject to the boundary conditions

$$\Phi(t_f) = (S - P_{ss})^{-1}. \qquad (3.124)$$

Equation (3.123) is the differential Lyapunov equation which can be solved in closed form which can subsequently be used to solve for $P(t)$ and the feedback gain in closed form. Appendix B presents a technique for solving the differential Lyapunov equation using a matrix exponential function.

Example 3.5: Design a finite-time LQR controller for the double integrator, where the cost function is

$$J = \frac{1}{2}\int_0^T \begin{Bmatrix} x \\ \dot{x} \end{Bmatrix}^T \begin{bmatrix} 1 & 0 \\ 0 & 1 \end{bmatrix} \begin{Bmatrix} x \\ \dot{x} \end{Bmatrix} + u^T[1]u \ dt \qquad (3.125)$$

and the state space equation of motion is

$$\left\{ \begin{matrix} \dot{x} \\ \ddot{x} \end{matrix} \right\} = \begin{bmatrix} 0 & 1 \\ 0 & 0 \end{bmatrix} \left\{ \begin{matrix} x \\ \dot{x} \end{matrix} \right\} + \left\{ \begin{matrix} 0 \\ 1 \end{matrix} \right\} u. \tag{3.126}$$

The differential Riccati equation is

$$- \begin{bmatrix} \dot{p}_{11} & \dot{p}_{12} \\ \dot{p}_{12} & \dot{p}_{22} \end{bmatrix} = \begin{bmatrix} 0 & 0 \\ 1 & 0 \end{bmatrix} \begin{bmatrix} p_{11} & p_{12} \\ p_{12} & p_{22} \end{bmatrix} +$$
$$\begin{bmatrix} p_{11} & p_{12} \\ p_{12} & p_{22} \end{bmatrix} \begin{bmatrix} 0 & 1 \\ 0 & 0 \end{bmatrix} + \begin{bmatrix} 1 & 0 \\ 0 & 1 \end{bmatrix} - \begin{bmatrix} p_{12}^2 & p_{12}p_{22} \\ p_{12}p_{22} & p_{22}^2 \end{bmatrix} \tag{3.127}$$

which is a nonlinear equation, subject to the boundary conditions

$$\begin{bmatrix} p_{11} & p_{12} \\ p_{12} & p_{22} \end{bmatrix} (T) = \begin{bmatrix} 0 & 0 \\ 0 & 0 \end{bmatrix} \tag{3.128}$$

since there is no terminal penalty on the states. The solution of the algebraic Riccati equation is:

$$P_{ss} = \begin{bmatrix} \sqrt{3} & 1 \\ 1 & \sqrt{3} \end{bmatrix} \tag{3.129}$$

Defining the solution of differential Riccati equation (3.127) as

$$P = \begin{bmatrix} p_{11} & p_{12} \\ p_{12} & p_{22} \end{bmatrix} = P_{ss} + V, \tag{3.130}$$

Equation (3.127) can be rewritten as

$$-\dot{V} = A^T V + VA - VBR^{-1}B^T V - P_{ss}BR^{-1}B^T V - VBR^{-1}B^T P_{ss}, \tag{3.131}$$

or

$$-\dot{V} = (A - BR^{-1}B^T P_{ss})^T V + V(A - BR^{-1}B^T P_{ss}) - VBR^{-1}B^T V. \tag{3.132}$$

subject to the boundary condition

$$V(T) = -P_{ss}. \tag{3.133}$$

Defining $\Phi = V^{-1}$, Equation (3.132) can be rewritten as

$$\dot{\Phi} = \Phi(A - BR^{-1}B^T P_{ss})^T + (A - BR^{-1}B^T P_{ss})\Phi - BR^{-1}B^T. \tag{3.134}$$

which is a differential Lyapunov equation subject to the boundary conditions

$$\Phi(T) = -P_{ss}^{-1}. \tag{3.135}$$

Equation (3.134) is

$$\dot{\Phi} = \Phi \begin{bmatrix} 0 & -1 \\ 1 & -\sqrt{3} \end{bmatrix} + \begin{bmatrix} 0 & 1 \\ -1 & -\sqrt{3} \end{bmatrix} \Phi - \begin{bmatrix} 0 & 0 \\ 0 & 1 \end{bmatrix} \tag{3.136}$$

which can be written in vector differential form as

$$\left\{ \begin{matrix} \dot{\phi}_{11} \\ \dot{\phi}_{12} \\ \dot{\phi}_{22} \end{matrix} \right\} = \underbrace{\begin{bmatrix} 0 & 2 & 0 \\ -1 & -\sqrt{3} & 1 \\ 0 & -2 & -2\sqrt{3} \end{bmatrix}}_{\mathscr{A}} \left\{ \begin{matrix} \phi_{11} \\ \phi_{12} \\ \phi_{22} \end{matrix} \right\} + \underbrace{\left\{ \begin{matrix} 0 \\ 0 \\ -1 \end{matrix} \right\}}_{\mathscr{B}} \tag{3.137}$$

subject to boundary conditions

$$\left\{ \begin{matrix} \phi_{11} \\ \phi_{12} \\ \phi_{22} \end{matrix} \right\} (T) = \left\{ \begin{matrix} -\frac{\sqrt{3}}{2} \\ \frac{1}{2} \\ -\frac{\sqrt{3}}{2} \end{matrix} \right\} \tag{3.138}$$

Closed form solution of Equation (3.137) is

$$\left\{ \begin{matrix} \phi_{11} \\ \phi_{12} \\ \phi_{22} \end{matrix} \right\} (t) = e^{\mathscr{A}t} \left\{ \begin{matrix} \phi_{11} \\ \phi_{12} \\ \phi_{22} \end{matrix} \right\} (0) + \mathscr{A}^{-1}(e^{\mathscr{A}t} - \mathscr{I})\mathscr{B}. \tag{3.139}$$

where $\mathscr{I}$ is the identity matrix. The initial conditions can be solved for from the terminal constraint (Equation 3.136) which results in the equation

$$\left\{ \begin{matrix} \phi_{11} \\ \phi_{12} \\ \phi_{22} \end{matrix} \right\} (0) = \left\{ \begin{matrix} -\frac{\sqrt{3}}{6}\left(e^{-T\sqrt{3}} + 2\right)e^{T\sqrt{3}} \\ 1/2\,e^{T\sqrt{3}} \\ -\frac{\sqrt{3}}{6}\left(e^{-T\sqrt{3}} + 2\right)e^{T\sqrt{3}} \end{matrix} \right\} \tag{3.140}$$

which can be used to arrive at a closed-form expression for the V matrix which is subsequently used to solve for the P matrix in closed-form:

$$P(t) = \begin{bmatrix} \dfrac{\sqrt{3}\sinh\left(\sqrt{3}(-t+T)\right)}{2+\cosh\left(\sqrt{3}(-t+T)\right)} & \dfrac{-1+\cosh\left(\sqrt{3}(-t+T)\right)}{2+\cosh\left(\sqrt{3}(-t+T)\right)} \\ \dfrac{-1+\cosh\left(\sqrt{3}(-t+T)\right)}{2+\cosh\left(\sqrt{3}(-t+T)\right)} & \dfrac{\sqrt{3}\sinh\left(\sqrt{3}(-t+T)\right)}{2+\cosh\left(\sqrt{3}(-t+T)\right)} \end{bmatrix} \tag{3.141}$$

and the feedback control gain is given as

$$K = R^{-1}B^{T}P(t) = \left\{ \dfrac{-1+\cosh\left(\sqrt{3}(-t+T)\right)}{2+\cosh\left(\sqrt{3}(-t+T)\right)} \quad \dfrac{\sqrt{3}\sinh\left(-\sqrt{3}(-t+T)\right)}{2+\cosh\left(\sqrt{3}(-t+T)\right)} \right\} \tag{3.142}$$

Example 3.6:

Consider the floating oscillator benchmark problem where the masses and the spring stiffness are unity. The state-space equation of motion is

$$\begin{Bmatrix} \dot{x}_1 \\ \dot{x}_2 \\ \ddot{x}_1 \\ \ddot{x}_2 \end{Bmatrix} = \underbrace{\begin{bmatrix} 0 & 0 & 1 & 0 \\ 0 & 0 & 0 & 1 \\ -1 & 1 & 0 & 0 \\ 1 & -1 & 0 & 0 \end{bmatrix}}_{A} \begin{Bmatrix} x_1 \\ x_2 \\ \dot{x}_1 \\ \dot{x}_2 \end{Bmatrix} + \underbrace{\begin{Bmatrix} 0 \\ 0 \\ 1 \\ 0 \end{Bmatrix}}_{B} u. \tag{3.143}$$

Design a LQR controller for the cost function

$$J = \frac{1}{2} \begin{Bmatrix} x_1 \\ x_2 \\ \dot{x}_1 \\ \dot{x}_2 \end{Bmatrix}^T \begin{bmatrix} 1 & 0 & 0 & 0 \\ 0 & 1 & 0 & 0 \\ 0 & 0 & 1 & 0 \\ 0 & 0 & 0 & 1 \end{bmatrix} \begin{Bmatrix} x_1 \\ x_2 \\ \dot{x}_1 \\ \dot{x}_2 \end{Bmatrix} + \frac{1}{2} \int_0^{t_f} \begin{Bmatrix} x_1 \\ x_2 \\ \dot{x}_1 \\ \dot{x}_2 \end{Bmatrix}^T \begin{bmatrix} 1 & 0 & 0 & 0 \\ 0 & 1 & 0 & 0 \\ 0 & 0 & 1 & 0 \\ 0 & 0 & 0 & 1 \end{bmatrix} \begin{Bmatrix} x_1 \\ x_2 \\ \dot{x}_1 \\ \dot{x}_2 \end{Bmatrix} + u^T[1]u \, dt$$
$$\tag{3.144}$$

For a final time of t_f of 15 seconds, the differential Riccati equation

$$-\dot{P} = A^T P + PA + Q - PBR^{-1}B^T P \tag{3.145}$$

with the boundary condition

$$P(15) = \begin{bmatrix} 1 & 0 & 0 & 0 \\ 0 & 1 & 0 & 0 \\ 0 & 0 & 1 & 0 \\ 0 & 0 & 0 & 1 \end{bmatrix}. \tag{3.146}$$

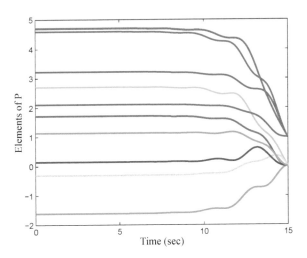

Figure 3.7: Evolution of the Terms of the Riccati Matrix

Figure 3.7 illustrates the variation of the terms of the Riccati matrix. It can be seen that the transients in the time response of the terms of the Riccati matrix exist towards the end of the maneuver.

The optimal control is given by the equation

$$u = -R^{-1}B^T P(t)x(t) \tag{3.147}$$

and is shown in Figure 3.8

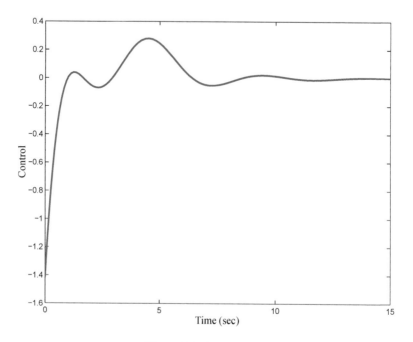

Figure 3.8: Control Input

Figure 3.9 illustrates the variation of the states of the system and it is clear that ten seconds into the controlled motion, the states are nearly zero. Figure 3.7 illustrates that the terms of the Riccati matrix are nearly constant for the first ten seconds of the maneuver. This prompts the use of the algebraic Riccati equation to determine the time invariant feedback gain vector which is:

$$K = \begin{bmatrix} 1.7212 & -0.3070 & 2.1077 & 1.1365 \end{bmatrix}. \tag{3.148}$$

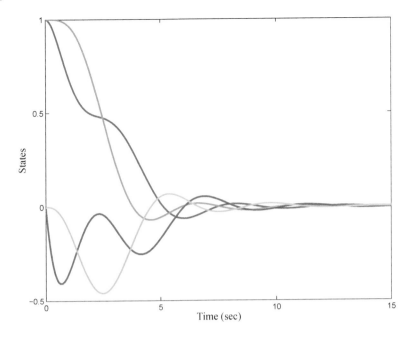

Figure 3.9: Evolution of System States

Example 3.7: Design a optimal control to minimize the cost function

$$J = x^T(T) \begin{bmatrix} \infty & 0 \\ 0 & \infty \end{bmatrix} x(T) + \int_0^T u^T R u \, dt \tag{3.149}$$

for the single spring-mass system

$$\begin{Bmatrix} \dot{x} \\ \ddot{x} \end{Bmatrix} = \begin{bmatrix} 0 & 1 \\ -1 & 0 \end{bmatrix} \begin{Bmatrix} x \\ \dot{x} \end{Bmatrix} + \begin{Bmatrix} 0 \\ 1 \end{Bmatrix} u. \tag{3.150}$$

while moving the system from any initial condition to the equilibrium point at the origin of the state space. This problem is the minimum power control problem when R is selected to be one. The infinite magnitude of the terminal penalty on the states is equivalent to requiring the terminal states to be zero for the LQR problem.

The differential Riccati equation for the system is

$$-\begin{bmatrix} \dot{p}_{11} & \dot{p}_{12} \\ \dot{p}_{12} & \dot{p}_{22} \end{bmatrix} = \begin{bmatrix} 0 & -1 \\ 1 & 0 \end{bmatrix} \begin{bmatrix} p_{11} & p_{12} \\ p_{12} & p_{22} \end{bmatrix} +$$
$$\begin{bmatrix} p_{11} & p_{12} \\ p_{12} & p_{22} \end{bmatrix} \begin{bmatrix} 0 & 1 \\ -1 & 0 \end{bmatrix} - \begin{bmatrix} p_{12}^2 & p_{12}p_{22} \\ p_{12}p_{22} & p_{22}^2 \end{bmatrix} \tag{3.151}$$

which is a nonlinear differential equation, subject to the boundary conditions

$$P(T) = \begin{bmatrix} p_{11} & p_{12} \\ p_{12} & p_{22} \end{bmatrix} (T) = \begin{bmatrix} \infty & 0 \\ 0 & \infty \end{bmatrix} \qquad (3.152)$$

Defining $\Phi = P^{-1}$, Equation (3.151) can be rewritten as

$$\dot{\Phi} = \Phi A^T + A\Phi - BR^{-1}B^T. \qquad (3.153)$$

which is a differential Lyapunov equation subject to the boundary conditions

$$\Phi(T) = -P^{-1}(T) = \begin{bmatrix} 0 & 0 \\ 0 & 0 \end{bmatrix}. \qquad (3.154)$$

Equation (3.153) is

$$\dot{\Phi} = \Phi \begin{bmatrix} 0 & -1 \\ 1 & 0 \end{bmatrix} + \begin{bmatrix} 0 & 1 \\ -1 & 0 \end{bmatrix} \Phi - \begin{bmatrix} 0 & 0 \\ 0 & 1 \end{bmatrix} \qquad (3.155)$$

which can be written in vector differential form as

$$\begin{Bmatrix} \dot{\phi}_{11} \\ \dot{\phi}_{12} \\ \dot{\phi}_{22} \end{Bmatrix} = \underbrace{\begin{bmatrix} 0 & 2 & 0 \\ -1 & 0 & 1 \\ 0 & -2 & 0 \end{bmatrix}}_{\mathscr{A}} \begin{Bmatrix} \phi_{11} \\ \phi_{12} \\ \phi_{22} \end{Bmatrix} + \underbrace{\begin{Bmatrix} 0 \\ 0 \\ -1 \end{Bmatrix}}_{\mathscr{B}} \qquad (3.156)$$

assuming that the Φ matrix is symmetric, subject to boundary conditions

$$\begin{Bmatrix} \phi_{11} \\ \phi_{12} \\ \phi_{22} \end{Bmatrix} (T) = \begin{Bmatrix} 0 \\ 0 \\ 0 \end{Bmatrix}. \qquad (3.157)$$

Closed form solution of Equation (3.156) is

$$\begin{Bmatrix} \phi_{11} \\ \phi_{12} \\ \phi_{22} \end{Bmatrix} (t) = \begin{bmatrix} \frac{1}{2}(1+\cos(2t)) & \sin(2t) & \frac{1}{2}(1-\cos(2t)) \\ -\frac{1}{2}\sin(2t) & \cos(2t) & \frac{1}{2}\sin(2t) \\ \frac{1}{2}(1-\cos(2t)) & -\sin(2t) & \frac{1}{2}(1+\cos(2t)) \end{bmatrix} \begin{Bmatrix} \phi_{11} \\ \phi_{12} \\ \phi_{22} \end{Bmatrix} (0)$$
$$+ \begin{Bmatrix} \frac{1}{4}\sin(2t) - \frac{1}{2}t \\ \frac{1}{4}\cos(2t) - \frac{1}{4} \\ -\frac{1}{4}\sin(2t) - \frac{1}{2}t \end{Bmatrix}. \qquad (3.158)$$

The initial conditions can be solved for from the terminal constraint (Equation (3.157)) which results in the equation

$$\begin{Bmatrix} \phi_{11} \\ \phi_{12} \\ \phi_{22} \end{Bmatrix} (0) = \begin{Bmatrix} -\frac{\sin(2T)}{4} + \frac{T}{2} \\ \frac{\cos(2T)}{4} - \frac{1}{4} \\ \frac{\sin(2T)}{4} + \frac{T}{2} \end{Bmatrix} \qquad (3.159)$$

which can used to arrive at a closed-form expression for the Φ matrix

$$\Phi(t) = \begin{bmatrix} \frac{1}{2}(T-t) - \frac{1}{4}\sin(2(T-t)) & \frac{1}{4}\cos(2(T-t)) - \frac{1}{4} \\ \frac{1}{4}\cos(2(T-t)) - \frac{1}{4} & \frac{1}{2}(T-t) + \frac{1}{4}\sin(2(T-t)) \end{bmatrix} \quad (3.160)$$

which is used to arrive at a closed-form expression for the matrix P,

$$P(t) = \begin{bmatrix} 2\frac{\sin(2(T-t))+2(T-t)}{2(t-T)^2+\cos(2(T-t))-1} & -2\frac{\cos(2(T-t))-1}{2(t-T)^2+\cos(2(T-t))-1} \\ -2\frac{\cos(2(T-t))-1}{2(t-T)^2+\cos(2(T-t))-1} & 2\frac{2(T-t)-\sin(2(T-t))}{2(t-T)^2+\cos(2(T-t))-1} \end{bmatrix} \quad (3.161)$$

and the feedback control gain is given as

$$K = R^{-1}B^T P(t) = \left\{ -2\frac{\cos(2(T-t))-1}{2(t-T)^2+\cos(2(T-t))-1}, \; 2\frac{-\sin(2(T-t))+2(T-t)}{2(t-T)^2+\cos(2(T-t))-1} \right\} \quad (3.162)$$

For a maneuver time of $T = 10$, the variation of the gain vector K as a function of time is plotted on a semilog plot in Figure 3.10 where it can be seen that the gain increases towards the end of the maneuver since the magnitude of the states are tending to zero.

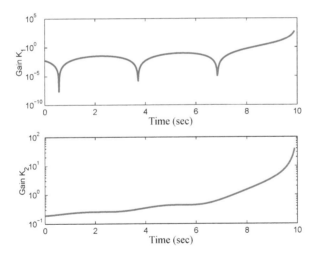

Figure 3.10: Variation of Feedback Gains

Figure 3.11 illustrates the variation of the states and the control as a function of time. The simulation is carried out using the time-varying state feedback gain matrix (Equation 3.162) for initial conditions which correspond to unit displacement and zero initial velocity.

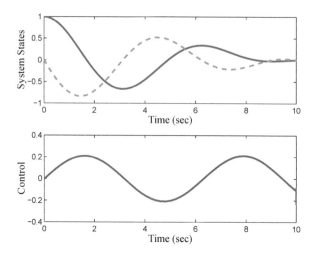

Figure 3.11: Evolution of States and Control

3.2.2   LQR without State Penalty

For the specific case of quadratic cost functions which are not a function of system states, a closed-form expression for the open-loop control profile can be derived to move the system from specified initial to final states in a specified time t_f. This section describes in detail the approach for designing the optimal control profile. Consider the cost function

$$J = \frac{1}{2} \int_0^{t_f} u^T R u \ dt \tag{3.163}$$

where the matrix R is positive definite, subject to the constraint

$$\dot{x} = Ax + Bu \tag{3.164}$$
$$x(0) = x_0 \tag{3.165}$$
$$x(t_f) = x_f. \tag{3.166}$$

which for convenience can be represented in its Jordan canonical form [6] using the similarity transformation $Vz = x$

$$\dot{z} = \underbrace{\bar{A}}_{V^{-1}AV} z + \underbrace{\bar{B}}_{V^{-1}B} u \tag{3.167}$$
$$z(0) = z_0 = V^{-1}x_0 \tag{3.168}$$
$$z(t_f) = z_f = V^{-1}x_f. \tag{3.169}$$

which can be used to solve for the states in closed form

$$z(t) = e^{\bar{A}t}z(0) + \int_0^t e^{\bar{A}(t-\tau)}\bar{B}u(\tau)\,d\tau \tag{3.170}$$

or at the final time,

$$e^{-\bar{A}t_f}z_f - z_0 = \int_0^{t_f} e^{-\bar{A}\tau}\bar{B}u(\tau)\,d\tau. \tag{3.171}$$

To solve for the optimal $u(t)$, calculus of variations is exploited. The cost function (Equation 3.163) is augmented with the terminal state constraint (Equation 3.171) resulting in the augmented cost

$$J_a = \frac{1}{2}\int_0^{t_f} u^T Ru\ dt + \lambda^T \left(e^{-\bar{A}t_f}z_f - z_0 - \int_0^{t_f} e^{-\bar{A}\tau}\bar{B}u(\tau)\,d\tau \right) \tag{3.172}$$

where λ is the Lagrange multiplier which is not a function of time. The first variation of the augmented cost J_a is

$$\delta J_a = \int_0^{t_f} \left\{ Ru - \lambda^T e^{-\bar{A}t}\bar{B} \right\} \delta u\ dt \tag{3.173}$$

which has to equal zero for the optimal control profile. This results in the equation

$$u(t) = R^{-1}\lambda^T e^{-\bar{A}t}\bar{B} \tag{3.174}$$

Substituting Equation (3.174) into Equation (3.171) results in n constraints for a nth order state space system in terms of the unknown variable λ which can be solved to result in a closed form expression for the optimal control profile.

Example 3.8: Consider the single spring-mass system and design an optimal control profile to minimize the cost function

$$J = \int_0^{10} u^2\ dt \tag{3.175}$$

subject to the constraints

$$\left\{ \begin{matrix} \dot{x}_1 \\ \ddot{x}_1 \end{matrix} \right\} = \begin{bmatrix} 0 & 1 \\ -1 & 0 \end{bmatrix} \left\{ \begin{matrix} x_1 \\ \dot{x}_1 \end{matrix} \right\} + \left\{ \begin{matrix} 0 \\ 1 \end{matrix} \right\} u \tag{3.176}$$

$$\left\{ \begin{matrix} x_1 \\ \dot{x}_1 \end{matrix} \right\}(0) = \left\{ \begin{matrix} 1 \\ 0 \end{matrix} \right\} \tag{3.177}$$

$$\left\{ \begin{matrix} x_1 \\ \dot{x}_1 \end{matrix} \right\}(10) = \left\{ \begin{matrix} 0 \\ 0 \end{matrix} \right\} \tag{3.178}$$

The optimal control profile is given by the Equation (3.174):

$$u(t) = \lambda^T e^{-At} B = \{\lambda_1 \ \lambda_2\} \begin{bmatrix} \cos(t) & -\sin(t) \\ \sin(t) & \cos(t) \end{bmatrix} \begin{Bmatrix} 0 \\ 1 \end{Bmatrix} \tag{3.179}$$

$$u(t) = -\lambda_1 \sin(t) + \lambda_2 \cos(t) \tag{3.180}$$

since $R = 1$. Substituting Equation (3.180) into the constraint (Equation 3.171), we have

$$\begin{bmatrix} \cos(10) & -\sin(10) \\ \sin(10) & \cos(10) \end{bmatrix} \begin{Bmatrix} 0 \\ 0 \end{Bmatrix} - \begin{Bmatrix} 1 \\ 0 \end{Bmatrix} =$$
$$\int_0^{10} \begin{bmatrix} \cos(t) & -\sin(t) \\ \sin(t) & \cos(t) \end{bmatrix} \begin{Bmatrix} 0 \\ 1 \end{Bmatrix} (-\lambda_1 \sin(t) + \lambda_2 \cos(t)) \, dt \tag{3.181}$$

or

$$-\begin{Bmatrix} 1 \\ 0 \end{Bmatrix} = \int_0^{10} \begin{bmatrix} \lambda_1 \sin^2(t) - \lambda_2 \sin(t) \cos(t) \\ -\lambda_1 \sin(t) \cos(t) + \lambda_2 \cos^2(t) \end{bmatrix} dt$$
$$= \begin{bmatrix} \frac{1}{4}(2t - \sin(2t)) & -\frac{1}{4}\cos(2t) \\ -\frac{1}{4}\cos(2t) & \frac{1}{4}(2t + \sin(2t)) \end{bmatrix} \Big|_0^{10} \begin{Bmatrix} \lambda_1 \\ \lambda_2 \end{Bmatrix} \tag{3.182}$$

which results in the closed-form solution

$$\begin{Bmatrix} \lambda_1 \\ \lambda_2 \end{Bmatrix} = -\begin{bmatrix} \frac{1}{4}(20 - \sin(20)) & \frac{1}{4}\cos(20) - \frac{1}{4} \\ \frac{1}{4}\cos(20) - \frac{1}{4} & \frac{1}{4}(20 + \sin(20)) \end{bmatrix}^{-1} \begin{Bmatrix} 1 \\ 0 \end{Bmatrix} = \begin{Bmatrix} -0.2098 \\ -0.0059 \end{Bmatrix} \tag{3.183}$$

Figure 3.12 illustrates the variation of the system states and control for the open-loop optimal control profile. Comparing Figure 3.12 to Figure 3.11 it is clear that the optimal solutions are identical.

3.2.3   Desensitized LQR Control

It is desirous to design controllers which are robust to errors in model parameters, i.e., they are insensitive to errors in estimated values of the parameters of the system model. The motivation comes from the fact that in real physical systems, the values which are assumed to be constant (e.g., natural frequency, damping ratio) are not known exactly. When this is the case, the control profile determined using estimated model parameters will not move the system from its initial states to the desired final states. To desensitize the control profile to errors in estimated model parameters, the sensitivity of the system states to the uncertain parameters are forced to zero at the end of the maneuver [7]. Consider the state-space model

$$\dot{x} = A(p)x + B(p)u \tag{3.184}$$

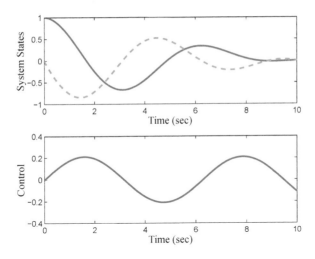

Figure 3.12: Evolution of States and Control

where p is the vector of uncertain parameters. The sensitivity of the states to the ith uncertain parameter p_i is given by the equation

$$\frac{d\dot{x}}{dp_i} = A(p)\frac{dx}{dp_i} + \frac{\partial A(p)}{\partial p_i}x + \frac{\partial B}{\partial p_i}u. \qquad (3.185)$$

The robustness constraint is imposed by requiring the sensitivity states to be forced to zero at the end of the maneuver, i.e., the initial and final sensitivity state constraints are

$$\frac{dx}{dp_i}(0) = 0 \qquad (3.186)$$

$$\frac{dx}{dp_i}(t_f) = 0. \qquad (3.187)$$

The augmented state-space model is given by the equation

$$\left\{\begin{array}{c} \dot{x} \\ \frac{d\dot{x}}{dp_i} \end{array}\right\} = \left[\begin{array}{cc} A(p) & 0 \\ \frac{\partial A(p)}{\partial p_i} & A(p) \end{array}\right]\left\{\begin{array}{c} x \\ \frac{dx}{dp_i} \end{array}\right\} + \left\{\begin{array}{c} B \\ \frac{\partial B}{\partial p_i} \end{array}\right\}u \qquad (3.188)$$

subject to the constraints

$$x(0) = x_0, \quad \frac{dx}{dp_i}(0) = 0 \qquad (3.189)$$

$$x(t_f) = x_f, \quad \frac{dx}{dp_i}(t_f) = 0. \qquad (3.190)$$

Example 3.9: Consider the single spring-mass system and design an optimal control profile to minimize the cost function

$$J = \int_0^{10} u^2 \, dt \tag{3.191}$$

subject to the constraints

$$\begin{Bmatrix} \dot{x} \\ \ddot{x} \end{Bmatrix} = \begin{bmatrix} 0 & 1 \\ -k & 0 \end{bmatrix} \begin{Bmatrix} x \\ \dot{x} \end{Bmatrix} + \begin{Bmatrix} 0 \\ 1 \end{Bmatrix} u \tag{3.192}$$

$$\begin{Bmatrix} x \\ \dot{x} \end{Bmatrix}(0) = \begin{Bmatrix} 1 \\ 0 \end{Bmatrix} \tag{3.193}$$

where the spring stiffness k is uncertain. The sensitivity of the states to k is given by the equation

$$\begin{Bmatrix} \frac{d\dot{x}}{dk} \\ \frac{d\ddot{x}}{dk} \end{Bmatrix} = \begin{bmatrix} 0 & 1 \\ -k & 0 \end{bmatrix} \begin{Bmatrix} \frac{dx}{dk} \\ \frac{d\dot{x}}{dk} \end{Bmatrix} + \begin{bmatrix} 0 & 0 \\ -1 & 0 \end{bmatrix} \begin{Bmatrix} x \\ \dot{x} \end{Bmatrix} \tag{3.194}$$

and the augmented state-space model is

$$\begin{Bmatrix} \dot{x} \\ \ddot{x} \\ \frac{d\dot{x}}{dk} \\ \frac{d\ddot{x}}{dk} \end{Bmatrix} = \underbrace{\begin{bmatrix} 0 & 1 & 0 & 0 \\ -k & 0 & 0 & 0 \\ 0 & 0 & 0 & 1 \\ -1 & 0 & -k & 0 \end{bmatrix}}_{A_a} \begin{Bmatrix} x \\ \dot{x} \\ \frac{dx}{dk} \\ \frac{d\dot{x}}{dk} \end{Bmatrix} + \underbrace{\begin{Bmatrix} 0 \\ 1 \\ 0 \\ 0 \end{Bmatrix}}_{B_a} u \tag{3.195}$$

subject to the initial and final state constraints

$$x(0) = x_0 \quad \frac{dx}{dk}(0) = 0 \tag{3.196}$$

$$x(t_f) = x_f, \quad \frac{dx}{dk}(t_f) = 0. \tag{3.197}$$

Assuming that the nominal estimate of the spring stiffness $k = 1$, the optimal control is given by the equation

$$u(t) = \qquad\qquad\qquad\qquad\qquad\qquad\qquad\qquad\qquad \lambda^T e^{-A_a t} B_a$$

$$u(t) = \begin{Bmatrix} \lambda_1 \\ \lambda_2 \\ \lambda_3 \\ \lambda_4 \end{Bmatrix}^T \begin{bmatrix} \cos(t) & -\sin(t) & 0 & 0 \\ \sin(t) & \cos(t) & 0 & 0 \\ -\frac{1}{2}t\sin(t) & -\frac{1}{2}t\cos(t) + \frac{1}{2}\sin(t) & \cos(t) & -\sin(t) \\ \frac{1}{2}t\cos(t) + \frac{1}{2}\sin(t) & -\frac{1}{2}t\sin(t) & \sin(t) & \cos(t) \end{bmatrix} \begin{Bmatrix} 0 \\ 1 \\ 0 \\ 0 \end{Bmatrix}$$

$$u(t) = -\lambda_1 \sin(t) + \lambda_2 \cos(t) + \lambda_3(-\frac{1}{2}t\cos(t) + \frac{1}{2}\sin(t)) - \lambda_4\frac{1}{2}t\sin(t)$$
$$\tag{3.198}$$

Substituting Equation (3.198) into Equation (3.171) results in the Lagrange multipliers

$$\begin{Bmatrix} \lambda_1 \\ \lambda_2 \\ \lambda_3 \\ \lambda_4 \end{Bmatrix} = \begin{Bmatrix} -0.97838108758535 \\ -0.27419663930798 \\ -0.08530392469518 \\ 0.29711406335070 \end{Bmatrix} \qquad (3.199)$$

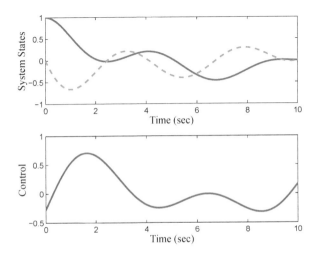

Figure 3.13: Evolution of States and Control

Figure 3.13 illustrates the variation of the system states and control profile for the robust minimum power control profile. The residual energies of the minimum power and the robust minimum power control profiles for variations in estimated spring stiffness is shown in Figure 3.14 where it is clear that around the nominal value of the spring stiffness, the robust control profile has increased the insensitivity of the controllers to errors in estimated model parameter.

One can conjecture that the increased robustness of the control trade-offs an increase in the performance index. The cost function corresponding to the optimal control in Example 3.8 is 0.2098 and the cost function of the insensitive control in Example 3.9 is 0.9784, a 366% increase which is a significant penalty for the resulting robustness.

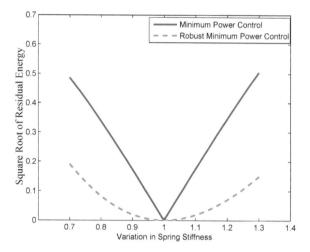

Figure 3.14: Controller Sensitivity

3.3   Minimum Power Control

Another important subset of the LQR problem is one where the state errors are not penalized. Minimizing the integral of the quadratic of the input is referred to as minimum power control. This section will describe the development of a closed-form expression for the open-loop minimum power, point-to-point motion control. The focus here is on the control of structures whose dynamics are represented by second-order matrix differential equations.

3.3.1   Minimum Power Control of Maneuvering Structures

This section deals with the design of rest-to-rest maneuver controllers which minimize the cost function

$$J = \frac{1}{2} \int_0^T u^T u \, dt \tag{3.200}$$

for multi-mode flexible structures with one rigid body mode and N flexible modes, modeled as

$$M\ddot{x} + \xi\dot{x} + Kx = Pu \tag{3.201}$$

where M is the mass matrix, ξ the damping matrix, K the stiffness matrix and the maneuver time T is specified. P is the control influence vector, and u and x are the scalar control input and state vector, respectively. It is assumed that the damping matrix ξ satisfies the Rayleigh damping model [8]

$$\xi = \alpha M + \beta K. \tag{3.202}$$

Defining $x = \Phi z$ where Φ is the eigenvector matrix resulting from the generalized eigenvalue problem of the matrix pair (K, M), the equations of motion can be rewritten as

$$\ddot{z} + \begin{bmatrix} 0 & 0 & 0 \\ & 0 & 2\zeta_1\omega_1 \\ & & 0 & 2\zeta_2\omega_2 & 0 \\ & & & & \ddots \end{bmatrix} \dot{z} + \begin{bmatrix} 0 & 0 & 0 \\ & 0 & \omega_1^2 \\ & & 0 & \omega_2^2 & 0 \\ & & & & \ddots \end{bmatrix} z = \left\{ \begin{matrix} \phi_0 \\ \phi_1 \\ \phi_2 \\ \vdots \end{matrix} \right\} u \tag{3.203}$$

which in state-space form is

$$\dot{w} = \begin{bmatrix} 0 & 1 & & & & 0 \\ 0 & 0 & & & & \\ & & 0 & 1 & & \\ & & -\omega_1^2 & -2\zeta_1\omega_1 & & \\ & & & & 0 & 1 \\ & & & & -\omega_2^2 & -2\zeta_2\omega_2 \\ 0 & & & & & \ddots \end{bmatrix} w + \begin{bmatrix} 0 \\ \phi_0 \\ 0 \\ \phi_1 \\ 0 \\ \phi_2 \\ \vdots \end{bmatrix} u. \tag{3.204}$$

Defining a similarity transformation

$$w = \underbrace{\begin{bmatrix} 1 & 0 & 0 & 0 & & & 0 \\ 0 & 1 & 0 & 0 & & & \\ 0 & 0 & -\frac{\zeta_1+j\sqrt{1-\zeta_1^2}}{\omega_1} & -\frac{\zeta_1-j\sqrt{1-\zeta_1^2}}{\omega_1} & & & \\ 0 & 0 & 1 & 1 & & 0 & 0 \\ & & 0 & 0 & -\frac{\zeta_2+j\sqrt{1-\zeta_2^2}}{\omega_2} & -\frac{\zeta_2-j\sqrt{1-\zeta_2^2}}{\omega_2} & \\ & & & & 1 & 1 & \\ 0 & & & & & & \ddots \end{bmatrix}}_{\Psi} y, \tag{3.205}$$

Equation (3.204) can be rewritten in the Jordan canonical form as

$$\dot{y} = \underbrace{\begin{bmatrix} 0 & 1 & & & & 0 \\ 0 & 0 & & & & \\ & & -p_1+jq_1 & 0 & & \\ & & 0 & -p_1-jq_1 & & \\ & & & & -p_2+jq_2 & 0 \\ & & & & 0 & -p_2-jq_2 \\ 0 & & & & & \ddots \end{bmatrix}}_{\mathcal{A}} y + \underbrace{\begin{bmatrix} 0 \\ \psi_2 \\ \psi_3 \\ \psi_4 \\ \psi_5 \\ \psi_6 \\ \vdots \end{bmatrix}}_{\mathcal{B}} u. \tag{3.206}$$

where

$$p_i = \zeta_i \omega_i \text{ and } q_i = \omega_i \sqrt{1 - \zeta_i^2}. \tag{3.207}$$

It can be easily shown that

$$exp(-\mathscr{A}t) = \begin{bmatrix} 1 & -t & 0 & & & \\ 0 & 1 & & & & \\ & & e^{p_1 t}(\cos(q_1 t) - j\sin(q_1 t)) & & 0 & \\ & & 0 & & e^{p_1 t}(\cos(q_1 t) + j\sin(q_1 t)) & \\ & 0 & & & & \ddots \end{bmatrix}. \tag{3.208}$$

Equation (3.206) can be solved in closed form

$$exp(-\mathscr{A}T)y(T) - y(0) = \int_0^T exp(-\mathscr{A}t)\mathscr{B}u(t)\,dt \tag{3.209}$$

Augmenting the cost function given by Equation (3.200) with the constraint Equation (3.209), we have

$$J_a = \frac{1}{2} \int_0^T u^T u\,dt + \lambda^T \left(exp(-\mathscr{A}T)y(T) - y(0) - \int_0^T exp(-\mathscr{A}t)\mathscr{B}u(t)\,dt \right) \tag{3.210}$$

where λ is a constant. The variation of Equation (3.210) results in the equation:

$$\delta J_a = \int_0^T \left(u - \lambda^T exp(-\mathscr{A}t)\mathscr{B} \right) \delta u(t)\,dt \tag{3.211}$$

which leads to the equation:

$$u(t) = \lambda^T exp(-\mathscr{A}t)\mathscr{B} \tag{3.212}$$

$$u(t) = \lambda^T \begin{Bmatrix} -t\psi_2 \\ \psi_2 \\ e^{p_1 t}(\cos(q_1 t) - j\sin(q_1 t))\psi_3 \\ e^{p_1 t}(\cos(q_1 t) + j\sin(q_1 t))\psi_4 \\ \vdots \end{Bmatrix} \tag{3.213}$$

$$u(t) = \kappa_1 + \kappa_2 t + \sum_{i=1}^{N} \kappa_{2i+1} e^{p_i t} \cos(q_i t) + \kappa_{2i+2} e^{p_i t} \sin(q_i t) \tag{3.214}$$

The constants κ_i which are functions of λ and ψ_i, are determined by solving the constraint Equation (3.209). It is clear that the minimum power control profile consists of a ramp input which corresponds to the rigid body mode and an exponentially growing harmonic part which corresponds to the underdamped flexible modes. For undamped systems, the harmonic part of the control profile does not grow with time.

Example 3.10: Design a minimum power control for the undamped floating oscillator benchmark problem where the maneuver time is 10 seconds and the objective is to move from an initial point of rest at unity to a final position of rest at the origin.

The equation of motion of the system is

$$\begin{bmatrix} 1 & 0 \\ 0 & 1 \end{bmatrix} \begin{Bmatrix} \ddot{y}_1 \\ \ddot{y}_2 \end{Bmatrix} + \begin{bmatrix} 1 & -1 \\ -1 & 1 \end{bmatrix} \begin{Bmatrix} y_1 \\ y_2 \end{Bmatrix} = \begin{bmatrix} 1 \\ 0 \end{bmatrix} u. \tag{3.215}$$

with the initial and final conditions

$$\begin{Bmatrix} y_1 \\ y_2 \\ \dot{y}_1 \\ \dot{y}_2 \end{Bmatrix} (0) = \begin{Bmatrix} 1 \\ 1 \\ 0 \\ 0 \end{Bmatrix} \quad \text{and} \quad \begin{Bmatrix} y_1 \\ y_2 \\ \dot{y}_1 \\ \dot{y}_2 \end{Bmatrix} (10) = \begin{Bmatrix} 0 \\ 0 \\ 0 \\ 0 \end{Bmatrix} \tag{3.216}$$

Using the similarity transformation

$$\begin{Bmatrix} y_1 \\ y_2 \end{Bmatrix} = \begin{bmatrix} -\frac{1}{\sqrt{2}} & -\frac{1}{\sqrt{2}} \\ -\frac{1}{\sqrt{2}} & \frac{1}{\sqrt{2}} \end{bmatrix} \begin{Bmatrix} x_1 \\ x_2 \end{Bmatrix}, \tag{3.217}$$

the modal equation of motion can be derived

$$\begin{bmatrix} 1 & 0 \\ 0 & 1 \end{bmatrix} \begin{Bmatrix} \ddot{x}_1 \\ \ddot{x}_2 \end{Bmatrix} + \begin{bmatrix} 0 & 0 \\ 0 & 2 \end{bmatrix} \begin{Bmatrix} x_1 \\ x_2 \end{Bmatrix} = - \begin{bmatrix} \frac{1}{\sqrt{2}} \\ \frac{1}{\sqrt{2}} \end{bmatrix} u. \tag{3.218}$$

with the corresponding boundary conditions

$$\begin{Bmatrix} x_1 \\ x_2 \\ \dot{x}_1 \\ \dot{x}_2 \end{Bmatrix} (10) = \begin{Bmatrix} 0 \\ 0 \\ 0 \\ 0 \end{Bmatrix} \quad \text{and} \quad \begin{Bmatrix} x_1 \\ x_2 \\ \dot{x}_1 \\ \dot{x}_2 \end{Bmatrix} (0) = \begin{Bmatrix} -1.4142 \\ 0 \\ 0 \\ 0 \end{Bmatrix} \tag{3.219}$$

which in state-space form is

$$\begin{Bmatrix} \dot{x}_1 \\ \ddot{x}_1 \\ \dot{x}_2 \\ \ddot{x}_2 \end{Bmatrix} = \begin{bmatrix} 0 & 1 & 0 & 0 \\ 0 & 0 & 0 & 0 \\ 0 & 0 & 0 & 1 \\ 0 & 0 & -2 & 0 \end{bmatrix} \begin{Bmatrix} x_1 \\ \dot{x}_1 \\ x_2 \\ \dot{x}_2 \end{Bmatrix} + \begin{bmatrix} 0 \\ -\frac{1}{\sqrt{2}} \\ 0 \\ -\frac{1}{\sqrt{2}} \end{bmatrix} u. \tag{3.220}$$

Based on Equation (3.214), the optimal control has the form

$$u(t) = \kappa_1 + \kappa_2 t + \kappa_3 \cos(\sqrt{2}t) + \kappa_4 \sin(\sqrt{2}t) \tag{3.221}$$

and the constraint equation for the determination of κ_i is

$$\begin{bmatrix} 1 & -10 & 0 & 0 \\ 0 & 1 & 0 & 0 \\ 0 & 0 & -0.005 & -0.7071 \\ 0 & 0 & 1.4142 & -0.0050 \end{bmatrix} \begin{Bmatrix} x_1 \\ x_2 \\ \dot{x}_1 \\ \dot{x}_2 \end{Bmatrix} (10) - \begin{Bmatrix} x_1 \\ x_2 \\ \dot{x}_1 \\ \dot{x}_2 \end{Bmatrix} (0) = \int_0^{10} \begin{Bmatrix} \frac{t}{\sqrt{2}} \\ -\frac{1}{\sqrt{2}} \\ \frac{1}{2}\sin(\sqrt{2}t) \\ \frac{1}{\sqrt{2}}\cos(\sqrt{2}t) \end{Bmatrix} u(t)\, dt$$

$$(3.222)$$

which reduces to

$$-\begin{Bmatrix} -1.4142 \\ 0 \\ 0 \\ 0 \end{Bmatrix} = \begin{bmatrix} 35.3553 & 235.7023 & 4.6446 & 0.3784 \\ -7.0711 & -35.3553 & -0.5000 & -0.5025 \\ 0.3553 & 0.2676 & 0.1768 & 2.5009 \\ -0.5000 & -4.6446 & -3.5343 & -0.2500 \end{bmatrix} \begin{Bmatrix} \kappa_1 \\ \kappa_2 \\ \kappa_3 \\ \kappa_4 \end{Bmatrix}$$

$$(3.223)$$

which can be solved for κ_i. Figure 3.15 illustrates the system response and the minimum power control profile.

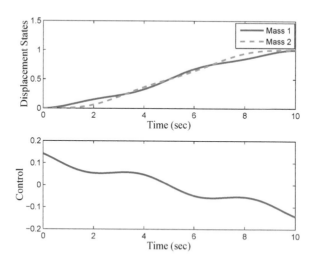

Figure 3.15: Evolution of States and Control

3.3.2 Robust Minimum Power Control of Maneuvering Structures

Consider the system

$$\ddot{z} + \begin{bmatrix} 0 & 0 & 0 \\ 0 & 2\zeta_1\omega_1 & \\ & 0 & 2\zeta_2\omega_2 & 0 \\ & & & \ddots \end{bmatrix} \dot{z} + \begin{bmatrix} 0 & 0 & 0 \\ 0 & \omega_1^2 & \\ & 0 & \omega_2^2 & 0 \\ & & & \ddots \end{bmatrix} z = \begin{Bmatrix} \phi_0 \\ \phi_1 \\ \phi_2 \\ \vdots \end{Bmatrix} u \qquad (3.224)$$

where the damping coefficient ζ_i or the natural frequency ω_i are uncertain. To add robustness to uncertainties in the system parameters, the sensitivity states can be determined and used to augment the equations of motion of the system. Enforcing the constraints that the sensitivity states be zero at the end of the maneuver will result in desensitized controllers as shown earlier. For illustrative purposes, assume that ω_1 is uncertain. From the equation

$$\ddot{z}_1 = -2\zeta_1\omega_1\dot{z}_1 - \omega_1^2 z_1 + \phi_1 u, \qquad (3.225)$$

the sensitivity of z_1 with respect to ω_1 is

$$\frac{d\ddot{z}_1}{d\omega_1} = -2\zeta_1\dot{z}_1 - 2\omega_1 z_1 - 2\zeta_1\omega_1 \frac{d\dot{z}_1}{d\omega_1} - \omega_1^2 \frac{dz_1}{d\omega_1}. \qquad (3.226)$$

The final augmented system model is

$$\ddot{z} + \begin{bmatrix} 0 & 0 & & 0 \\ 0 & 2\zeta_1\omega_1 & 0 & \\ 0 & 2\zeta_1 & 2\zeta_1\omega_1 & \\ 0 & 0 & 2\zeta_2\omega_2 & 0 \\ & & & \ddots \end{bmatrix} \dot{z} + \begin{bmatrix} 0 & 0 & & 0 \\ 0 & \omega_1^2 & 0 & \\ 0 & 2\omega_1 & \omega_1^2 & \\ 0 & 0 & \omega_2^2 & 0 \\ & & & \ddots \end{bmatrix} z = \begin{Bmatrix} \phi_0 \\ \phi_1 \\ 0 \\ \phi_2 \\ \vdots \end{Bmatrix} u \qquad (3.227)$$

where the state vector has been augmented with the sensitivity states [7]. The second-order model of the system can be rewritten in Jordan canonical form as

$$\dot{y} = \underbrace{\begin{bmatrix} 0 & 1 & & & & & & 0 \\ 0 & 0 & & & & & & \\ & & -p_1-jq_1 & 1 & & & & \\ & & 0 & -p_1-jq_1 & 0 & & & \\ & & 0 & 0 & -p_1+jq_1 & 1 & & \\ & & 0 & 0 & 0 & -p_1+jq_1 & & \\ & & 0 & 0 & 0 & 0 & -p_2+jq_2 & 0 \\ & & 0 & 0 & 0 & 0 & 0 & -p_2-jq_2 \\ 0 & & & & & & & \ddots \end{bmatrix}}_{\mathscr{A}} y + \underbrace{\begin{bmatrix} 0 \\ \psi_2 \\ \psi_3 \\ \psi_4 \\ \psi_5 \\ \psi_6 \\ \vdots \end{bmatrix}}_{\mathscr{B}} u.$$

$$(3.228)$$

where

$$p_i = \zeta_i \omega_i \text{ and } q_i = \omega_i \sqrt{1 - \zeta_i^2}. \tag{3.229}$$

It can be seen from Equation (3.228) that the addition of the sensitivity state equation to the original system model results in repeated poles of the augmented system at the nominal location of the uncertain poles. Representing $\cos(.) + j\sin(.)$ as $\mathcal{CS}^+(.)$, and $\cos(.) - j\sin(.)$ as $\mathcal{CS}^-(.)$ it can be easily shown that the matrix exponential is:

$$exp(-\mathscr{A}t) = \begin{bmatrix} \mathscr{A}_0 & 0 & 0 & 0 & \cdots \\ 0 & \mathscr{A}_1^- & 0 & 0 & \cdots \\ 0 & 0 & \mathscr{A}_1^+ & 0 & \cdots \\ 0 & 0 & 0 & \mathscr{A}_2^- & \cdots \\ 0 & 0 & 0 & 0 & \ddots \end{bmatrix} \tag{3.230}$$

where

$$\mathscr{A}_0 = \begin{bmatrix} 1 & -t \\ 0 & 1 \end{bmatrix} \tag{3.231}$$

$$\mathscr{A}_i^- = \begin{bmatrix} e^{p_i t}(\mathcal{CS}^-(q_i t)) & te^{p_i t}(\mathcal{CS}^-(q_i t)) \\ 0 & e^{p_i t}(\mathcal{CS}^-(q_i t)) \end{bmatrix} \tag{3.232}$$

$$\mathscr{A}_i^+ = \begin{bmatrix} e^{p_i t}(\mathcal{CS}^+(q_i t)) & te^{p_i t}(\mathcal{CS}^+(q_i t)) \\ 0 & e^{p_i t}(\mathcal{CS}^+(q_i t)) \end{bmatrix} \tag{3.233}$$

and the optimal control can be shown to have the form

$$u(t) = \kappa_1 + \kappa_2 t + t e^{p_1 t}(\kappa_3 \cos(q_1 t) + \kappa_4 j \sin(q_1 t))$$
$$+ \sum_{i=1}^{N} \kappa_{2i+3} e^{p_i t} \cos(q_i t) + \kappa_{2i+4} j e^{p_i t} \sin(q_i t) \tag{3.234}$$

when robustness is desired with respect to uncertainties in the damping and natural frequency of the first mode. If robustness with respect to additional modes is required, terms of the form $t e^{p_i t}(\kappa_m \cos(q_i t) + \kappa_{m+1} j \sin(q_i t))$ are included into $u(t)$ to desensitize the controller to uncertainties in ζ_i or ω_i.

Example 3.11: The decoupled equations of motion of the two-mass floating oscillator problem are

$$\begin{bmatrix} 1 & 0 \\ 0 & 1 \end{bmatrix} \begin{Bmatrix} \ddot{x}_1 \\ \ddot{x}_2 \end{Bmatrix} + \begin{bmatrix} 0 & 0 \\ 0 & 2 \end{bmatrix} \begin{Bmatrix} x_1 \\ x_2 \end{Bmatrix} = -\begin{bmatrix} \frac{1}{\sqrt{2}} \\ \frac{1}{\sqrt{2}} \end{bmatrix} u \tag{3.235}$$

with a nominal natural frequency ω of $\sqrt{2}$. To design a controller which is robust to uncertainties in the natural frequency of the system, the sensitivity state equations are derived and the augmented dynamical model of the system is

$$
\left\{\begin{array}{c} \ddot{x}_1 \\ \ddot{x}_2 \\ \frac{d\ddot{x}_2}{d\omega} \end{array}\right\} + \begin{bmatrix} 0 & 0 & 0 \\ 0 & 2 & 0 \\ 0 & 2\sqrt{2} & 2 \end{bmatrix} \left\{\begin{array}{c} x_1 \\ x_2 \\ \frac{dx_2}{d\omega} \end{array}\right\} = - \begin{bmatrix} \frac{1}{\sqrt{2}} \\ \frac{1}{\sqrt{2}} \\ 0 \end{bmatrix} u \tag{3.236}
$$

which can be represented in Jordan canonical form as

$$
\left\{\begin{array}{c} \dot{y}_1 \\ \dot{y}_2 \\ \dot{y}_3 \\ \dot{y}_4 \\ \dot{y}_5 \\ \dot{y}_6 \end{array}\right\} = \begin{bmatrix} 0 & 1 & 0 & 0 & 0 & 0 \\ 0 & 0 & 0 & 0 & 0 & 0 \\ 0 & 0 & \sqrt{2}j & 1 & 0 & 0 \\ 0 & 0 & 0 & \sqrt{2}j & 0 & 0 \\ 0 & 0 & 0 & 0 & -\sqrt{2}j & 1 \\ 0 & 0 & 0 & 0 & 0 & -\sqrt{2}j \end{bmatrix} \left\{\begin{array}{c} y_1 \\ y_2 \\ y_3 \\ y_4 \\ y_5 \\ y_6 \end{array}\right\} + \begin{bmatrix} 0 \\ -\frac{1}{\sqrt{2}} \\ -\frac{\sqrt{2}}{4} \\ \frac{1}{2}j \\ -\frac{\sqrt{2}}{4} \\ -\frac{1}{2}j \end{bmatrix} u \tag{3.237}
$$

with the boundary conditions

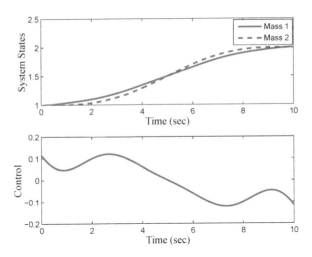

Figure 3.16: System Response and Control Input

$$
\left\{\begin{array}{c} y_1 \\ y_2 \\ y_3 \\ y_4 \\ y_5 \\ y_6 \end{array}\right\}(0) = \left\{\begin{array}{c} 0 \\ 0 \\ 0 \\ 0 \\ 0 \\ 0 \end{array}\right\} \text{ and } \left\{\begin{array}{c} y_1 \\ y_2 \\ y_3 \\ y_4 \\ y_5 \\ y_6 \end{array}\right\}(10) = \left\{\begin{array}{c} -1.4142 \\ 0 \\ 0 \\ 0 \\ 0 \\ 0 \end{array}\right\} \tag{3.238}
$$

The optimal control is given as

$$u(t) = \kappa_1 + \kappa_2 t + \kappa_3 \cos(\sqrt{2}t) + j\kappa_4 \sin(\sqrt{2}t) + \kappa_5 t \cos(\sqrt{2}t) + j\kappa_6 t \sin(\sqrt{2}t)$$
$$(3.239)$$

where the optimal κ_i are

$$\begin{Bmatrix} \kappa_1 \\ \kappa_2 \\ \kappa_3 \\ \kappa_4 \\ \kappa_5 \\ \kappa_5 \end{Bmatrix} = \begin{Bmatrix} 0.1666 \\ -0.0333 \\ -0.0506 \\ 0.0952i \\ 0.0146 \\ -0.0146i \end{Bmatrix}. \qquad (3.240)$$

Figure 3.16 illustrates the displacement response and the robust minimum power control profile.

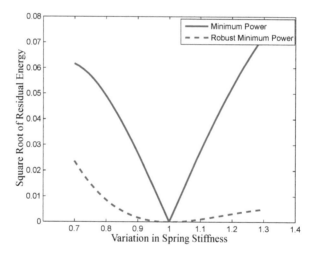

Figure 3.17: Controller Sensitivity

It is patent from Figure 3.17 which plots the residual energy of the system at the end of the maneuver, that the robust minimum power control profile provides robustness to uncertainties of the spring stiffness around the nominal estimate of the system model. The cost trade-off corresponds to an increase in the performance index from 0.0504 (Example 3.10) to 0.0666 (Example 3.11), a modest 32% increase in the power consumed.

3.3.3 Minimum Time/Power Control

The minimum power optimal control formulation requires the maneuver time to be specified. In application such as disk drives for laptops, minimizing the consumed power is clearly desirable. However, if the tradeoff of power versus maneuver time is large, the resulting controller is not realistic. Minimization of a weighted combination of maneuver time and power consumed can lead to appealing control strategies which permits users to dial up or down, the penalty associated with the maneuver time and power consumed. This section will present the optimal control formulation and illustrate it with two examples.

The cost function which minimizes a convex combination of the maneuver time and the power consumed is:

$$J = \int_0^{t_f} \left((1 - \alpha) + \alpha \frac{1}{2} u^T R u \right) dt \tag{3.241}$$

where α is a scalar which lies between zero and one and permits the cost function to span the spectrum of minimum time control and minimum power control problems. The positive definite matrix R permits relatively weighting inputs in a multi-input control system. The linear model of the controlled system is:

$$\dot{x} = Ax + Bu \tag{3.242}$$

and the controller is subject to the boundary conditions:

$$x(0) = x_0, \quad x(t_f) = x_f \tag{3.243}$$

The Hamiltonian of the system is:

$$\mathcal{H} = (1 - \alpha) + \alpha \frac{1}{2} u^T R u + \lambda^T (Ax + Bu). \tag{3.244}$$

The necessary conditions for optimality are:

$$\dot{x} = \frac{\partial \mathcal{H}}{\partial \lambda} = Ax + Bu \tag{3.245a}$$

$$\dot{\lambda} = -\frac{\partial \mathcal{H}}{\partial x} = -A^T \lambda \tag{3.245b}$$

$$\frac{\partial \mathcal{H}}{\partial u} = 0 = \alpha R u + B^T \lambda \tag{3.245c}$$

$$x(0) = x_0 \text{ and } x(t_f) = x_f \tag{3.245d}$$

The costate equation can be solved resulting in the equation

$$\lambda(t) = exp(-A^T t)\lambda(0) \tag{3.246}$$

which results in the control

$$u = -\frac{1}{\alpha} R^{-1} B^T \lambda(t) = -\frac{1}{\alpha} R^{-1} B^T exp(-A^T t)\lambda(0) \tag{3.247}$$

Substituting the expression for the control into the state equation leads to:

$$x(t_f) = exp(At_f)x(0) + \int_0^{t_f} exp(A(t_f - \tau))Bu(\tau)\, d\tau \qquad (3.248)$$

which can be rewritten as:

$$\alpha\left(exp(-At_f)x_f - x_0\right) = -\int_0^{t_f} exp(-A\tau)BR^{-1}B^T exp(-A^T\tau)\lambda(0)\, d\tau \qquad (3.249)$$

Define the Gramian matrix

$$P = \int_0^{t_f} exp(-A\tau)BR^{-1}B^T exp(-A^T\tau)\, d\tau \qquad (3.250)$$

which is related to the controllability Gramian $W_c(t_f)$ by the equation

$$W_c(t_f) = exp(At_f)Pexp(A^T t_f). \qquad (3.251)$$

Therefore, for a controllable system, P is a nonsingular matrix. Differentiating Equation (3.250) with respect to time leads to:

$$\dot{P} = \qquad BR^{-1}B^T - A\int_0^{t_f} exp(-A\tau)BR^{-1}B^T exp(-A^T\tau)\, d\tau -$$
$$\int_0^{t_f} exp(-A\tau)BR^{-1}B^T exp(-A^T\tau)\, d\tau A^T \qquad (3.252)$$

or

$$\dot{P} = BR^{-1}B^T - AP - PA^T, \quad P(0) = 0 \qquad (3.253)$$

which is the differential Lyapunov equation which can solved using the technique presented in Appendix B.

Equation (3.249) can now be solved for $\lambda(0)$

$$\lambda(0) = -P^{-1}\alpha\left(exp(-At_f)x_f - x_0\right) \qquad (3.254)$$

which can be used to rewrite the cost function as:

$$J = (1 - \alpha)t_f + \int_0^{t_f} \alpha\frac{1}{2}u^T Ru\, dt$$
$$= (1 - \alpha)t_f + \int_0^{t_f} \frac{1}{2\alpha}\lambda(0)^T exp(-A^T t)^T BR^{-1}B^T exp(-A^T t)\lambda(0)\, dt \qquad (3.255)$$

Define:

$$S = \int_0^{t_f} exp(-A^T t)^T BR^{-1}B^T exp(-A^T t)\, dt \qquad (3.256)$$

which leads to the differential Lyapunov equation:

$$\dot{S} = BR^{-1}B^T - AS - SA^T \qquad (3.257)$$

which is identical to Equation (3.253). Solving Equation (3.257), leads to the cost function:

$$J = (1 - \alpha)t_f + \frac{1}{2\alpha}\lambda(0)^T S\lambda(0) \tag{3.258}$$

where $\lambda(0)$ and S are implicit functions of t_f. Since S and P are identical, the cost function can be rewritten as:

$$J = (1 - \alpha)t_f + \frac{\alpha}{2}\left(\exp(-At_f)x_f - x_0\right)^T (P^{-1})^T SP^{-1}\left(\exp(-At_f)x_f - x_0\right) \tag{3.259}$$

which reduces to:

$$J = (1 - \alpha)t_f + \frac{\alpha}{2}\left(\exp(-At_f)x_f - x_0\right)^T (P^{-1})^T \left(\exp(-At_f)x_f - x_0\right) \tag{3.260}$$

A single-dimension parameter optimization problem can now be solved to determine t_f.

Example 3.12:

Consider the problem:

$$J = \int_0^{t_f} \left(1 + \frac{1}{2}u^2\right) dt \tag{3.261}$$

for the system:

$$\begin{Bmatrix} \dot{x}_1 \\ \dot{x}_2 \end{Bmatrix} = \begin{bmatrix} 0 & 1 \\ 0 & 0 \end{bmatrix}\begin{Bmatrix} x_1 \\ x_2 \end{Bmatrix} + \begin{bmatrix} 0 \\ 1 \end{bmatrix}u \tag{3.262}$$

subject to the boundary conditions:

$$x_1(0) = x_2(0) = 0, \ x_1(t_f) = 1, x_2(t_f) = 0. \tag{3.263}$$

The optimal control is given by the equation:

$$u = \begin{bmatrix} t & -1 \end{bmatrix}\lambda(0) \tag{3.264}$$

where $\lambda(0)$ is given by the equation:

$$\lambda(0) = -P^{-1}\left(\exp(-At_f)x_f - x_0\right) \tag{3.265}$$

where P is the solution of the equation:

$$\begin{bmatrix} \dot{p}_1 & \dot{p}_2 \\ \dot{p}_2 & \dot{p}_3 \end{bmatrix} = \begin{bmatrix} 0 & 0 \\ 0 & 1 \end{bmatrix} - \begin{bmatrix} 0 & 1 \\ 0 & 0 \end{bmatrix}\begin{bmatrix} p_1 & p_2 \\ p_2 & p_3 \end{bmatrix} - \begin{bmatrix} p_1 & p_2 \\ p_2 & p_3 \end{bmatrix}\begin{bmatrix} 0 & 0 \\ 1 & 0 \end{bmatrix} \tag{3.266}$$

which can be solved resulting in the equation:

$$P = \begin{bmatrix} p_1 & p_2 \\ p_2 & p_3 \end{bmatrix} = \begin{bmatrix} \frac{t_f^3}{3} & -\frac{t_f^2}{2} \\ -\frac{t_f^2}{2} & t_f \end{bmatrix} \tag{3.267}$$

Thus, $\lambda(0)$ can be solved as:

$$\lambda(0) = -\begin{bmatrix} 12t_f^{-3} & 6t_f^{-2} \\ 6t_f^{-2} & 4t_f^{-1} \end{bmatrix}\begin{bmatrix} 1 \\ 0 \end{bmatrix} = \begin{bmatrix} -12t_f^{-3} \\ -6t_f^{-2} \end{bmatrix} \tag{3.268}$$

The final cost function can now be written as:

$$J = t_f + \frac{6}{t_f^3} \tag{3.269}$$

Forcing the derivative of J with respect to t_f to zero, leads to the optimal solution:

$$t_f = \sqrt[4]{18} \tag{3.270}$$

Example 3.13:

Consider the problem of minimizing:

$$J = \int_0^{t_f}\left((1-\alpha) + \frac{1}{2}\alpha u^2\right) dt \tag{3.271}$$

for the undamped benchmark floating oscillator:

$$\begin{Bmatrix} \dot{x}_1 \\ \dot{x}_2 \\ \dot{x}_3 \\ \dot{x}_4 \end{Bmatrix} = \begin{bmatrix} 0 & 0 & 1 & 0 \\ 0 & 0 & 0 & 1 \\ -1 & 1 & 0 & 0 \\ 1 & -1 & 0 & 0 \end{bmatrix}\begin{Bmatrix} x_1 \\ x_2 \\ x_3 \\ x_4 \end{Bmatrix} + \begin{bmatrix} 0 \\ 0 \\ 1 \\ 0 \end{bmatrix} u \tag{3.272}$$

subject to the boundary conditions:

$$x_1(0) = x_2(0) = 0, x_3(0) = x_4(0) = 0, \ x_1(t_f) = 1, x_2(t_f) = 1, x_3(t_f) = 0, x_4(t_f) = 0. \tag{3.273}$$

α is constrained to:

$$0 \le \alpha \le 1. \tag{3.274}$$

To simplify the solution process, the system model is represented in the Jordan canonical form as:

$$\begin{Bmatrix} \dot{y}_1 \\ \dot{y}_2 \\ \dot{y}_3 \\ \dot{y}_4 \end{Bmatrix} = \begin{bmatrix} 0 & 1 & 0 & 0 \\ 0 & 0 & 0 & 0 \\ 0 & 0 & \sqrt{2}i & 0 \\ 0 & 0 & 0 & -\sqrt{2}i \end{bmatrix}\begin{Bmatrix} y_1 \\ y_2 \\ y_3 \\ y_4 \end{Bmatrix} + \begin{bmatrix} 0 \\ 1 \\ 1 \\ 1 \end{bmatrix} u \tag{3.275}$$

subject to the boundary conditions:

$$y_1(0) = y_2(0) = 0, y_3(0) = y_4(0) = 0, \ y_1(t_f) = 2, y_2(t_f) = 0, y_3(t_f) = 0, y_4(t_f) = 0. \tag{3.276}$$

The optimal control is given by the equation:

$$u = -\frac{1}{\alpha}\left[-t,\ 1,\ \cos(\sqrt{2}t) - i\sin(\sqrt{2}t),\ \cos(\sqrt{2}t) + i\sin(\sqrt{2}t)\right]\lambda(0) \qquad (3.277)$$

where $\lambda(0)$ is given by the equation:

$$\lambda(0) = -\alpha P^{-1}\left(exp(-At_f)x_f - x_0\right) \qquad (3.278)$$

where P is the solution of the equation:

$$
\begin{bmatrix} \dot{p}_1 & \dot{p}_2 & \dot{p}_3 & \dot{p}_4 \\ \dot{p}_2 & \dot{p}_5 & \dot{p}_6 & \dot{p}_7 \\ \dot{p}_3 & \dot{p}_6 & \dot{p}_8 & \dot{p}_9 \\ \dot{p}_4 & \dot{p}_7 & \dot{p}_9 & \dot{p}_{10} \end{bmatrix} = \begin{bmatrix} 0&0&0&0 \\ 0&1&1&1 \\ 0&1&1&1 \\ 0&1&1&1 \end{bmatrix} - \begin{bmatrix} 0&1&0&0 \\ 0&0&0&0 \\ 0&0&\sqrt{2}i&0 \\ 0&0&0&-\sqrt{2}i \end{bmatrix}\begin{bmatrix} p_1 & p_2 & p_3 & p_4 \\ p_2 & p_5 & p_6 & p_7 \\ p_3 & p_6 & p_8 & p_9 \\ p_4 & p_7 & p_9 & p_{10} \end{bmatrix} \qquad (3.279)
$$

$$
- \begin{bmatrix} p_1 & p_2 & p_3 & p_4 \\ p_2 & p_5 & p_6 & p_7 \\ p_3 & p_6 & p_8 & p_9 \\ p_4 & p_7 & p_9 & p_{10} \end{bmatrix}\begin{bmatrix} 0&0&0&0 \\ 1&0&0&0 \\ 0&0&\sqrt{2}i&0 \\ 0&0&0&-\sqrt{2}i \end{bmatrix} \qquad (3.280)
$$

which results in:

$$
\underbrace{\begin{bmatrix} p_1 & p_2 & p_3 & p_4 \\ p_2 & p_5 & p_6 & p_7 \\ p_3 & p_6 & p_8 & p_9 \\ p_4 & p_7 & p_9 & p_{10} \end{bmatrix}}_{P}(t) = \begin{bmatrix} \frac{t^3}{3} - \frac{t^2}{2} & \frac{1}{2} - e^{-it\sqrt{2}}\left(\frac{1}{2}i\sqrt{2}t + \frac{1}{2}\right) & \frac{1}{2} - e^{it\sqrt{2}}\left(-\frac{1}{2}i\sqrt{2}t + \frac{1}{2}\right) \\ * & t & \frac{1}{2}i\sqrt{2}\left(-1 + e^{-it\sqrt{2}}\right) & -\frac{1}{2}i\sqrt{2}\left(-1 + e^{it\sqrt{2}}\right) \\ * & * & \frac{1}{4}i\sqrt{2}\left(-1 + e^{-2i\sqrt{2}t}\right) & t \\ * & * & * & -\frac{1}{4}i\sqrt{2}\left(-1 + e^{2i\sqrt{2}t}\right) \end{bmatrix}
$$

$$ \qquad (3.281)$$

The cost function can now be represented as:

$$J = (1-\alpha)t_f + \frac{\alpha}{2}\{2\ 0\ 0\ 0\}\,P^{-1}(t_f)\begin{Bmatrix} 2 \\ 0 \\ 0 \\ 0 \end{Bmatrix}. \qquad (3.282)$$

Figure 3.18 illustrates the variation of the cost J as a function of the maneuver time t_f for $\alpha = 0.5$. The optimal maneuver time can be determined by a line search algorithm and results in $t_f = 4.5315$.

Figure 3.19 illustrates the variation of the maneuver time and the consumed power as a function of the weighting parameter α. The variation of power consumed (dashed line) is plotted on a log scale which implies that the power consumed rapidly reduces as α is increased from 0. It is also clear that the maneuver time (solid line) varies slowly till α is close to 1.

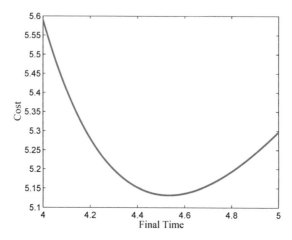

Figure 3.18: Variation of Cost Function

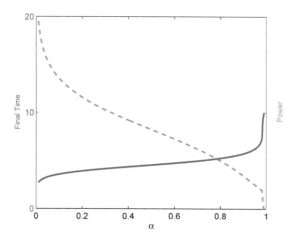

Figure 3.19: Variation of Cost Function

3.4 Frequency-Shaped LQR Controller

The LQR design approach assumes that the model describing the system is valid at all input frequencies. However, most models are inherently erroneous with both structured and unstructured uncertainties. The structured uncertainties refers to the errors in model parameters and unstructured uncertainties corresponds to unmodeled dynamics. If one can include a penalty function which is a function of frequency, one can then compensate for errors in system models in specific frequency regions. Gupta [9] proposed an elegant approach to combine frequency domain penalty functions into the classical LQR design process. This section describes the frequency shaped LQR design process and illustrates it on the floating oscillator benchmark problem.

The standard LQR design process is related to the determination of a controller which minimizes the cost function

$$J = \frac{1}{2} \int_0^\infty \left(x^T Q x + u^T R u \right) dt \tag{3.283}$$

for the linear system

$$\dot{x} = Ax + Bu. \tag{3.284}$$

where $x \in \mathcal{R}^n$ is the state vector, and $u \in \mathcal{R}^m$ is the control vector. The optimal controller for the infinite-time horizon problem is given by the equation

$$u = -Kx = -R^{-1} B^T P x \tag{3.285}$$

where P is the solution of the algebraic Riccati equation

$$A^T P + PA + Q - PBR^{-1} B^T P = 0. \tag{3.286}$$

The Q and R matrices are the state and control penalty matrices, respectively, which are generally constant matrices. The Parseval's theorem [10] can be exploited to rewrite the quadratic cost function (Equation 3.283) in the frequency domain as

$$J = \frac{1}{4\pi} \int_{-\infty}^{\infty} \left(x^T(-j\omega)Qx(j\omega) + u^T(-j\omega)Ru(j\omega) \right) d\omega. \tag{3.287}$$

In this formulation, it can be seen that the states and control are equally penalized at all frequencies. In some systems, the actuators, sensors, or the system might have poor characteristics at specific frequencies. Increasing the penalty of the states and control at these frequencies permits shaping the frequency response characteristics of the closed-loop system. The frequency-shaped cost function can be represented as

$$J = \frac{1}{4\pi} \int_{-\infty}^{\infty} \left(x^T(-j\omega)Q(j\omega)x(j\omega) + u^T(-j\omega)R(j\omega)u(j\omega) \right) d\omega. \tag{3.288}$$

where Q and R are Hermitian matrices, i.e., $Q_{ij} = conj(Q_{ji})$ and $R_{ij} = conj(R_{ji})$.

Assuming that the penalty matrices $Q(j\omega)$ and $R(j\omega)$ can be rewritten as

$$Q(j\omega) = \Phi(-j\omega)\Phi(j\omega) \tag{3.289}$$

$$R(j\omega) = \Psi(-j\omega)\Psi(j\omega) \tag{3.290}$$

where $\Phi(j\omega)$ and $\Psi(j\omega)$ are rational matrices. The filtered states

$$x^f = \Phi(j\omega)x \tag{3.291}$$

$$u^f = \Psi(j\omega)u \tag{3.292}$$

can be represented in state-space form as

$$\dot{w}_x = A_x w_x + B_x x$$

$$x^f = C_x w_x + D_x x \tag{3.293}$$

$$\dot{w}_u = A_u w_u + B_u u$$

$$u^f = C_u w_u + D_u u. \tag{3.294}$$

Augmenting the system model (Equation 3.284) with Equations (3.293) and (3.294), results in the augmented state-space system

$$\underbrace{\begin{Bmatrix} \dot{x} \\ \dot{w}_x \\ \dot{w}_u \end{Bmatrix}}_{\dot{X}} = \underbrace{\begin{bmatrix} A & 0 & 0 \\ B_x & A_x & 0 \\ 0 & 0 & A_u \end{bmatrix}}_{\mathcal{A}} \underbrace{\begin{Bmatrix} x \\ w_x \\ w_u \end{Bmatrix}} + \underbrace{\begin{bmatrix} B \\ 0 \\ B_u \end{bmatrix}}_{\mathcal{B}} u \tag{3.295}$$

and the modified cost function

$$J = \frac{1}{2} \int_0^\infty X^T \underbrace{\begin{bmatrix} D_x^T D_x & D_x^T C_x & 0 \\ C_x^T D_x & C_x^T C_x & 0 \\ 0 & 0 & C_u^T C_u \end{bmatrix}}_{\mathcal{Q}} X +$$

$$2X^T \underbrace{\begin{bmatrix} 0 \\ 0 \\ C_u^T D_u \end{bmatrix}}_{\mathcal{N}} u + u^T \underbrace{D_u^T D_u}_{\mathcal{R}} u \, dt. \tag{3.296}$$

The Hamiltonian of the system is

$$\mathcal{H} = \frac{1}{2}\left(X^T \mathcal{Q} X + 2X^T \mathcal{N} u + u^T \mathcal{R} u\right) + \lambda^T \left(\mathcal{A} X + \mathcal{B} u\right). \tag{3.297}$$

The Euler-Lagrange equations are

$$\dot{X} = \frac{\partial \mathcal{H}}{\partial \lambda} = \mathcal{A} X + \mathcal{B} u \tag{3.298}$$

$$\dot{\lambda} = -\frac{\partial \mathcal{H}}{\partial X} = -\mathcal{N} u - \mathcal{Q} X - \mathcal{A}^T \lambda \tag{3.299}$$

$$0 = \frac{\partial \mathcal{H}}{\partial u} = \mathcal{R} u + \mathcal{B}^T \lambda + \mathcal{N}^T X \tag{3.300}$$

The control can be solved from Equation (3.300) resulting in the equation

$$u = -\mathcal{R}^{-1}\left(\mathcal{B}^T \lambda + \mathcal{N}^T X\right). \tag{3.301}$$

Substituting Equation (3.301) into the state and costate equations (Equations 3.298 and 3.299), the resulting coupled equations are

$$\dot{X} = \mathcal{A}X - \mathcal{B}\mathcal{R}^{-1}\left(\mathcal{B}^T \lambda + \mathcal{N}^T X\right) \tag{3.302}$$

$$\dot{\lambda} = \mathcal{N}\mathcal{R}^{-1}\left(\mathcal{B}^T \lambda + \mathcal{N}^T X\right) - \mathcal{Q}X - \mathcal{A}^T \lambda. \tag{3.303}$$

Assuming the relationship

$$\lambda = PX \Rightarrow \dot{\lambda} = \dot{P}X + P\dot{X} \tag{3.304}$$

and substituting Equations (3.302) and (3.303) into Equation (3.304), we have the following matrix differential equation

$$-\dot{P} = P\left(\mathcal{A} - \mathcal{B}\mathcal{R}^{-1}\mathcal{N}^T\right) + \left(\mathcal{A} - \mathcal{B}\mathcal{R}^{-1}\mathcal{N}^T\right)^T P$$
$$+ \left(\mathcal{Q} - \mathcal{N}\mathcal{R}^{-1}\mathcal{N}^T\right) - P\mathcal{B}\mathcal{R}^{-1}\mathcal{B}^T P \tag{3.305}$$

which for the infinite time horizon problem reduces to the algebraic Riccati equation

$$P\left(\mathcal{A} - \mathcal{B}\mathcal{R}^{-1}\mathcal{N}^T\right) + \left(\mathcal{A} - \mathcal{B}\mathcal{R}^{-1}\mathcal{N}^T\right)^T P$$
$$+ \left(\mathcal{Q} - \mathcal{N}\mathcal{R}^{-1}\mathcal{N}^T\right) - P\mathcal{B}\mathcal{R}^{-1}\mathcal{B}^T P = 0. \tag{3.306}$$

Equation 3.306 can be interpreted as the algebraic Riccati equation for the design of a LQR controller for the *prefeedback* system [11]

$$\dot{X} = \left(\mathcal{A} - \mathcal{B}\mathcal{R}^{-1}\mathcal{N}^T\right)X + \mathcal{B}u. \tag{3.307}$$

which can be solved using a variety of commercially available software tools.

Figure 3.20 is a schematic of the frequency-shaped LQR controller where the blocks within the dashed box correspond to the system being controlled. This structure presumes that the states are available for feedback control. For output feedback control, the structure of controller is modified by including the output matrices to permit the output of the system to be filtered instead of the system states.

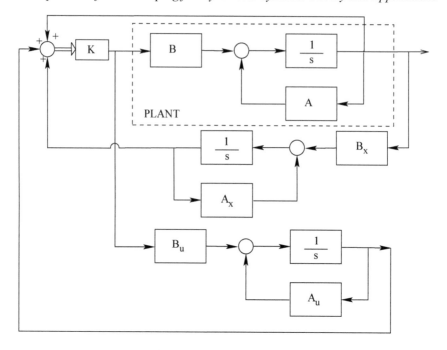

Figure 3.20: Control Structure

Example 3.14:

Consider the undamped floating oscillator benchmark problem whose model in state-space form is

$$
\begin{Bmatrix} \dot{x}_1 \\ \dot{x}_2 \\ \ddot{x}_1 \\ \ddot{x}_2 \end{Bmatrix} = \begin{bmatrix} 0 & 0 & 1 & 0 \\ 0 & 0 & 0 & 1 \\ -1 & 1 & 0 & 0 \\ 1 & -1 & 0 & 0 \end{bmatrix} \begin{Bmatrix} x_1 \\ x_2 \\ \dot{x}_1 \\ \dot{x}_2 \end{Bmatrix} + \begin{Bmatrix} 0 \\ 0 \\ 1 \\ 0 \end{Bmatrix} u. \tag{3.308}
$$

The optimal control which minimizes the cost function

$$
J = \frac{1}{2} \int_0^\infty \begin{Bmatrix} x_1 \\ x_2 \\ \dot{x}_1 \\ \dot{x}_2 \end{Bmatrix}^T \begin{bmatrix} 1 & 0 & 0 & 0 \\ 0 & 0 & 0 & 0 \\ 0 & 0 & 0 & 0 \\ 0 & 0 & 0 & 0 \end{bmatrix} \begin{Bmatrix} x_1 \\ x_2 \\ \dot{x}_1 \\ \dot{x}_2 \end{Bmatrix} + u^T [1] u \; dt \tag{3.309}
$$

is given by the solution of the algebraic Riccati equation and is given as

$$
u = - \begin{bmatrix} 0.7673 & 0.2327 & 1.2388 & 0.6413 \end{bmatrix} \begin{Bmatrix} x_1 \\ x_2 \\ \dot{x}_1 \\ \dot{x}_2 \end{Bmatrix} \tag{3.310}
$$

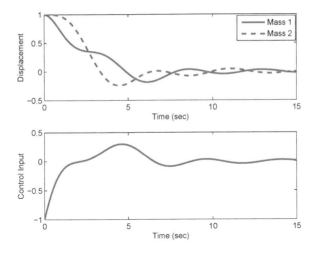

Figure 3.21: System Response and Control Input

Figure 3.21 illustrates the displacement response of the two masses of the closed-loop system and the corresponding control input. It is clear that an underdamped oscillatory response dominates the response of the system. This corresponds to a closed-loop frequency of $\omega = 1.3899$, which corresponds to the undamped mode of frequency $\omega = 1.4142$ of the open-loop system. It is therefore warranted to penalize the frequency content of the states at frequencies around the undamped natural frequency of the uncontrolled system. A frequency domain quadratic cost function which penalizes the velocity of the first mass is proposed

$$J = \frac{1}{2} \int_{-\infty}^{\infty} \begin{Bmatrix} x_1 \\ x_2 \\ \dot{x}_1 \\ \dot{x}_2 \end{Bmatrix}^T \begin{bmatrix} 1 & 0 & 0 & 0 \\ 0 & 0 & 0 & 0 \\ 0 & 0 & \frac{1}{(\omega^2-2)^2} & 0 \\ 0 & 0 & 0 & 0 \end{bmatrix} \begin{Bmatrix} x_1 \\ x_2 \\ \dot{x}_1 \\ \dot{x}_2 \end{Bmatrix} + u^T[1]u \, d\omega \tag{3.311}$$

Defining a filtered state

$$x_1^f = \frac{1}{s^2+2}\dot{x}_1, \tag{3.312}$$

which in state-space form is

$$\begin{Bmatrix} \dot{w}_1 \\ \dot{w}_2 \end{Bmatrix} = \begin{bmatrix} 0 & -2 \\ 1 & 0 \end{bmatrix} \begin{Bmatrix} w_1 \\ w_2 \end{Bmatrix} + \begin{bmatrix} 1 \\ 0 \end{bmatrix} \dot{x}_1 \tag{3.313}$$

$$x_1^f = \begin{bmatrix} 0 & 1 \end{bmatrix} \begin{Bmatrix} w_1 \\ w_2 \end{Bmatrix} + [0]\dot{x}_1. \tag{3.314}$$

The frequency-domain LQR cost can be rewritten in time domain as

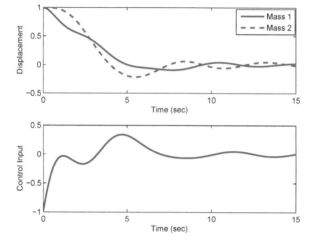

Figure 3.22: System Response and Control Input

$$J = \frac{1}{2}\int_{-\infty}^{\infty} \left\{ \begin{matrix} x_1 \\ x_2 \\ \dot{x}_1 \\ \dot{x}_2 \\ w_1 \\ w_2 \end{matrix} \right\}^T \begin{bmatrix} 1 & 0 & 0 & 0 & 0 & 0 \\ 0 & 0 & 0 & 0 & 0 & 0 \\ 0 & 0 & 0 & 0 & 0 & 0 \\ 0 & 0 & 0 & 0 & 0 & 0 \\ 0 & 0 & 0 & 0 & 0 & 0 \\ 0 & 0 & 0 & 0 & 0 & 1 \end{bmatrix} \left\{ \begin{matrix} x_1 \\ x_2 \\ \dot{x}_1 \\ \dot{x}_2 \\ w_1 \\ w_2 \end{matrix} \right\} + u^T[1]u \; dt \qquad (3.315)$$

and the feedback controller is

$$u = -\begin{bmatrix} 0.9862, & 0.0138, & 1.7910, & 0.1656, & 0.6176, & 0.4869 \end{bmatrix} \left\{ \begin{matrix} x_1 \\ x_2 \\ \dot{x}_1 \\ \dot{x}_2 \\ w_1 \\ w_2 \end{matrix} \right\} \qquad (3.316)$$

or

$$u = -\begin{bmatrix} 0.9862 & 0.0138 & 1.7910 & 0.1656 \end{bmatrix} \left\{ \begin{matrix} x_1 \\ x_2 \\ \dot{x}_1 \\ \dot{x}_2 \end{matrix} \right\} - \frac{0.6176s + 0.4869}{s^2 + 2}\dot{x}_1 \qquad (3.317)$$

Figure 3.22 illustrates the displacement response of the two masses of the frequency shaped closed-loop system and the corresponding control input. It can be seen that the peak control magnitude does not change compared to the standard LQR control. To illustrate the benefit of shaping the frequency content of the states, the potential energy in the spring of the benchmark problem is plotted in Figure 3.23 for both the LQR and frequency-shaped LQR controllers. It is clear that there is a significant reduction in the peak potential energy of the controlled system which corresponds to minimizing the excitation of the flexible mode of the system.

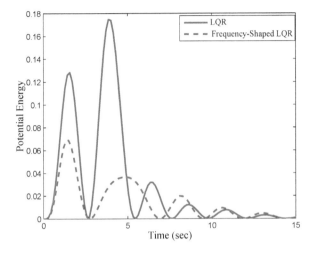

Figure 3.23: Potential Energy Evolution

The magnitude plot of the bode diagram of the LQR and frequency-shaped LQR controller is shown in Figure 3.24 where it is clear that the frequency shaping results in a notch at a frequency of $\sqrt{2}$. This also implies that if there are errors in estimating the natural frequency of the benchmark problem, the frequency shaped LQR controller can accommodate small errors and it reduces the magnitude plot in the vicinity of the estimated natural frequency.

3.5   LQR Control with Noisy Input

Real-world applications include situations when the system is subject to random disturbance inputs. For instance, random wind gusts act as disturbances to the controllers of the *Segway scooter*, helicopters, airplanes, the effect of waves on a shipborne crane, besides others. The effect of the disturbances in the presence of modeling uncertainties has been an active area of research over the past decade [12]. This

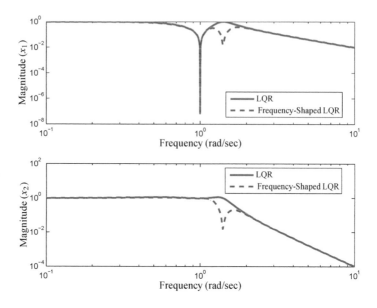

Figure 3.24: Bode Diagram: Magnitude Plot

section will describe controller design for systems subject to Gaussian random input.

Consider the linear system

$$\dot{x} = Ax + Bu + Gw \tag{3.318}$$

where $w(t)$ is a zero-mean Gaussian noise process with covariance

$$\mathscr{E}\left\{w(t)w^T(\tau)\right\} = W(t)\delta(t-\tau). \tag{3.319}$$

$\mathscr{E}$ is the expectation operator. To design a feedback controller which minimizes the cost function

$$J = \mathscr{E}\left[\lim_{T\to\infty}\frac{1}{T}\int_0^T \left(x^T Qx + u^T Ru\right) dt\right] \tag{3.320}$$

assume that the controller can be expressed as

$$u = -Kx. \tag{3.321}$$

The resulting cost function can be represented as

$$J = \mathcal{E}\left[\lim_{T \to \infty} \frac{1}{T}\int_0^T \left(x^T(Q + K^T R K)x\right)dt\right] \tag{3.322}$$

$$\Rightarrow \mathcal{E}\left[\lim_{T \to \infty} \frac{1}{T}\int_0^T \mathrm{tr}\left(xx^T(Q + K^T R K)\right)dt\right] \tag{3.323}$$

where tr is the trace operator. The closed-loop dynamics are given as

$$\dot{x} = (A - BK)x + Gw \tag{3.324}$$

$$x(t) = \Phi(t,0)x(0) + \int_0^t \Phi(t,\tau)Gw(\tau)\,d\tau \tag{3.325}$$

where $\Phi(t,\tau)$ is the transition matrix. Define the state covariance matrix as

$$P(t) \equiv \mathcal{E}\left\{x(t)x^T(t)\right\} \tag{3.326}$$

which can be rewritten using Equation (3.325) as

$$P(t) = \Phi(t,0)P(0)\Phi^T(t,0) + \int_0^t \Phi(t,\tau)GWG^T\Phi^T(t,\tau)\,d\tau \tag{3.327}$$

The time derivative of Equation (3.327) is

$$\dot{P}(t) = \frac{\partial \Phi(t,0)}{\partial t}P(0)\Phi^T(t,0) + \Phi(t,0)P(0)\frac{\partial \Phi^T(t,0)}{\partial t} + \Phi(t,t)GWG^T\Phi^T(t,t)$$
$$+ \int_0^t \frac{\partial \Phi(t,\tau)}{\partial t}GWG^T\Phi^T(t,\tau)\,d\tau + \int_0^t \Phi(t,\tau)GWG^T\frac{\partial \Phi^T(t,\tau)}{\partial t}\,d\tau \tag{3.328}$$

which can be simplified to

$$\dot{P}(t) = (A - BK)\Phi(t,0)P(0)\Phi^T(t,0) + \Phi(t,0)P(0)\Phi^T(t,0)(A - BK)^T + GWG^T$$
$$+ (A - BK)\int_0^t \Phi(t,\tau)GWG^T\Phi^T(t,\tau)\,d\tau + \int_0^t \Phi(t,\tau)GWG^T\Phi^T(t,\tau)\,d\tau\,(A - BK)^T \tag{3.329}$$

or

$$\dot{P}(t) = (A - BK)P + P(A - BK)^T + GWG^T \tag{3.330}$$

which is a differential Lyapunov equation. At steady state, P can be solved from the algebraic Lyapunov function

$$(A - BK)P + P(A - BK)^T + GWG^T = 0 \tag{3.331}$$

and the cost J is given by the equation

$$J = tr(P(Q + K^T RK)) \tag{3.332}$$

Since the cost function J is a function of the feedback gain, one can formulate an optimization problem to minimize J with respect to the feedback gain matrix K. Gradient optimization algorithm determine the gradients numerically by finite difference or analytical expressions could be provided for the gradients which reduce the numerical inaccuracies of a finite difference gradient. The gradient of J with respect to k_i, the ith term of the feedback gain matrix is

$$
\begin{aligned}
\frac{\partial J}{\partial k_i} &= tr(\frac{\partial P}{\partial k_i}(Q + K^T RK) + P\frac{\partial (Q + K^T RK)}{\partial k_i}) \\
&= tr(\frac{\partial P}{\partial k_i}(Q + K^T RK) + P(\frac{\partial K^T}{\partial k_i} RK + K^T R\frac{\partial K}{\partial k_i}))
\end{aligned}
\tag{3.333}
$$

where $\frac{\partial P}{\partial k_i}$ is the solution of the Lyapunov equation

$$(A - BK)\frac{\partial P}{\partial k_i} + \frac{\partial P}{\partial k_i}(A - BK)^T - B\frac{\partial K}{\partial k_i}P - P\frac{\partial K^T}{\partial k_i}B^T = 0 \tag{3.334}$$

Example 3.15: Design a feedback controller which minimizes the cost function

$$J = \mathcal{E}\left[\lim_{T \to \infty} \frac{1}{T} \int_0^T \left(x^T \begin{bmatrix} q_{11} & q_{12} \\ q_{12} & q_{22} \end{bmatrix} x + u^T Ru \right) dt \right] \tag{3.335}$$

for the rigid body system

$$\dot{x} = \begin{bmatrix} 0 & 1 \\ 0 & 0 \end{bmatrix} x + \begin{bmatrix} 0 \\ 1 \end{bmatrix} u + \begin{bmatrix} 0 \\ 1 \end{bmatrix} w \tag{3.336}$$

where w is a zero-mean white noise process with a covariance

$$W = \mathcal{E}\{ww^T\}. \tag{3.337}$$

The Lyapunov equation to be solved is

$$\begin{bmatrix} 0 & 1 \\ -k_1 & -k_2 \end{bmatrix} \begin{bmatrix} p_{11} & p_{12} \\ p_{12} & p_{22} \end{bmatrix} + \begin{bmatrix} p_{11} & p_{12} \\ p_{12} & p_{22} \end{bmatrix} \begin{bmatrix} 0 & -k_1 \\ 1 & -k_2 \end{bmatrix} + \begin{bmatrix} 0 & 0 \\ 0 & W \end{bmatrix} = \begin{bmatrix} 0 & 0 \\ 0 & 0 \end{bmatrix} \tag{3.338}$$

which results in the solution

$$P = \begin{bmatrix} p_{11} & p_{12} \\ p_{12} & p_{22} \end{bmatrix} = \begin{bmatrix} \frac{W}{2k_1 k_2} & 0 \\ 0 & \frac{W}{2k_2} \end{bmatrix}. \tag{3.339}$$

The cost function can be written in terms of the P matrix as

$$J = \frac{W(q_{11} + Rk_1^2)}{2k_1 k_2} + \frac{W(q_{22} + Rk_2^2)}{2k_2}. \tag{3.340}$$

The optimal gains are determined by forcing the gradient of J with respect to k_1 and k_2 to zero, resulting in the feedback gain matrix

$$K = \left\{ \sqrt{\frac{q_{11}}{R}} \quad \sqrt{2\sqrt{\frac{q_{11}}{R}} + \frac{q_{22}}{R}} \right\}. \tag{3.341}$$

It is clear that the feedback gain is not a function of the covariance of the noise.
Solving the algebraic Riccati equation

$$\begin{bmatrix} 0 & 0 \\ 1 & 0 \end{bmatrix} \begin{bmatrix} p_{11} & p_{12} \\ p_{12} & p_{22} \end{bmatrix} + \begin{bmatrix} p_{11} & p_{12} \\ p_{12} & p_{22} \end{bmatrix} \begin{bmatrix} 0 & 1 \\ 0 & 0 \end{bmatrix} + \begin{bmatrix} q_{11} & q_{12} \\ q_{12} & q_{22} \end{bmatrix} -$$

$$\begin{bmatrix} p_{11} & p_{12} \\ p_{12} & p_{22} \end{bmatrix} \begin{bmatrix} 0 \\ 1 \end{bmatrix} R^{-1} \begin{bmatrix} 0 & 1 \end{bmatrix} \begin{bmatrix} p_{11} & p_{12} \\ p_{12} & p_{22} \end{bmatrix} = \begin{bmatrix} 0 & 0 \\ 0 & 0 \end{bmatrix} \tag{3.342}$$

which results in the optimal feedback controller for the infinite time horizon LQR problem leads to the feedback gain matrix

$$K = R^{-1} \begin{bmatrix} 0 & 1 \end{bmatrix} \begin{bmatrix} p_{11} & p_{12} \\ p_{12} & p_{22} \end{bmatrix} = \left\{ \sqrt{\frac{q_{11}}{R}} \quad \sqrt{2\sqrt{\frac{q_{11}}{R}} + \frac{q_{22}}{R}} \right\} \tag{3.343}$$

which is identical to the solution given by Equation (3.341).

Burl [13] presents a simple development to prove that the optimal feedback gain for the LQR problem is identical to the stochastic regulator problem (LQR with noisy input). The benefit of using the development presented in this section is that it provides P, the covariance of the states. One can pose constrained optimization problems where the variance of a state can be required to be less than a designer specified limit.

Example 3.16: The LQR closed-loop dynamics of the system illustrated in Example 3.15 is:

$$\dot{x} = \begin{bmatrix} 0 & 1 \\ -10.00 & -10.96 \end{bmatrix} x + \begin{bmatrix} 0 \\ 1 \end{bmatrix} w \tag{3.344}$$

for the weighting matrices:

$$Q = \begin{bmatrix} 1 & 0 \\ 0 & 1 \end{bmatrix} \text{ and } R = 0.01. \tag{3.345}$$

The time evolution of the covariance matrix is given by the equation:

$$\begin{bmatrix} \dot{p}_{11} & \dot{p}_{12} \\ \dot{p}_{12} & \dot{p}_{22} \end{bmatrix} = \begin{bmatrix} 0 & 1 \\ -10.0 & -10.96 \end{bmatrix} \begin{bmatrix} p_{11} & p_{12} \\ p_{12} & p_{22} \end{bmatrix}$$
$$+ \begin{bmatrix} p_{11} & p_{12} \\ p_{12} & p_{22} \end{bmatrix} \begin{bmatrix} 0 & -10.0 \\ 1 & -10.96 \end{bmatrix} + \begin{bmatrix} 0 & 0 \\ 0 & 1 \end{bmatrix} \qquad (3.346)$$

when the variance of the Gaussian white noise input is $W=1$. Figure 3.25 illustrates the time evolution of the displacement and velocity states by the solid line. The 1σ variation of the states is illustrated by the dashed lines. The initial conditions of the states are assumed to be known and are $x = \{1\ 0\}$. It is clear that the states lie within the bounds provided by the differential Lyapunov equation.

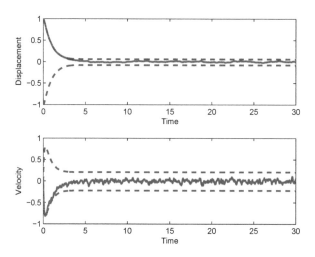

Figure 3.25: System Response

3.6   Summary

   This chapter reviewed calculus of variations and derived the necessary conditions for optimality. The Hamiltonian formulation for the design of optimal controllers was presented which was subsequently used to derive the optimal state feedback controllers for the LQR problem. A simple technique to solve the finite time horizon problem was described and benchmark examples were used to illustrate the technique. The LQR problem has been extensively studied and some of the extensions to the classic LQR problems were presented. These include the minimum power control problem, the desensitized LQR problem, and the frequency-shaped LQR problem, which provided a means to include frequency weighting in the design of the state feedback controller.

Exercises

3.1 The shape of a cable under its own weight when supported at its end can be shown to be a *catenary* using *calculus of variations*. The inverted *catenary* is a near perfect shape of an arch which supports its own weight and is the shape of the *Gateway Arch* in St. Louis, Missouri.

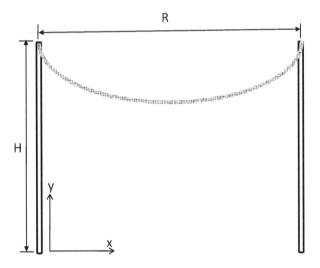

Figure 3.26: Catenary

The equation for catenary (Figure 3.26) can be developed by minimizing the potential energy of a cable under it own weight:

$$J = \int_0^L \rho g y \, ds$$

subject to the integral constraint:

$$\int_0^L ds = L$$

where L is the length of the cable. Determine the solution $y(x)$ when $R = 1$, and $L = 3$ and the height of the pole from the frame of reference $= H = 2$.

3.2   A pipeline needs to be constructed connecting a source of oil to a port which lies across a mountain range illustrated in Figure 3.27. The coordinates of the source and port are:

$$\text{Source coordinates } \begin{bmatrix} x \\ y \\ z \end{bmatrix} = \begin{bmatrix} -1 \\ -1 \\ 0.2 \end{bmatrix}$$

$$\text{Port coordinates } \begin{bmatrix} x \\ y \\ z \end{bmatrix} = \begin{bmatrix} 1 \\ 1 \\ 0.2 \end{bmatrix}.$$

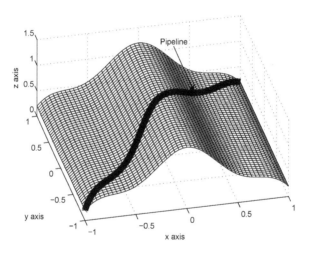

Figure 3.27: Pipeline Path

To save on the cost of material, the shortest pipeline has to be constructed. Determine the coordinates of the pipeline which minimize the cost function:

$$\min L = \int_{-1}^{1} \sqrt{1 + (\frac{dy}{dx})^2 + (\frac{dz}{dx})^2}\, dx$$

which is the length of the pipeline, subject to:

$$z = 1 - x^2 + 0.2\cos(2\pi x),$$

the model of the mountain range for any value of y.

3.3 For the benchmark problem,

$$\begin{bmatrix} 1 & 0 \\ 0 & 1 \end{bmatrix} \ddot{x} + \begin{bmatrix} 1 & -1 \\ -1 & 1 \end{bmatrix} x = \begin{bmatrix} 1 \\ 0 \end{bmatrix} u,$$

design a controller which minimizes the cost function

$$J = \frac{1}{2} \int_0^{4\pi} \dot{u}^2 \, dt$$

so as to satisfy the boundary conditions:

$y_1(0) = y_2(0) = 0$, and $\dot{y}_1(0) = \dot{y}_2(0) = 0 y_1(t_f) = y_2(t_f) = y_f$, and $\dot{y}_1(t_f) = \dot{y}_2(t_f) = 0$

3.4 Numerically solve the following optimal control problem:

$$\min \qquad x^2(\pi) + 0.1\dot{x}^2(\pi) + \int_0^\pi (x^2 + \dot{x}^2 + u^2)dt$$

subject to

$$\ddot{x} + x = u$$

and

$$x(0) = \dot{x}(0) = 1$$

Plot the evolution of the states, control and the terms of the Riccati matrix.

3.5 A linearized model of a helicopter in hover is [14]:

$$\dot{x} = \underbrace{\begin{bmatrix} -0.0257 & 0.013 & -0.322 & 0 \\ 1.26 & -1.765 & 0 & 0 \\ 0 & 1 & 0 & 0 \\ 1 & 0 & 0 & 0 \end{bmatrix}}_{A} x + \underbrace{\begin{bmatrix} 0.086 \\ -7.408 \\ 0 \\ 0 \end{bmatrix}}_{B} u + \underbrace{\begin{bmatrix} 0.0257 \\ -1.26 \\ 0 \\ 0 \end{bmatrix}}_{G} w$$

Design a linear quadratic regulator assuming no process noise. Assume a state weighting matrix Q and control weighting matrix R to be identity. Design a controller when the process noise has a variance $R_w = 1$ and plot the system response to an initial condition of $x(0) = [1, 0, 0, 0]$ and plot the 1σ variation of the first two states from the covariance dynamics.

3.6 Design a LQR controller for the flexible transmission system shown in Figure 3.28. The transfer function of the system is

$$\frac{\phi(s)}{\tau(s)} = \frac{1.491s^3 - 98.46s^2 + 704.8s + 1.746e5}{s^4 + 1.036s^3 + 1292s^2 + 422.3s + 1.796e5}$$

and the cost function is:

$$J = \int_0^\infty \left(\phi(t)^2 + \tau(t)^2 \right) dt$$

Simulate the closed-loop system response to an initial condition of $\phi(0) = 1$, $\dot{\phi}(0) = 0$, $\ddot{\phi}(0) = 0$, and $\dddot{\phi}(0) = 0$.

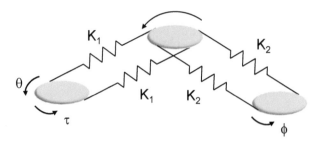

Figure 3.28: Flexible Transmission Problem

3.7   Design a desensitized controller for the system shown in Figure 3.29 for a rest-to-rest maneuver which minimize the cost function

$$J = \frac{1}{2} \int_0^{10} u^2 \, dt$$

Assume that the springs stiffness K is uncertain. The initial displacement of the mass and inertia are zero and the final rotation of the inertia is required to be 1 radian.

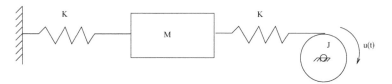

Figure 3.29: Two-Degrees of Freedom System

The equations of motion of the system are:

$$M\ddot{x} + Kx + K(x - R\theta) = 0$$
$$J\ddot{\theta} - KR(x - R\theta) = u$$

where M is the mass of the block, J is the inertia of the pulley whose radius is R. $u(t)$ is the torque input to the pulley. Assume $M = 1$, $J = 1$, $R = 1$ and the nominal stiffness coefficient is $K = 1$.

3.8   Design a controller which minimizes the cost function:

$$J = \int_0^{t_f} (1 + u^2) \, dt$$

subject to the state equations:

$$\begin{Bmatrix} \dot{x}_1 \\ \dot{x}_2 \end{Bmatrix} = \begin{bmatrix} 0 & 1 \\ -\omega^2 & 0 \end{bmatrix} \begin{Bmatrix} x_1 \\ x_2 \end{Bmatrix} + \begin{bmatrix} 0 \\ 1 \end{bmatrix} u$$

The initial and final conditions for the states are:

$$x(0) = 0, \dot{x}(0) = 0, x(t_f) = 1, \dot{x}(t_f) = 0.$$

References

[1] E. W. Weisstein. Brachistochrone problem. A Wolfram Web resource. http://mathworld.wolfram.com/BrachistochroneProblem.html.

[2] G. Hansjorg. Christiaan huygens and contact geometry. *Nieuw Arch. Wiskd*, 5:117, 2005.

[3] A. P. Sage and C. C. White III. *Optimum Systems Control*. Prentice Hall, 1977.

[4] N. Wiener. *Extrapolation, Interpolation, and Smoothing of Stationary Time Series*. Technology Press, Cambridge, MA, 1949.

[5] R. E. Kalman and R. W. Koepcke. Optimal synthesis of linear sampling control systems using generalized performance indices. *Transactions ASME Series D (J. of Basic Engineering)*, 80:1820–1826, 1958.

[6] S. P. Bhat and D. K. Miu. Minimum power and minimum jerk control and its application in computer disk drives. *IEEE Trans. on Magnetics*, 27(6):4471–4476, 1991.

[7] T. Hindle and T. Singh. Robust minimum power/jerk control of maneuvering structures. *J. of Guid., Cont. and Dyn.*, 24(4):816–826, 2001.

[8] W. T. Thomson and M. D. Dahleh. *Theory of Vibration with Applications*. Prentice Hall, 1998.

[9] N. K. Gupta. Frequency-shaped cost functionals: Extension of linear-quadratic-Gaussian design methods. *J. of Guid., Cont. and Dyn.*, 3(6):529–535, 1980.

[10] G. C. Goodwin anad S. F. Graebe and M. E. Salgado. *Control System Design*. Prentice Hall, 2001.

[11] Y. Kim and J. L. Junkins. *Introduction to Dynamics and Control of Flexible Structures*. AIAA Education Series, 1993.

[12] A. E. Bryson and R. A. Mills. Linear-quadratic-Gaussian controllers with specified parameter robustness. *J. of Guid., Cont. and Dyn.*, 21(1):11–18, February 1998.

[13] J.B. Burl. *Linear Optimal Control: $\mathcal{H}_2$ and $\mathcal{H}_\infty$ Methods*. Addison-Wesley Longman, 1999.

[14] L. El Ghaoui, A. Carrier, and A. E. Bryson. Linear quadratic minimax controllers. *IEEE Trans. on Cont. Systems Tech.*, 15(4):953–961, 1992.

4

Saturating Control

We know very little, and yet it is astonishing that we know so much, and still more astonishing that so little knowledge can give us so much power.

Bertrand Russell (1872–1970); English philosopher, mathematician

O PTIMAL control is a phrase used to describe the field of study where one endeavors to minimize or maximize a cost function subject to the dynamic model, state and control constraints and has been extensively studied over the past 50 years. Contributions by Pontryagin and Bellman laid the foundation on which the field had grown rapidly and matured. Table 4.1 lists a few applications in the first column and three parameters which are of interest in the optimal control formulation for the listed applications. A check mark is used to highlight the parameters of interest. These applications have been selected to motivate the cost functions which will be presented in this chapter.

Consider a race car pit crew for a Formula 1 racing team. They would like to minimize the time for their car to complete the racetrack in minimum time, subject to the constraint that the available fuel is constrained. On the other hand, a subway railcar that starts and stops frequently requires a control strategy that moves people as rapidly as possible subject to constraints on the rate of change of acceleration (jerk), which influences passenger comfort. It is clear that different applications demand different performance metrics. In this chapter we will address the problem of

	Time	Budget/Fuel	Jerk/Jolt
Race Car	X	X	
Desktop Hard Disk Drive	X		X
Airline Industry	X	X	
Hubble Telescope	X		X
High-Speed Elevator	X		X
Burt Rutan Voyager		X	
Subway Railcar	X		X

Table 4.1: Applications and Metrics of Interest

rest-to-rest maneuvers of systems where the control input is subject to saturation constraints. Four cost functions will be considered which include: (1) Minimum-Time,
(2) Minimum Fuel/Time, (3) Fuel Limited Time Optimal, and (4) Jerk Limited Time
Optimal Control problems. All the optimal controllers presented in this chapter are
open-loop. For real-world applications, models which are used for the design of
controllers invariably include modeling errors, process and sensor noise, and initial condition uncertainties. In such situations, open-loop controllers cannot reliably
satisfy terminal state constraint. A hybrid control scheme where the open-loop controllers are used to move the states to the vicinity of desired states and subsequently
a feedback controller is used to asymptotically reach the final states, is a potential
approach for obtaining the benefits of open-loop and closed-loop controllers. These
open-loop controllers can also be used to generate nominal control and state profiles
which in conjunction with a tracking feedback controller can influence the states
such that the tracking error asymptotically tends to zero. All of the aforementioned
problems will be solved for the rest-to-rest maneuver of a double integrator which
represents the motion of a rigid body and the floating oscillator benchmark problem
which captures the characteristics of a maneuvering flexible structure [1].

4.1   Benchmark Problem

Figure 4.1: Floating Oscillator

The equations of motion of the benchmark floating oscillator problem illustrated
in Figure 4.1 are

$$\begin{bmatrix} m_1 & 0 \\ 0 & m_2 \end{bmatrix} \begin{bmatrix} \ddot{y}_1 \\ \ddot{y}_2 \end{bmatrix} + \begin{bmatrix} k & -k \\ -k & k \end{bmatrix} \begin{bmatrix} y_1 \\ y_2 \end{bmatrix} = \begin{bmatrix} 1 \\ 0 \end{bmatrix} u. \tag{4.1}$$

The state constraint for all the optimization problems considered in this section are

$$y_1(0) = y_2(0) = 0, \text{ and } \dot{y}_1(0) = \dot{y}_2(0) = 0$$
$$y_1(t_f) = y_2(t_f) = y_f, \text{ and } \dot{y}_1(t_f) = \dot{y}_2(t_f) = 0 \tag{4.2}$$

and the normalized control is subject to the constraint

$$-1 \le u \le 1. \tag{4.3}$$

The nominal values of the parameters of the system are

$$m_1 = m_2 = k = 1 \tag{4.4}$$

The equations of motion can be decoupled by the similarity transformation

$$\theta = \frac{1}{2}(y_1 + y_2) \text{ and } q = \frac{1}{2}(y_2 - y_1) \tag{4.5}$$

resulting in the equations of motion

$$\ddot{\theta} = \frac{1}{2}u = \phi_0 u \tag{4.6a}$$

$$\ddot{q} + 2q = -\frac{1}{2}u = \phi_1 u, \tag{4.6b}$$

and the corresponding boundary conditions are

$$\theta(0) = q(0) = 0 \text{ and } \dot{\theta}(0) = \dot{q}(0) = 0$$
$$\theta(t_f) = 1, q(t_f) = 0 \text{ and } \dot{\theta}(tf) = \dot{q}(t_f) = 0. \tag{4.7}$$

The decoupled equations are used to derive the constraint equations for the optimiza-tion problem since this approach can be generalized for any number of modes.

Every section in this chapter will study three problems. The first corresponds to the rigid body model given by Equation (4.6a). The next problem studied is the undamped benchmark problem and finally, the results of the effect of damping on the solution to the benchmark problem will be presented.

4.2 Minimum-Time Control

The minimum-time control problem can be stated as [2–4]:

$$\min J = \int_0^{t_f} dt \tag{4.8a}$$

subject to

$$\dot{x} = Ax + Bu \tag{4.8b}$$
$$x(0) = x_0 \text{ and } x(t_f) = x_f \tag{4.8c}$$
$$-1 \le u \le 1 \forall t \tag{4.8d}$$

Defining the Hamiltonian as:

$$H = 1 + \lambda^T (Ax + Bu), \tag{4.9}$$

the necessary conditions for optimality can be derived using calculus of variations, resulting in the equations

$$\dot{x} = \frac{\partial H}{\partial \lambda} = Ax + Bu \tag{4.10a}$$

$$\dot{\lambda} = -\frac{\partial H}{\partial x} = -A^T \lambda \tag{4.10b}$$

$$u = -sgn\left(B^T \lambda\right) \tag{4.10c}$$

$$x(0) = x_0 \text{ and } x(t_f) = x_f \tag{4.10d}$$

$$H = 0 \text{ at } t = 0. \tag{4.10e}$$

Equation (4.10c) is derived using *Pontryagin's minimum principle (PMP)*. The *Pontryagin's minimum principle (PMP)* states that the Hamiltonian (H) should be minimized over all possible control input which lie in the set of permissible control $\mathcal{U}$. This permits including constraints on the control input.

$$H(x^*(t), \lambda^*(t), u^*(t), t) \leq H(x^*(t), \lambda^*(t), u(t), t) \quad \forall u(t) \in \mathcal{U} \tag{4.11}$$

for all time $t \in \begin{bmatrix} 0 & t_f \end{bmatrix}$. $(.)^*$ refer to optimal quantities. The form of control defined by Equation (4.10c) is referred to as a bang-bang control since the control switches between its limits.

4.2.1 Singular Time-Optimal Control

Suppose that there exists an interval of time where the switching function $q(t)$ is exactly zero, i.e.,

$$q(t) = B^T \lambda(t) = 0 \quad \forall t \in \begin{bmatrix} T_1 & T_2 \end{bmatrix}, \tag{4.12}$$

the Hamiltonian H is not affected by any selection of the control $u(t)$, and consequently the interval $t \in [T_1 \quad T_2]$ is called a singular interval and the resulting problem is called a *singular time-optimal* problem. Figure 4.2 schematically illustrates a singular time-optimal control profile.

Since, the switching function is identically zeros in the interval $[T_1 \quad T_2]$, this implies

$$\frac{dq}{dt} = B^T A^T \exp(-A^T t)\lambda(0) = 0 \tag{4.13a}$$

$$\frac{d^2q}{dt^2} = B^T (A^T)^2 \exp(-A^T t)\lambda(0) = 0 \tag{4.13b}$$

$$\frac{d^3q}{dt^3} = B^T (A^T)^3 \exp(-A^T t)\lambda(0) = 0 \tag{4.13c}$$

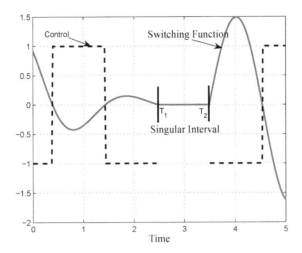

Figure 4.2: Switching Function for a Singular Time-Optimal Control

$$\ldots \tag{4.13d}$$

$$\frac{d^{n-1}q}{dt^{n-1}} = B^T (A^T)^{n-1} \exp(-A^T t)\lambda(0) = 0, \tag{4.13e}$$

since the costate equation can be written as

$$\lambda(t) = \exp(-A^T t)\lambda(0). \tag{4.14}$$

Equation (4.13) can be concisely written as

$$\left[B \; AB \; A^2 B \; \ldots \; A^{n-1} B \right]^T \exp(-A^T t)\lambda(0) = 0. \tag{4.15}$$

Since $\exp(-A^T t)$ is nonsingular and $\lambda(0)$ cannot be zero, since it results in a trivial solution, a singular solution requires that the *controllability matrix* be singular which corresponds to an uncontrollable linear time invariant system. Thus, a controllable system is normal and the resulting time-optimal control is bang-bang.

4.2.2   Rigid Body

With the knowledge that the time-optimal control is bang-bang, the control profile can be represented as the output of a time-delay filter subject to a step input which is parameterized to generate a bang-bang control profile. For example, the time-optimal control profile for a rest-to-rest maneuver of double integrator,

$$\ddot{\theta} = \phi_0 u \tag{4.16}$$

is:

$$u(t) = 1 - 2\mathcal{H}(t - T) + \mathcal{H}(t - 2T) \tag{4.17a}$$

$$u(s) = \frac{1}{s}(1 - 2\exp(-sT) + \exp(-2sT)) = \frac{1}{s}G_c(s) \tag{4.17b}$$

which is a single switch bang-bang control profile which satisfies the constraint that the control has to lie in the range

$$-1 \le u(t) \le 1. \tag{4.18}$$

$\mathcal{H}$ is the Heaviside function.

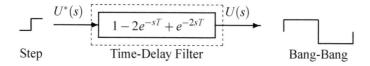

Step   Time-Delay Filter   Bang-Bang

Figure 4.3: Time-Delay Filter Structure

The transfer function $G_c(s)$ of the time-delay filter (Figure 4.3) that generates the bang-bang control profile can be rewritten as

$$G_c(s) = (1 - 2\exp(-sT) + \exp(-2sT)) = (1 - \exp(-sT))^2 \tag{4.19}$$

which has zeros at

$$s = 0, 0, \pm j\frac{2n\pi}{T}, \text{ where, } n = 1, 2, 3, \dots \tag{4.20}$$

It should be noted that the two poles at the origin of the rigid body model, are canceled by the two zeros of the time-delay filter, which are also located at the origin.

The switch time is a function of the rigid body displacement θ_f, and can be shown to be

$$T = \sqrt{\frac{\theta_f}{\phi_0}}. \tag{4.21}$$

Since the Hamiltonian of the system is

$$H = 1 + \lambda_1 \dot{\theta} + \lambda_2 \phi_0 u, \tag{4.22}$$

the costates satisfy

$$\dot{\lambda}_1 = -\frac{\partial H}{\partial \theta} = 0 \tag{4.23a}$$

$$\dot{\lambda}_2 = -\frac{\partial H}{\partial \dot{\theta}} = -\lambda_1 \tag{4.23b}$$

which can be integrated to give

$$\lambda_1(t) = E \tag{4.24a}$$

$$\lambda_2(t) = -Et + F. \tag{4.24b}$$

Since the Hamiltonian is zero at the initial time, we have

$$H(0) = 1 + \phi_0 \lambda_2(0) = 0 \Rightarrow \lambda_2(0) = -\frac{1}{\phi_0}. \tag{4.25}$$

Further, since the switching function $\phi_0 \lambda_2$ should be zero at the switch time T, we have

$$\phi_0 \lambda_2(T) = \phi_0 \left(-ET - \frac{1}{\phi_0} \right) = 0 \tag{4.26}$$

which results in the closed form expression for the control

$$u = -sgn(\phi_0 \lambda_2(t)) = -sgn \left(\frac{1}{\sqrt{\phi_0 \theta_f}} t - \frac{1}{\phi_0} \right). \tag{4.27}$$

The time-optimal control of any linear controllable time invariant system can be stated as the design of a time-delay filter so that a set of zeros of the transfer function of the time-delay filter cancel all the poles of the system.

4.2.3 Time-Optimal Rest-to-Rest Maneuvers

Many researchers have noted the antisymmetric characteristic of the minimum-time control profiles designed for rest-to-rest maneuvers for flexible structures without damping [5–7]. The term *antisymmetry* refers to profiles which can be generated by rotating the first half of the profile about the mid-maneuver time, by π radians. For example, a single period of a sinusoid $(\sin(\omega t))$ is antisymmetric since the profile from $t=0$ to $\frac{\pi}{\omega}$ when rotated about the point $t=\frac{\pi}{\omega}$, which is half the period of oscillation, by π radians, generates one cycle of the sinusoid.

We first show that a control profile that is antisymmetric about the mid-maneuver time leads to a transfer function of a time-delay filter with two zeros at the origin of the s plane. Figure 4.5 illustrates the minimum-time control profile for an undamped system with the time delays selected to ensure antisymmetry. The transfer function of a time-delay filter (Figure 4.4) containing $2n+2$ time-delays ($2n+1$ switches) is

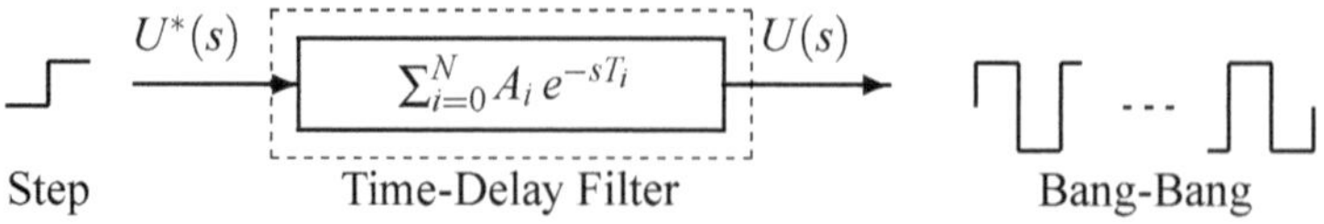

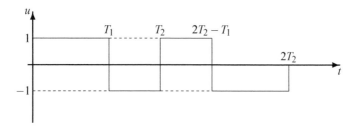

Figure 4.4: Time-Delay Filter Structure

$$G_c(s) = 1 + 2\sum_{i=1}^{n}(-1)^i \exp(-sT_i) + 2(-1)^{n+1}\exp(-sT_{n+1})$$

$$+ 2\sum_{i=1}^{n}(-1)^i \exp(-s(2T_{n+1} - T_i)) + \exp(-2sT_{n+1}). \tag{4.28}$$

Figure 4.5: Three-Switch Antisymmetric Control Profile

A zero of the transfer function is located at $s = 0$, as Equation (4.28) goes to zero at $s = 0$. The derivative of Equation (4.28) with respect to s is

$$\frac{dG_c(s)}{ds} = -2\sum_{i=1}^{n} T_i(-1)^i \exp(-sT_i) - 2(-1)^{n+1}T_{n+1}\exp(-sT_{n+1})$$

$$- 2\sum_{i=1}^{n}(2T_{n+1} - T_i)(-1)^i \exp(-s(2T_{n+1} - T_i)) - 2T_{n+1}\exp(-2sT_{n+1}). \tag{4.29}$$

which is also zero at $s = 0$. Thus, we conclude that Equation (4.28) has at least two zeros at the origin that cancel the rigid body poles, thus satisfying the velocity boundary conditions for a rest-to-rest maneuver.

To arrive at equations to cancel the imaginary poles of the system, we substitute

$$s = \sigma + j\omega \tag{4.30}$$

into Equation (4.28), and equating the real and imaginary parts, we have,

$$\Re(G_c(s)) = 1 + 2\sum_{i=1}^{n}(-1)^i \exp(-\sigma T_i)\cos(\omega T_i) + 2(-1)^{n+1}\exp(-\sigma T_{n+1})\cos(\omega T_{n+1})$$

$$+ 2\sum_{i=1}^{n}(-1)^i \exp(-\sigma(2T_{n+1} - T_i))\cos(\omega(2T_{n+1} - T_i))$$

$$+ \exp(-2\sigma T_{n+1})\cos(2\omega T_{n+1}) = 0. \tag{4.31}$$

$$\Im(G_c(s)) = 2\sum_{i=1}^{n}(-1)^i \exp(-\sigma T_i)\sin(\omega T_i) + 2(-1)^{n+1}\exp(-\sigma T_{n+1})\sin(\omega T_{n+1})$$

$$+ 2\sum_{i=1}^{n}(-1)^i \exp(-\sigma(2T_{n+1} - T_i))\sin(\omega(2T_{n+1} - T_i))$$

$$+ \exp(-2\sigma T_{n+1})\sin(2\omega T_{n+1}) = 0 \tag{4.32}$$

where $\Re(.)$ and $\Im(.)$ refer to the real and imaginary parts of the argument. As we require the undamped poles of the system to be canceled by the zeros of the time-delay filter transfer function, we substitute $\sigma = 0$, in Equations (4.31) and (4.32) and rewrite them as

$$\cos(\omega T_{n+1})\left(2\sum_{i=1}^{n}(-1)^i \cos(\omega(T_{n+1} - T_i)) + (-1)^{n+1} + \cos(\omega T_{n+1})\right) = 0. \tag{4.33}$$

$$\sin(\omega T_{n+1})\left(2\sum_{i=1}^{n}(-1)^i \cos(\omega(T_{n+1} - T_i)) + (-1)^{n+1} + \cos(\omega T_{n+1})\right) = 0. \tag{4.34}$$

To cancel poles at $\omega = \pm j\omega_1, \pm j\omega_2, ..., \pm j\omega_n$, we substitute ω_i into the coefficient of $\sin(\omega T_{n+1})$ in Equation (4.34), which is the same as the coefficient of $\cos(\omega T_{n+1})$ in Equation (4.33), which leads to n equations in $n+1$ unknowns $T_1, T_2, ..., T_{n+1}$.

The final constraint is derived from the system transfer function

$$\frac{y(s)}{u(s)} = G(s) \tag{4.35}$$

so as to satisfy the rigid body boundary condition. The constraint can be written as:

$$y_f = \lim_{s \to 0} G(s)G_c(s). \tag{4.36}$$

Since the transfer function of the time-delay filter $G_c(s)$ has zeros at the locations of the rigid body poles of the system, the limit value is indeterminate. Repeatedly apply L'Hospital's rule until the limit is determinate. That is the final constraint for the parameter optimization problem which can be stated as

$$\min \ J = T_{n+1} \tag{4.37a}$$

subject to

$$2\sum_{i=1}^{n}(-1)^i\cos(\omega_j(T_{n+1}-T_i))+(-1)^{n+1}+\cos(\omega_j T_{n+1}) = 0, \forall\ \omega_j$$

$$\tag{4.37b}$$

$$y_f = \lim_{s\to 0} G(s)G_c(s) \tag{4.37c}$$

$$0 < T_1 < T_2 < \ldots < T_{n+1}. \tag{4.37d}$$

4.2.4   Implications of Pole-Zero Cancelation

To illustrate that the parameter optimization problem given by Equation 4.37 is the same as one derived by requiring the final state conditions to be satisfied, consider a second-order model of an undamped system

$$M\ddot{x} + Kx = Du \tag{4.38}$$

where the mass matrix M is positive definite, and K is the stiffness matrix which is positive definite or positive semidefinite. The stiffness matrix is positive semi-definite when the system includes rigid body modes. The second-order system model given by Equation (4.38) can be represented in a modal form as:

$$\ddot{z}+\begin{bmatrix}0 & 0 & 0 \\ 0 & \omega_1^2 \\ & 0 & \omega_2^2 & 0 \\ & & & \ddots\end{bmatrix}z=\begin{Bmatrix}\phi_0 \\ \phi_1 \\ \phi_2 \\ \vdots\end{Bmatrix}u \tag{4.39}$$

by defining a transformation $x = \Phi z$ where Φ is the eigenvector matrix resulting from the generalized eigenvalue problem of the matrix pair (K,M). The modal equations

$$\ddot{z}_0 = \phi_0 u \tag{4.40}$$

$$\ddot{z}_j + \omega_j^2 z_j = \phi_j u \text{ for } j=1,2,\ldots \tag{4.41}$$

can be represented in the frequency domain, assuming zero initial conditions, as:

$$Z_0(s) = \phi_0\frac{1}{s^3}G_c(s) \tag{4.42}$$

$$Z_j(s) = \phi_j\frac{1}{s(s^2+\omega_j^2)}G_c(s) \text{ for } j=1,2,\ldots \tag{4.43}$$

for the control parameterization given by Equation (4.28). Symbolically integrating the modal equations of motion result in the equations:

$$z_0(t) = \phi_0 \left(\frac{t^2}{2} + 2 \sum_{i=1}^{n} (-1)^i \frac{(t - T_i)^2}{2} \mathcal{H}(t - T_i) \right.$$

$$+ 2 \sum_{i=1}^{n} (-1)^i \frac{(t - (2T_{n+1} - T_i))^2}{2} \mathcal{H}(t - (2T_{n+1} - T_i)) \quad (4.44)$$

$$\left. + (-1)^{n+1} (t - T_{n+1})^2 \mathcal{H}(t - T_{n+1}) + \frac{(t - 2T_{n+1})^2}{2} \mathcal{H}(t - 2T_{n+1}) \right)$$

$$z_j(t) = \phi_i \left(\frac{1 - \cos(\omega_j t)}{\omega_j^2} + 2 \sum_{i=1}^{n} (-1)^i \frac{1 - \cos(\omega_j(t - T_i))}{\omega_j^2} \mathcal{H}(t - T_i) \right.$$

$$+ 2 \sum_{i=1}^{n} (-1)^i \frac{1 - \cos(\omega_j(t - (2T_{n+1} - T_i)))}{\omega_j^2} \mathcal{H}(t - (2T_{n+1} - T_i))$$

$$+ 2(-1)^{n+1} \frac{1 - \cos(\omega_j(t - T_{n+1}))}{\omega_j^2} \mathcal{H}(t - T_{n+1}) \quad (4.45)$$

$$\left. + \frac{1 - \cos(\omega_j(t - 2T_{n+1}))}{\omega_j^2} \mathcal{H}(t - 2T_{n+1}) \right) \quad \text{for } j{=}1,2,\dots.$$

Evaluating the modal states at the final time, result in the equations:

$$z_0(2T_{n+1}) = \phi_0 \left(2T_{n+1}^2 + 2 \sum_{i=1}^{n} (-1)^i \frac{(2T_{n+1} - T_i)^2}{2} + \right.$$

$$\left. (-1)^{n+1} (T_{n+1})^2 + 2 \sum_{i=1}^{n} (-1)^i \frac{T_i^2}{2} \right) \quad (4.46)$$

$$z_j(2T_{n+1}) = \phi_i \left(\frac{-\cos(\omega_j 2T_{n+1})}{\omega_j^2} + 2 \sum_{i=1}^{n} (-1)^i \frac{-\cos(\omega_j(2T_{n+1} - T_i))}{\omega_j^2} \right.$$

$$\left. + 2 \sum_{i=1}^{n} (-1)^i \frac{-\cos(\omega_j T_i)}{\omega_j^2} + 2(-1)^{n+1} \frac{-\cos(\omega_j T_{n+1})}{\omega_j^2} - \frac{1}{\omega_j^2} \right)$$

$$\text{for } j{=}1,2,\dots. \quad (4.47)$$

Since we require $z_j(2T_{n+1})$ to be zero, we have:

$$z_j(2T_{n+1}) = 0 = -\cos(\omega_j 2T_{n+1}) - 2 \sum_{i=1}^{n} (-1)^i \cos(\omega_j(2T_{n+1} - T_i)) \quad (4.48)$$

$$- 2 \sum_{i=1}^{n} (-1)^i \cos(\omega_j T_i) - 2(-1)^{n+1} \cos(\omega_j T_{n+1}) - 1 \text{ for } j{=}1,2,\dots.$$

which is identical to Equation (4.31) when $\sigma = 0$ and $\omega = \omega_j$.

The time derivative of $z_i(t)$ evaluated at the final time is:

$$\dot{z}_i(2T_{n+1}) = \frac{\phi_i}{\omega_i}\left(\sin(\omega_i 2T_{n+1}) + 2\sum_{i=1}^{n}(-1)^i \sin(\omega_i(T_{n+1} - T_i)) + \right.$$ (4.49)

$$\left. 2\sum_{i=1}^{n}(-1)^i \sin(\omega_i T_i) + 2(-1)^{n+1}\sin(T_{n+1}) \right) \text{ for i=1,2,...}$$

which is required to be zero at the final time for a rest-to-rest maneuver. Equation (4.49) is identical to Equation (4.32) when $\sigma = 0$ and $\omega = \omega_i$. Since Equations (4.31) and (4.32) which are derived by forcing zeros of the time-delay filter to cancel the undamped poles of the system are identical to Equations (4.48) and (4.49), it can be seen that the pole-zero cancelation constraint is equivalent to forcing the displacement and velocity of the flexible modes to zero at the final time. Finally, Equation (4.46) can be shown to be the same as Equation (4.36) which illustrates that the boundary conditions for the states of a rest-to-rest maneuver are satisfied by designing a time-delay filter whose zeros cancel the poles of the system and which is required to satisfy the final value constraint given by Equation (4.36).

4.2.5   Sufficiency Condition

The parameter optimization problem to be solved includes nonlinear constraints and may have multiple solutions. To verify the optimality of the switch times arrived at from the parameter optimization problem, the satisfaction of the costate equations and the optimality conditions which results in the switching control, need to be verified. Since the costate equation for the minimum-time optimal control problem for the system

$$\dot{x} = Ax + Bu$$ (4.50)

is

$$\dot{\lambda} = -A^T\lambda.$$ (4.51)

The optimal control is given by the switching function

$$B^T\lambda(t) = B^T \exp(-A^T t)\lambda(0)$$ (4.52)

which is equal to zero at the $(2n+1)$ switch times. Collecting the switching time constraints into one equation, we have:

$$\underbrace{\begin{bmatrix} B^T \exp(-A^T T_1) \\ B^T \exp(-A^T T_2) \\ B^T \exp(-A^T T_3) \\ \cdots \\ B^T \exp(-A^T (2T_{n+1} - T_2)) \\ B^T \exp(-A^T (2T_{n+1} - T_1)) \end{bmatrix}}_{P} \lambda(0) = \begin{bmatrix} 0 \\ 0 \\ 0 \\ \cdots \\ 0 \\ 0 \end{bmatrix}.$$ (4.53)

Thus $\lambda(0)$ has to lie in the null space of the $(2n+1) \times (2n+2)$ matrix P. We can determine the null space of P and use that vector as $\lambda(0)$, to determine $\lambda(t)$. Since, the parameter optimization permits multiple solutions that satisfy the boundary conditions, the control profile determined from

$$u(t) = -sgn\left(B^T \lambda(t)\right) \qquad (4.54)$$

must switch at the predetermined switch times, to be optimal.

4.2.6 Benchmark Problem

The transfer function of the benchmark problem where the displacement of the second mass is the output, is

$$G(s) = \frac{Y_2(s)}{U(s)} = \frac{1}{s^2(s^2+2)} \qquad (4.55)$$

which has poles at

$$s = 0,0,\pm j\sqrt{2}. \qquad (4.56)$$

The control profile for a rest-to-rest maneuver with a final displacement of y_f of the benchmark problem masses, is parameterized as

$$u(t) = 1 - 2\mathcal{H}(t - T_1) + 2\mathcal{H}(t - T_2)) - 2\mathcal{H}(t - (2T_2 - T_1)) + \mathcal{H}(t - 2T_2) \qquad (4.57a)$$

$$u(s) = \frac{1}{s}(1 - 2\exp(-sT_1) + 2\exp(-sT_2) - 2\exp(-s(2T_2 - T_1)) + \exp(-2sT_2)) \qquad (4.57b)$$

$$= \frac{1}{s}G_c(s). \qquad (4.57c)$$

The parameter optimization problem to be solved for the rest-to-rest maneuver of the undamped floating oscillator problem

$$\min J = T_2 \qquad (4.58a)$$

subject to

$$-2\cos(\omega(T_2 - T_1)) + 1 + \cos(\omega T_2) = 0 \qquad (4.58b)$$

$$\frac{-2T_1^2 + 2T_2^2 - 2(2T_2 - T_1)^2 + (2T_2)^2}{4} = y_f \qquad (4.58c)$$

$$T_2 > T_1 > 0 \qquad (4.58d)$$

where ω, the natural frequency of the system is $\sqrt{2}$, and y_f represents the final displacement of the two masses.

Figure 4.6 illustrates the variations of the switch times and the maneuver time, as a function of the rigid body displacement. It can be seen that for specific maneuvers, the three switches collapse into a single switch, which is given by the equation

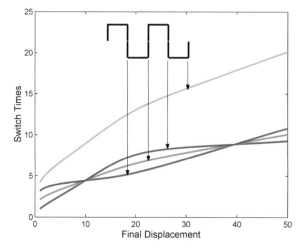

Figure 4.6: Controller Parameters vs. Displacement

$$y_f = \frac{T_1^2}{2}. \tag{4.59}$$

This single switch control profile cancels the poles corresponding to the rigid body and flexible modes. This can be easily explained by studying Equation (4.20) which lists the location of zeros of a single-switch bang-bang control profile. Substituting the switch time given by Equation (4.59) into Equation (4.20), the time-delay filter poles are located at

$$s = 0, 0, \pm j \frac{2n\pi}{\sqrt{2y_f}} \tag{4.60}$$

It can be seen that for a system with a natural frequency of ω_n, we require

$$\frac{2n\pi}{\sqrt{2y_f}} = \omega_n \Rightarrow y_f = 2 \left(\frac{n\pi}{\omega_n} \right)^2 \tag{4.61}$$

to cancel the poles at $\pm \omega_n j$. For the benchmark problem, the system natural frequency is $\sqrt{2}$, and the rigid body displacement has to satisfy the equation

$$y_f = 2 \left(\frac{n\pi}{\sqrt{2}} \right)^2 = (n\pi)^2. \tag{4.62}$$

Equation (4.61) states that for a system with a natural frequency of ω_n, there are numerous maneuvers for which the single switch profile is time optimal.

4.2.7 Effect of Damping

The time-optimal control profile for systems with damping are no longer antisymmetric about the mid-maneuver time [8]. With this in mind, the transfer function of a time-delay filter which generates the optimal control profile can be parameterized as:

$$G_c(s) = 1 + 2\sum_{i=1}^{N}(-1)^i \exp(-sT_i) + \exp(-sT_{N+1}), \qquad (4.63)$$

where, N is an odd integer. This transfer function has one zero at $s = 0$. To force the transfer function to have an additional zero at $s = 0$, we require

$$2\sum_{i=1}^{N} T_i(-1)^i \exp(-sT_i) + T_{N+1}\exp(-sT_{N+1}) = 0, \qquad (4.64)$$

which is the derivative of Equation (4.63) with respect to s. To have a second zero at $s = 0$, the constraint equation to be be satisfied is

$$2\sum_{i=1}^{N} T_i(-1)^i + T_{N+1} = 0. \qquad (4.65)$$

To cancel the damped poles

$$s = \sigma_k \pm j\omega_k, \text{ where } k = 1,2,3,... \qquad (4.66)$$

of the system, we require

$$1 + 2\sum_{i=1}^{N}(-1)^i \exp(-\sigma_k T_i)\cos(\omega_k T_i) + \exp(-\sigma_k T_{N+1})\cos(\omega_k T_{N+1}) = 0, \forall\, k \quad (4.67)$$

$$2\sum_{i=1}^{N}(-1)^i \exp(-\sigma_k T_i)\sin(\omega_k T_i) + \exp(-\sigma_k T_{N+1})\sin(\omega_k T_{N+1}) = 0, \forall\, k \quad (4.68)$$

The final constraint is arrived at from the rigid body boundary condition which can be written as:

$$y_f = \lim_{s \to 0} G(s)G_c(s). \qquad (4.69)$$

4.2.8 Example

The transfer function for the damped floating oscillator is:

$$G(s) = \frac{cs + k}{(m^2 s^4 + 2mcs^3 + 2mks^2)} \qquad (4.70)$$

The parameter optimization problem for the damped floating oscillator is

$$\min\ J = T_4 \tag{4.71a}$$

subject to

$$-2T_1 + 2T_2 - 2T_3 + T_4 = 0 \tag{4.71b}$$

$$1 - 2\exp(-\sigma T_1)\cos(\omega T_1) + 2\exp(-\sigma T_2)\cos(\omega T_2)$$
$$-2\exp(-\sigma T_3)\cos(\omega T_3) + \exp(-\sigma T_4)\cos(\omega T_4) = 0 \tag{4.71c}$$

$$-2\exp(-\sigma T_1)\sin(\omega T_1) + 2\exp(-\sigma T_2)\sin(\omega T_2)$$
$$-2\exp(-\sigma T_3)\sin(\omega T_3) + \exp(-\sigma T_4)\sin(\omega T_4) = 0 \tag{4.71d}$$

$$\frac{k(-2T_1^2 + 2T_2^2 - 2T_3^2 + T_4^2)}{4mk} = y_f \tag{4.71e}$$

$$T_4 > T_3 > T_2 > T_1 > 0 \tag{4.71f}$$

where

$$s = \sigma \pm j\omega = -\frac{c}{m} \pm j\sqrt{\frac{2k}{m} - \frac{c^2}{m^2}}. \tag{4.72}$$

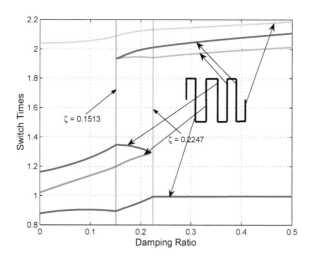

Figure 4.7: Controller Parameters vs. Damping Ratio

To study the effect of the damping ratio on the structure of the time-optimal control profile, the parameter optimization problem is solved for the damping ratio varying from zero to one [9, 10]. The spring stiffness is assumed to be $k = 50$ and the rigid body displacement $y_f = 0.5$. Figure 4.7 plots the variation of the switch times and the maneuver time as a function of the damping ratio. It is clear that for small and large magnitude of the damping ratio, the time-optimal control includes three switches. However, for a certain region of the damping ratio, the time-optimal control profile includes five switches. Figure 4.8 illustrates the switching function and the corresponding time-optimal control profile for three values of the damping ratios. It can be conjectured that as the damping ratio is increased, two switches are introduced towards the end of the maneuver. As the damping ratio is further increased, two switches near the middle of the maneuver coalesce resulting in a three-switch control profile.

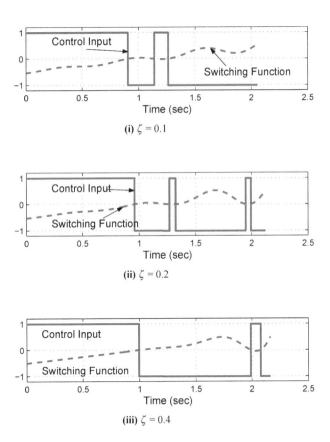

Figure 4.8: Typical Solutions

The time-optimal control profile for the damped benchmark problem is given by the equation

$$u = -sgn(B^T \lambda(t)) \tag{4.73}$$

which indicates that the control input changes sign when the switching function $B^T \lambda(t)$ passes through zero. With the variation of certain parameters in the control problem, switches can be introduced or eliminated from the control structure. If there exists a control profile transition created by switches appearing at the initial or final time of the control input, the equation that has to be satisfied is

$$B^T \lambda(t_{tran}) = 0 \tag{4.74}$$

at the initial or final time respectively, since only one switch can enter from either end of the time boundaries. The variation of the costates with respect to the system parameters is assumed to be smooth. Since the costates are smooth functions of time, if the switches appear in between the initial and final time, then two switches are introduced simultaneously and the transition occurs when the equations [10]

$$B^T \lambda(t) = 0 \tag{4.75a}$$

$$\frac{d\left(B^T \lambda(t)\right)}{dt} = 0 \tag{4.75b}$$

are satisfied. Equation (4.75a) is the constraint that forces the value of the switching function to be zero at the transition point, and Equation (4.75b) indicates that the slope of the switching curve at the transition point should also be zero.

A parameter optimization problem is solved where the parameters to be solved for are the damping ratio and the time which corresponds to when Equations (4.75a) and (4.75b) are simultaneously satisfied. Figure 4.9 illustrates the transition of the time-optimal control profile from a three-switch to a five-switch control profile as the damping is increased. Figure 4.10 corresponds to the transition from a five switch to a three switch control profile as the damping is increased.

Defining fuel consumed as $\int_0^{t_f} |u| \, dt$, it is clear that the bang-bang control profile corresponds to a large consumption of fuel. In applications where the available fuel is limited such as satellites, laptop computers, hybrid or plug-in cars, and so forth, control policies which maximize the operation life or range of the application between recharge would be the optimization objective. A judicious balance between fuel consumed and maneuver time can be posed as a *fuel/time optimal control* problem. The next section addresses this problem.

4.3   Fuel/Time Optimal Control

The minimum fuel/time optimal control problem can be stated as [11–14]:

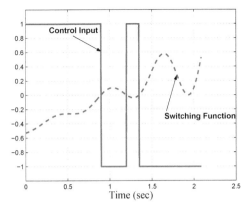

Figure 4.9: Transition Control Profile

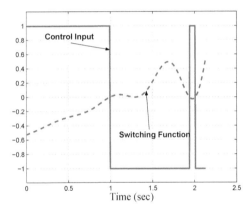

Figure 4.10: Transition Control Profile

$$\min \; J = \int_0^{t_f} (1 + \alpha |u|) \, dt \tag{4.76a}$$

subject to

$$\dot{x} = Ax + Bu \tag{4.76b}$$

$$x(0) = x_0 \text{ and } x(t_f) = x_f \tag{4.76c}$$

$$-1 \le u \le 1 \; \forall t \tag{4.76d}$$

where α is a weighting parameter which prescribe a penalty which has to be imposed on the fuel consumed relative to the maneuver time. $\alpha = 0$, corresponds to the time-optimal control problem and when $\alpha \to \infty$, the problem corresponds to a fuel optimal control. Defining the Hamiltonian as

$$H = 1 + \alpha|u| + \lambda^T (Ax + Bu), \qquad (4.77)$$

the necessary conditions for optimality can be derived using calculus of variations, resulting in the equations

$$\dot{x} = \frac{\partial H}{\partial \lambda} = Ax + Bu \qquad (4.78a)$$

$$\dot{\lambda} = -\frac{\partial H}{\partial x} = -A^T \lambda \qquad (4.78b)$$

$$u = -dez\left(\frac{B^T \lambda}{\alpha}\right) \qquad (4.78c)$$

$$x(0) = x_0 \text{ and } x(t_f) = x_f \qquad (4.78d)$$

$$H(0) = 0. \qquad (4.78e)$$

where *dez* is the deadzone function (Figure 4.11) defined as

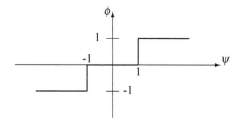

Figure 4.11: Deadzone Function

$$\phi = dez(\psi) \Rightarrow \begin{cases} \phi = 0 & \text{if } |\psi| < 1 \\ \phi = sgn(\psi) & \text{if } |\psi| > 1 \\ 0 \leq \phi \leq 1 & \text{if } \psi = 1 \\ -1 \leq \phi \leq 0 & \text{if } \psi = -1 \end{cases} \qquad (4.79)$$

Equation (4.78c) is derived using *Pontryagin's minimum principle (PMP)* and results in a bang-off-bang control profile since the control is constrained to its limiting values and zero.

4.3.1 Singular Fuel/Time Optimal Control

Suppose that there exists an interval of time where the switching function $q(t)$ is exactly $\pm\alpha$, i.e.,

$$q(t) = B^T \lambda(t) = \pm\alpha \ \forall t \in [T_1 \ T_2], \tag{4.80}$$

the control input does not change the value of the Hamiltonian in the interval $[T_1 \ T_2]$, resulting in a singular fuel/time optimal control profile. Figure 4.12 schematically illustrates a singular fuel/time optimal control profile for a weighting parameter of $\alpha = 0.5$.

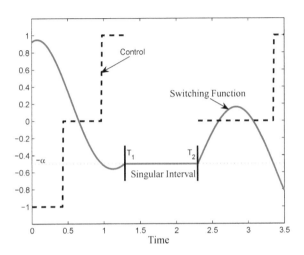

Figure 4.12: Switching Function for a Singular Fuel/Time Optimal Control

Since, the switching function is identically equal to $\pm\alpha$ in the interval $[T_1 \ T_2]$, this implies

$$\frac{dq}{dt} = B^T A^T \exp(-A^T t)\lambda(0) = 0 \tag{4.81a}$$

$$\frac{d^2q}{dt^2} = B^T (A^T)^2 \exp(-A^T t)\lambda(0) = 0 \tag{4.81b}$$

$$\frac{d^3q}{dt^3} = B^T (A^T)^3 \exp(-A^T t)\lambda(0) = 0 \tag{4.81c}$$

$$\dots \tag{4.81d}$$

$$\frac{d^{n-1}q}{dt^{n-1}} = B^T (A^T)^{n-1} \exp(-A^T t)\lambda(0) = 0 \tag{4.81e}$$

since the costate equation can be written as

$$\lambda(t) = \exp(-A^T t)\lambda(0). \tag{4.82}$$

Equation (4.81) can be concisely written as

$$\begin{bmatrix} B & AB & A^2B & A^3B \end{bmatrix}^T A^T \exp(-A^T t)\lambda(0) = 0. \tag{4.83}$$

Since $\exp(-A^T t)$ is nonsingular and $\lambda(0)$ cannot be zero, since it results in a trivial solution, a singular solution requires that the *controllability matrix* be singular which corresponds to an uncontrollable system, or the A matrix be singular. Since the class of problem being studied here include rigid body modes, the A matrix is singular.

To preclude the existence of a singular interval for a linear system with multiple damped flexible modes and a rigid body modes, consider the modal form of the system:

$$\dot{x} = Ax + Bu \tag{4.84}$$

where

$$A = \begin{bmatrix} 0 & 1 & . & . & . & . & . & . & . & . & . & . \\ 0 & 0 & . & . & . & . & . & . & . & . & . & . \\ . & . & 0 & & 1 & & . & . & . & . & . & . \\ . & . & -\omega_1^2 & -2\zeta_1\omega_1 & . & . & . & . & . & . & . \\ . & . & . & . & . & . & 0 & & 1 & & . & . \\ . & . & . & . & . & . & -\omega_2^2 & -2\zeta_2\omega_2 & . & . \\ . & . & . & . & . & . & . & . & . & & \ddots \end{bmatrix}, B = \begin{bmatrix} 0 \\ \phi_0 \\ 0 \\ \phi_1 \\ 0 \\ \phi_2 \\ \vdots \end{bmatrix} \tag{4.85}$$

and the costate equation is

$$\begin{bmatrix} \dot{\lambda}_1 \\ \dot{\lambda}_2 \\ \dot{\lambda}_3 \\ \dot{\lambda}_4 \\ \vdots \end{bmatrix} = - \begin{bmatrix} 0 & 0 & . & . & . & . & . & . & . & . \\ 1 & 0 & . & . & . & . & . & . & . & . \\ . & . & 0 & -\omega_1^2 & . & . & . & . & . \\ . & . & 1 & -2\zeta_1\omega_1 & . & . & . & . & . \\ . & . & . & . & . & 0 & -\omega_2^2 & . & . \\ . & . & . & . & . & 1 & -2\zeta_2\omega_2 & . & . \\ . & . & . & . & . & . & . & . & \ddots \end{bmatrix} \begin{bmatrix} \lambda_1 \\ \lambda_2 \\ \lambda_3 \\ \lambda_4 \\ \vdots \end{bmatrix} \tag{4.86}$$

From Equation (4.86), we have

$$\lambda_1 = C \tag{4.87a}$$

$$\lambda_2 = -Ct + D \tag{4.87b}$$

$$\lambda_4 = A_1 e^{\zeta_1\omega_1 t} \sin(\omega_1 t) + B_1 e^{\zeta_1\omega_1 t} \cos(\omega_1 t) \tag{4.87c}$$

$$\lambda_6 = A_2 e^{\zeta_2\omega_2 t} \sin(\omega_2 t) + B_2 e^{\zeta_2\omega_2 t} \cos(\omega_2 t) \tag{4.87d}$$

$$\vdots \tag{4.87e}$$

and the switching function is

$$B^T \lambda = \phi_0 \lambda_2 + \phi_1 \lambda_4 + \phi_2 \lambda_6 + \dots \tag{4.88}$$

which can be rewritten as

$$B^T \lambda = \phi_0(-Ct + D) + \phi_1(A_1 e^{\zeta_1 \omega_1 t} \sin(\omega_1 t) + B_1 e^{\zeta_1 \omega_1 t} \cos(\omega_1 t))$$
$$+ \phi_2(A_2 e^{\zeta_2 \omega_2 t} \sin(\omega_2 t) + B_2 e^{\zeta_2 \omega_2 t} \cos(\omega_2 t)) + \dots \tag{4.89}$$

For a singular control to exist, the switching function should be

$$B^T \lambda = \pm \alpha \ \forall t \in \begin{bmatrix} T_1 & T_2 \end{bmatrix} \tag{4.90}$$

which, assuming $\phi_i \neq 0$, can only be satisfied by the following set of parameters

$$C = A_1 = B_1 = A_2 = B_2 = \dots = 0, \text{ and } D = \frac{\pm \alpha}{\phi_0} \tag{4.91}$$

which results in a trivial solution. Thus a multi-mode controllable linear system cannot support a singular fuel/time optimal control. Thus, we have shown that the necessary conditions for optimality conflicts with the condition for the existence of singular intervals, thus, precluding singular intervals.

4.3.2 Rigid Body

With the knowledge that the time-optimal control is bang-off-bang, the control profile can be represented as the output of a time-delay filter subject to a step input which is parameterized to generate a bang-off-bang control profile. For example, the fuel/time optimal control profile for a double integrator, for a rest-to-rest maneuver is:

$$u(t) = 1 - \mathcal{H}(t - T_1) - \mathcal{H}(t - (2T_2 - T_1)) + \mathcal{H}(t - 2T_2) \tag{4.92a}$$

$$u(s) = \frac{1}{s}(1 - \exp(-sT_1) - \exp(-s(2T_2 - T_1)) + \exp(-2sT_2)) = \frac{1}{s} G_c(s) \tag{4.92b}$$

which is a two-switch antisymmetric bang-off-bang control profile which satisfies the constraint that the control has to lie in the range

$$-1 \leq u(t) \leq 1. \tag{4.93}$$

The rigid body boundary condition leads to

$$y_f = \frac{\phi_0}{2}(-2T_1^2 + 4T_1 T_2) \tag{4.94}$$

and the resulting cost function is

$$J = 2\alpha T_1 + \frac{1}{2T_1}\left(\frac{2y_f}{\phi_0} + 2T_1^2\right) \tag{4.95}$$

The parameter T_1, which optimizes the cost function is given by the equation

$$\frac{dJ}{dT_1} = 0 \qquad (4.96)$$

which results in

$$T_1^2 = \frac{y_f}{\phi_0 \, (2\alpha + 1)}. \qquad (4.97)$$

Since the Hamiltonian of the system is

$$H = 1 + \alpha|u| + \lambda_1 \dot{\theta} + \lambda_2 \phi_0 u, \qquad (4.98)$$

the costate equations are

$$\dot{\lambda}_1 = -\frac{\partial H}{\partial \theta} = 0 \qquad (4.99a)$$

$$\dot{\lambda}_2 = -\frac{\partial H}{\partial \dot{\theta}} = -\lambda_1. \qquad (4.99b)$$

which can be integrated to give

$$\lambda_1(t) = E \qquad (4.100a)$$
$$\lambda_2(t) = -Et + F. \qquad (4.100b)$$

Since the Hamiltonian is zero at the initial time, we have

$$H(0) = 1 + \alpha + \phi_0 \lambda_2(0) = 0 \Rightarrow \lambda_2(0) = -\frac{1+\alpha}{\phi_0} = F. \qquad (4.101)$$

Further, since the switching function $\phi_0 \lambda_2$ should be $-\alpha$ at the switch time T_1, we have

$$\phi_0 \lambda_2(T_1) = \phi_0 \left(-ET_1 - \frac{1+\alpha}{\phi_0} \right) = -\alpha \qquad (4.102)$$

which can be solved for E,

$$E = -\sqrt{\frac{2\alpha + 1}{\phi_0 y_f}} \qquad (4.103)$$

which results in the closed form expression for the control

$$u = -dez(\frac{\phi_0 \lambda_2(t)}{\alpha}) = -dez \left(\frac{\phi_0}{\alpha} \sqrt{\frac{2\alpha + 1}{\phi_0 y_f}} t - \frac{1+\alpha}{\alpha} \right). \qquad (4.104)$$

Figure 4.13 illustrates the tradeoff between fuel consumed and maneuver time for varying α and the final displacement y_f. It is clear that as α increases which corresponds to an increase on the penalty on the fuel consumed, the maneuver time increases. This graph can be used to locate the *knee of the curve* for the selection of α. Moving to left of the *knee of the curve* results in an rapid increase in the fuel consumed.

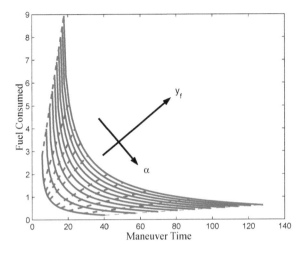

Figure 4.13: Fuel/Time Trade-Off

4.3.3 Fuel/Time Optimal Rest-to-Rest Maneuver

As in the case of time-optimal control profiles for undamped systems, the fuel/time optimal control profile for undamped flexible structures with a rigid body mode is antisymmetric about the mid-maneuver time (Figure 4.14) and this information is exploited in the parameterization of the transfer function of a time-delay filter which generates a bang-off-bang profile when subject to a unit step input [15]. The transfer function is

$$G_c(s) = 1 + \sum_{i=1}^{n} \pm \exp(-sT_i)$$
$$+ \sum_{i=1}^{n} \pm \exp(-s(2T_{n+1} - T_i)) + \exp(-2sT_{n+1}). \tag{4.105}$$

The transfer function has two zeros at $s = 0$ since

$$G_c(s = 0) = \left.\frac{dG_c}{ds}\right|_{s=0} = 0, \tag{4.106}$$

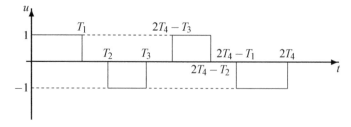

Figure 4.14: Fuel/Time Optimal Control Profile

which cancel the rigid body poles of the system. To cancel the modes corresponding
to the flexible modes at

$$s = \sigma + j\omega \tag{4.107}$$

we have,

$$
\begin{aligned}
\Re(G_c(s)) =&\, 1 + \sum_{i=1}^{n} \pm \exp(-\sigma T_i)\cos(\omega T_i) \\
&+ \sum_{i=1}^{n} \pm \exp(-\sigma(2T_{n+1} - T_i))\cos(\omega(2T_{n+1} - T_i)) \\
&+ \exp(-2\sigma T_{n+1})\cos(2\omega T_{n+1}) = 0.
\end{aligned}
\tag{4.108}
$$

$$
\begin{aligned}
\Im(G_c(s)) =&\, + \sum_{i=1}^{n} \pm \exp(-\sigma T_i)\sin(\omega T_i) \\
&+ \sum_{i=1}^{n} \pm \exp(-\sigma(2T_{n+1} - T_i))\sin(\omega(2T_{n+1} - T_i)) \\
&+ \exp(-2\sigma T_{n+1})\sin(2\omega T_{n+1}) = 0.
\end{aligned}
\tag{4.109}
$$

For undamped modes, we substitute $\sigma = 0$ into Equations (4.108), (4.109), and simplify to arrive at the constraints

$$2\cos(\omega T_{n+1})\left(\sum_{i=1}^{n} \pm \cos(\omega(T_{n+1} - T_i)) + \cos(\omega T_{n+1}) \right) = 0 \tag{4.110}$$

$$2\sin(\omega T_{n+1})\left(\sum_{i=1}^{n} \pm \cos(\omega(T_{n+1} - T_i)) + \cos(\omega T_{n+1}) \right) = 0 \tag{4.111}$$

The final constraint to satisfy the rigid body boundary conditions is derived as in
the case of time-optimal control, resulting in the equation

$$y_f = \lim_{s \to 0} G(s) G_c(s). \tag{4.112}$$

where $G(s)$ is the system transfer function.

4.3.4 Sufficiency Conditions

As in the case of the design of time-optimal control, the parameter optimization problem for the fuel/time optimal control problem can result in multiple solutions, since the constraints are nonlinear. To corroborate the optimality of the control profile, the rest of the necessary conditions for optimality have to be verified. This includes the deadzone function which defines the optimal control in terms of the optimal costates. The costates for the system

$$\dot{x} = Ax + Bu \tag{4.113}$$

are

$$\dot{\lambda} = -A^T \lambda. \tag{4.114}$$

The switching function can therefore be represented as

$$B^T \lambda = B^T \exp(-A^T t) \lambda(0), \tag{4.115}$$

which has to equal $\pm \alpha$ at each switch time. Thus $\lambda(0)$ can be solved from the equation

$$\begin{bmatrix} B^T \exp(-A^T T_1) \\ B^T \exp(-A^T T_2) \\ B^T \exp(-A^T T_3) \\ \cdots \\ B^T \exp(-A^T (2T_{n+1} - T_2)) \\ B^T \exp(-A^T (2T_{n+1} - T_1)) \end{bmatrix} \lambda(0) = \begin{bmatrix} \pm \alpha \\ \pm \alpha \\ \pm \alpha \\ \cdots \\ \pm \alpha \\ \pm \alpha \end{bmatrix} \tag{4.116}$$

We can determine $\lambda(0)$ from Equation (4.116) and thus $\lambda(t)$. If the control profile determined from

$$u = -dez \left(\frac{B^T \lambda(t)}{\alpha} \right) \tag{4.117}$$

switches at the times determined from the parameter optimization problem, then the solution is optimal.

4.3.5 Benchmark Problem

With the knowledge that the fuel/time optimal control profile is bang-off-bang, we construct a time-delay filter whose output is bang-off-bang when it is subject to a step-input. The amplitudes of the time-delayed signals are constrained by the bounds on the control, requiring the determination of the time-delays to completely specify the time-delay filter.

To define the time-delay filter completely, one can formulate a constrained parameter optimization problem, where the parameters to be determined are the time-delays. To eliminate any residual vibration of the system, the complex poles located at $\pm j\omega$ have to be canceled by a complex conjugate pair of zeros of the time-delay filter leading to the first constraint. The next constraint, to satisfy the boundary conditions, is arrived at by integrating the rigid body equation of motion. Singh and Vadali [16] have illustrated that the transfer function of a time-delay filter that generates a control profile which is antisymmetric about its mid-maneuver time, has a minimum of two zeros at the origin of the complex plane, which cancel the rigid body poles of the system. This fact is exploited to reduce the dimension of the parameter search space.

We consider the problem of reorientation, i.e., a rest-to-rest maneuver with boundary conditions

$$x_1(0) = x_2(0) = 0, \quad x_1(t_f) = x_2(t_f) = \frac{\theta_f}{2\phi_0}$$

$$\dot{x}_1(0) = \dot{x}_2(0) = 0, \quad \dot{x}_1(t_f) = \dot{x}_2(t_f) = 0 \qquad (4.118)$$

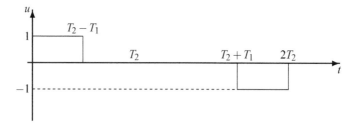

Figure 4.15: Two-Switch Fuel/Time Optimal Control Profile

The two-switch control profile for a fuel/time optimal (Figure 4.15) problem is parameterized in frequency domain as

$$u = \frac{1}{s}(1 - e^{-s(T_2-T_1)} - e^{-s(T_2+T_1)} + e^{-2sT_2}) \qquad (4.119)$$

where T_2 is the mid-maneuver time and $T_2 - T_1$ is the first switch time. The transfer function of the time-delay filter that generates the bang-off-bang control profile is

$$G(s) = (1 - e^{-s(T_2-T_1)} - e^{-s(T_2+T_1)} + e^{-2sT_2}). \qquad (4.120)$$

From the Laplace transform of the decoupled rigid-body mode (Equation 4.6a), we have

$$s^2\theta(s) = \frac{\phi_0}{s}(1 - e^{-s(T_2-T_1)} - e^{-s(T_2+T_1)} + e^{-2sT_2})$$ (4.121)

The inverse Laplace transform leads to

$$\theta(t) = \phi_0(\frac{t^2}{2} - \mathcal{H}(t-(T_2-T_1))\frac{(t-(T_2-T_1))^2}{2}$$ (4.122)

$$-\mathcal{H}(t-(T_2+T_1))\frac{(t-(T_2+T_1))^2}{2} - \mathcal{H}(t-2T_2)\frac{(t-2T_2)^2}{2})$$

where $\mathcal{H}(t-T_i)$ is the Heaviside's unit step function. The final state of θ is $\theta(2T_2)$ $= \theta_f$, which leads to

$$\theta_f = \phi_0(T_2^2 - T_1^2)$$ (4.123)

which can be rewritten as

$$\frac{\theta_f}{\phi_0} = (T_2 - T_1)(T_2 + T_1).$$ (4.124)

To cancel the vibration mode, we substitute $s = j\omega$ into Equation (4.120) and equating the real and imaginary parts to zero, we have

$$1 - \cos(\omega(T_2 - T_1)) - \cos(\omega(T_2 + T_1)) + \cos(2\omega T_2) = 0$$ (4.125)

and

$$-\sin(\omega(T_2 - T_1)) - \sin(\omega(T_2 + T_1)) + \sin(2\omega T_2) = 0$$ (4.126)

respectively. Equations (4.125) and (4.126) can be simplified to

$$2\cos(\omega T_2)(-\cos(\omega T_1) + \cos(\omega T_2)) = 0$$ (4.127)

and

$$2\sin(\omega T_2)(-\cos(\omega T_1) + \cos(\omega T_2)) = 0$$ (4.128)

respectively, leading to the constraint equation

$$-\cos(\omega T_1) + \cos(\omega T_2) = 0.$$ (4.129)

Solving Equation (4.129), we have

$$\omega T_2 = \pm\omega T_1 + 2n\pi, \quad n = \pm 1, \pm 2, ..., \pm\infty$$ (4.130)

substitute Equation (4.130) into Equation (4.124), and we have

$$T_1 = \begin{cases} \frac{\omega\theta_f}{4n\pi\phi_0} - \frac{n\pi}{\omega} & \text{for } \omega T_2 = \omega T_1 + 2n\pi \\ \frac{n\pi}{\omega} - \frac{\omega\theta_f}{4n\pi\phi_0} & \text{for } \omega T_2 = -\omega T_1 + 2n\pi \end{cases}$$ (4.131)

The next step includes determining the integer n. The cost function to be minimized is

$$J = \int_0^{2T_2} (1 + \alpha|u|)\, dt \tag{4.132}$$

which can be rewritten as

$$J = 2T_2 + 2\alpha(T_2 - T_1) \tag{4.133}$$

or

$$J = \begin{cases} \frac{\omega\theta_f}{2n\pi\phi_0} + (2 + 4\alpha)\frac{n\pi}{\omega} & \text{for } \omega T_2 = \omega T_1 + 2n\pi \\ (1 + 2\alpha)\frac{\omega\theta_f}{2n\pi\phi_0} + \frac{2n\pi}{\omega} & \text{for } \omega T_2 = -\omega T_1 + 2n\pi \end{cases} \tag{4.134}$$

Since the cost J is a unimodal function for positive n (negative n are not valid as they imply a negative cost J), we can solve for n_{opt} that minimizes J, assuming n is continuous. The integer values that flank n_{opt} are used to determine the optimum n. Assuming n is continuous, we require

$$\frac{dJ}{dn} = 0 \tag{4.135}$$

for optimality, which leads to

$$n^2 = \begin{cases} (\frac{\omega}{\pi})^2 \frac{\theta_f}{(2+4\alpha)\phi_0} & \text{for } \omega T_2 = \omega T_1 + 2n\pi \\ (\frac{\omega}{\pi})^2 \frac{(1+2\alpha)\theta_f}{2\phi_0} & \text{for } \omega T_2 = -\omega T_1 + 2n\pi \end{cases} \tag{4.136}$$

Since n is an integer, the value of J corresponding to the integer values of n that flank the value of n estimated by Equation (4.136) are evaluated to determine the optimum n. A point to note is that the optimal control profile leads to displacement of the rigid body mode at mid-maneuver time which is half of the total maneuver and the displacement of the vibratory mode which is zero at the mid-maneuver time.

Since, the structure of the control profile for the rigid body and the benchmark problem are identical, it is informative to plot the variation of the switch times and the maneuver time as a function of α. Figure 4.16 illustrates the variation of the first switch time for the rigid body (dashed line) and the benchmark problem. It can be seen that the switch time remains constant for a range of α. Figure 4.17 illustrates the variation of the maneuver time versus α. It can be seen that for some values of α, the maneuver time of the benchmark problem is smaller than the rigid body system. However, as is shown in Figure 4.18, the variation of the cost function of the rigid body system is either equal to or smaller than the cost of the benchmark problem. Figure 4.18 also illustrates the variation of the parameter n as a function of α.

It was shown in Section 4.2 that the time-optimal control profile, which corresponds to $\alpha=0$, is a three-switch bang-bang profile, which is antisymmetric about the mid-maneuver time. Since, the fuel/time optimal control profile should in the limit tend toward the time-optimal control profile ($\alpha \to 0$), it is clear that a transition control profile should exist which can generate the two-switch bang-off-bang,

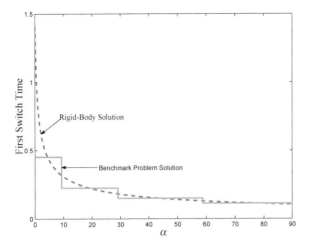

Figure 4.16: Switch Time Variation

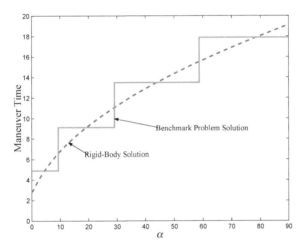

Figure 4.17: Maneuver Time Variation

and the three-switch bang-bang control profile. Figure 4.14 which illustrates a six-switch control profile, which is antisymmetric about the mid-maneuver time, satisfies the requirement of the transition control profile.

For $\alpha = 0$, $T_1 = T_2$, and $T_3 = 2T_4 - T_3$, resulting in the time-optimal control structure (Figure 4.5). For α greater than α_{cr}, the switch times $T_2 = T_3$, generating the control profile of Figure 4.15. α_{cr} is the critical value of α at which the structure of the

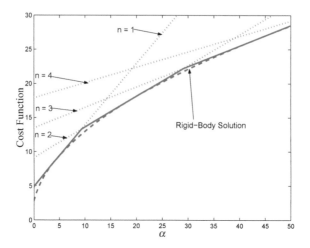

Figure 4.18: Cost Variation

control profile changes from that of Figure 4.14 to that represented in Figure 4.15.

The antisymmetric fuel/time optimal control profile for small α is parameterized as

$$u = \frac{1}{s}(1 - e^{-s(T_4-T_3)} - e^{-s(T_4-T_2)} + e^{-s(T_4-T_1)} + e^{-s(T_4+T_1)} - e^{-s(T_4+T_2)} - e^{-s(T_4+T_3)} + e^{-2sT_4})$$

(4.137)

where T_4 is the mid-maneuver time. Integrating the rigid-body mode with the control given by Equation (4.137), the constraints for the parameter optimization problem can be shown to be

$$\theta_f = \phi_0(T_4^2 - T_3^2 + T_2^2 - T_1^2)$$

(4.138)

which is derived from the rigid-body boundary condition, and

$$-\cos(\omega T_3) + \cos(\omega T_2) - \cos(\omega T_1) + \cos(\omega T_4) = 0$$

(4.139)

which forces the transfer function of the time-delay filter to cancel the frequency corresponding to the vibratory mode of the system.

A parameter optimization problem is formulated to solve for the time delays. The cost function to be minimized is

$$J = \int_0^{t_f} (1 + \alpha|u|)\, dt = 2T_4 + 2\alpha((T_4 - T_3) + (T_2 - T_1))$$

(4.140)

subject to constraints given by Equations (4.138) and (4.139). Since the constraints are nonlinear and there are more parameters than constraints, there exist a potential for the existence of multiple solutions. To prove optimality of the control profile, the control profile predicted by the switching function should coincide with that predicted by the parameter optimization. To determine the switching function, we need to solve for the initial costates. We know that

$$\lambda(t) = \exp(-A^T t)\lambda(0) \tag{4.141}$$

We also know that the switching function has to satisfy

$$\begin{bmatrix} B^T \exp(-A^T(T_4 - T_3)) \\ B^T \exp(-A^T(T_4 - T_2)) \\ B^T \exp(-A^T(T_4 + T_2)) \\ B^T \exp(-A^T(T_4 + T_3)) \end{bmatrix} \lambda(0) = \begin{Bmatrix} -\alpha \\ \alpha \\ -\alpha \\ \alpha \end{Bmatrix} \tag{4.142}$$

Solving for $\lambda(0)$ from Equation (4.142) and substituting in

$$u = -dez(B^T \exp(-A^T t)\lambda(0)/\alpha) \tag{4.143}$$

we can arrive at the control profile which should coincide with the one predicted by the parametrization problem, for optimality.

4.3.6 Determination of α_{cr}

The vector differential equation representing the equations of motion of the two-mass-spring system is

$$\dot{x} = Ax + Bu \quad |u| \le 1. \tag{4.144}$$

where $x \in \mathfrak{R}^4$, u is a scalar and the matrices A and B are

$$A = \begin{bmatrix} 0 & 1 & 0 & 0 \\ 0 & 0 & 0 & 0 \\ 0 & 0 & 0 & 1 \\ 0 & 0 & -\omega^2 & 0 \end{bmatrix} \tag{4.145}$$

and

$$B = \begin{bmatrix} 0 \\ \phi_0 \\ 0 \\ \phi_1 \end{bmatrix} \tag{4.146}$$

respectively. The optimal control is given by

$$u(t) = -dez(B^T \lambda(t)/\alpha) \tag{4.147}$$

where $\lambda(t)$ is the costate vector which is given by the equation

$$\lambda(t) = \exp(-A^T t)\lambda(0) \tag{4.148}$$

We know that the control structure of Figure 4.19 is generated by the switching function $B^T \lambda(t)$ also shown in Figure 4.19.

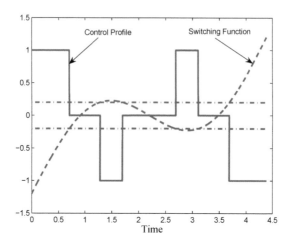

Figure 4.19: Six-Switch Control Profile

α_{cr} corresponds to the α when the switching function $B^T \lambda(t)$ equals values of α and $-\alpha$ at times t_{cr} and $T_f - t_{cr}$ respectively (Figure 4.20), i.e.,

$$B^T \lambda(t_{cr}) = B^T \exp(-A^T t_{cr}) \lambda(0) = \alpha_{cr} \qquad (4.149)$$

$$B^T \lambda(T_f - t_{cr}) = B^T \exp(-A^T (T_f - t_{cr})) \lambda(0) = -\alpha_{cr} \qquad (4.150)$$

and to ensure that the switching function touches the lines parallel to the abscissa with ordinate intercepts of $\pm\alpha_{cr}$, tangentially, we require the slope of $B^T \lambda(t)$ at t_{cr} and $T_f - t_{cr}$ be equal to zero (Figure 4.20), i.e.,

$$\frac{d}{dt}(B^T \lambda(t))\bigg|_{t=t_{cr}} = -B^T A^T \exp(-A^T t_{cr}) \lambda(0) = 0 \qquad (4.151)$$

and

$$\frac{d}{dt}(B^T \lambda(t))\bigg|_{t=T_f - t_{cr}} = -B^T A^T \exp(-A^T (T_f - t_{cr})) \lambda(0) = 0 \qquad (4.152)$$

Solving Equations (4.149) through (4.152), which are rewritten as

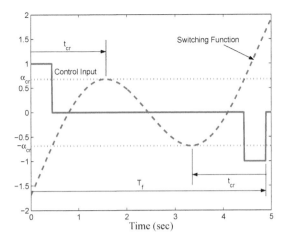

Figure 4.20: Large α Control Profile

$$
\begin{bmatrix}
B^T \exp(-A^T t_{cr}) \\
B^T \exp(-A^T (T_f - t_{cr})) \\
-B^T A^T \exp(-A^T t_{cr}) \\
-B^T A^T \exp(-A^T (T_f - t_{cr}))
\end{bmatrix}
\lambda(0) =
\left\{
\begin{array}{c}
\alpha_{cr} \\
-\alpha_{cr} \\
0 \\
0
\end{array}
\right\}
\tag{4.153}
$$

we arrive at t_{cr} and α_{cr}. To determine $\lambda(0)$ we consider the following equations:

$$
B^T \exp(-A^T (T_2 - T_1))\lambda(0) = -\alpha
\tag{4.154}
$$

and

$$
B^T \exp(-A^T (T_2 + T_1))\lambda(0) = \alpha
\tag{4.155}
$$

which are satisfied at the switch time $(T_2 - T_1)$ and $(T_2 + T_1)$ respectively. Equations (4.154) and (4.155) in conjunction with the switching function evaluated at time T_2 which can be shown to be [15],

$$
B^T \exp(-A^T T_2)\lambda(0) = 0
\tag{4.156}
$$

and the Hamiltonian evaluated at time = 0:

$$
B^T \lambda(0) = -1 - \alpha
\tag{4.157}
$$

can be used to solve for $\underline{\lambda}(0)$.

Figures 4.16 and 4.17 indicate that the switch times T_1 and T_2 remain constant for small α, i.e., for the range of α greater than α_{cr} and less than $\alpha = 9.87$ (for a maneuver of $\theta_f = 1$). Equations (4.154) through (4.157) can now be represented as

$$P\lambda(0) = \begin{Bmatrix} -\alpha \\ \alpha \\ 0 \\ -1-\alpha \end{Bmatrix} \tag{4.158}$$

where

$$P = \begin{bmatrix} B^T \exp(-A^T(T_2 - T_1)) \\ B^T \exp(-A^T(T_2 + T_1)) \\ B^T \exp(-A^T T_2) \\ B^T \end{bmatrix} \tag{4.159}$$

Since T_1 and T_2 are constant for α greater than α_{cr}, and less than 9.87, P is a constant matrix. Thus, α_{cr} and t_{cr} are solved from the equation

$$\begin{bmatrix} B^T \exp(-A^T t_{cr})P^{-1} \\ B^T \exp(-A^T(T_f - t_{cr}))P^{-1} \\ -B^T A^T \exp(-A^T t_{cr})P^{-1} \\ -B^T A^T \exp(-A^T(T_f - t_{cr}))P^{-1} \end{bmatrix} \begin{Bmatrix} -\alpha_{cr} \\ \alpha_{cr} \\ 0 \\ -1 - \alpha_{cr} \end{Bmatrix} = \begin{Bmatrix} \alpha_{cr} \\ -\alpha_{cr} \\ 0 \\ 0 \end{Bmatrix} \tag{4.160}$$

For $\theta_f = 1$, Figure 4.21 illustrates the variation of the structure of the control profile. As expected, for $\alpha = 0$, the control profile is bang-bang. Increasing α, the penalty on fuel consumed results in the three switches of the bang-bang control splitting into six switches of a bang-off-bang profile. Further increase in α leads to a transition occurring at $\alpha_{cr} = 0.6824$, where the two pulses near the mid-maneuver time collapse, resulting in a two-switch bang-off-bang control profile.

4.3.7   Effect of Damping

As in the case of the time-optimal control, the structure of the fuel/time optimal control profile is not antisymmetric about the mid-maneuver time. This implies that as the weighting parameter α of the fuel/time cost function is varied, there will be a transition when two switches of the bang-off-bang control profile collapse. However, unlike the undamped case where four switches collapsed simultaneously, in the damped case two switches will collapse at a time. To confirm this conjecture, the damped benchmark problem is studied with a damping ratio of $\zeta = 0.1$.

Two switches of the bang-off-bang control profile collapse when

$$f(x) = \begin{Bmatrix} B^T \exp(-A^T t_{cr}) + \alpha_{cr} \\ -B^T A^T \exp(-A^T t_{cr}) \end{Bmatrix} = 0 \tag{4.161}$$

To solve for the vector of parameters $x = [t_{cr} \quad \alpha_{cr}]^T$, the Newton-Rhapson method is given by:

$$x_{i+1} = x_i - (\nabla_{x_i} f)^{-1} f(x_i) \tag{4.162}$$

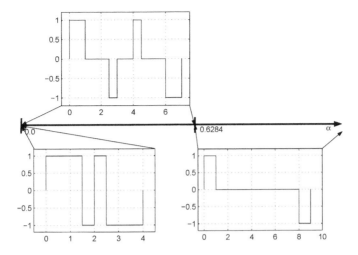

Figure 4.21: Spectrum of Fuel/Time Optimal Control

is used, which converges rapidly to the transition point. Figure 4.22 illustrates the transition of the structure of the control profile with the transition occurring at α_{cr} = 0.5268, and the corresponding t_{cr} is 3.4169 seconds. It can be seen that only the second positive pulse collapses.

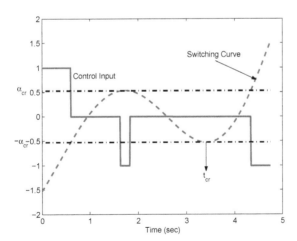

Figure 4.22: Transition of the Control Profile

The variation of the switch times of the fuel/time optimal control profile as a function of the weighting parameter α is shown in Figure 4.23. It can been seen that at the dashed line there is a transition from a six-switch control profile to a four-switch control profile.

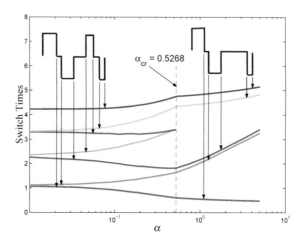

Figure 4.23: Switch Time Variation

Consider the problem of planning a control strategy for a race car to complete a race circuit in minimum time, with a finite amount of fuel. The *fuel/time optimal control* problem where a soft constraint is imposed on the fuel consumed would not be the appropriate problem formulation. One approach is to search for the value of α which solves the fuel/time optimal control problem with exactly the permitted fuel usage. But this requires solving the fuel/time problem many times. Another approach is to minimize time while setting a hard constraint on the fuel. To impose hard constraints on the fuel consumed, an integral constraint has to be imposed. The next section will develop the *fuel limited, minimum-time control* problem and illustrates that the optimal solution corresponds to a specific penalty imposed on fuel consumed in the *fuel/time optimal control* formulation.

4.4   Fuel Limited Minimum/Time Control

The minimum time control problem with constraints on the fuel, can be stated as:

$$\min \ J = \int_0^{t_f} dt \tag{4.163a}$$

subject to

$$\dot{x} = Ax + Bu \tag{4.163b}$$

$$\int_0^{t_f} |u| \, dt \le U \tag{4.163c}$$

$$x(0) = x_0 \text{ and } x(t_f) = x_f \tag{4.163d}$$

$$-1 \le u \le 1 \forall t \tag{4.163e}$$

where U represents the maximum available fuel for the maneuver [17]. For maneuvers where U is less than or equal to the fuel used for the unconstrained time-optimal control problem, the fuel consumed for the maneuver will be the maximum permitted. The fuel constraint can be rewritten as a differential equation

$$\dot{\psi} = |u| \tag{4.164}$$

with the boundary conditions

$$\psi(0) = 0, \psi(t_f) = U. \tag{4.165}$$

Defining the Hamiltonian as:

$$H = 1 + \lambda^T (Ax + Bu) + \mu|u|, \tag{4.166}$$

the necessary conditions for optimality can be derived using calculus of variations, resulting in the equations

$$\dot{x} = \frac{\partial H}{\partial \lambda} = Ax + Bu \tag{4.167a}$$

$$\dot{\psi} = \frac{\partial H}{\partial \mu} = |u| \tag{4.167b}$$

$$\dot{\lambda} = -\frac{\partial H}{\partial x} = -A^T \lambda \tag{4.167c}$$

$$\dot{\mu} = -\frac{\partial H}{\partial \psi} = 0 \tag{4.167d}$$

$$u = -dez\left(\frac{B^T \lambda}{\mu}\right) \tag{4.167e}$$

$$x(0) = x_0 \text{ and } x(t_f) = x_f \tag{4.167f}$$

$$\psi(0) = 0 \text{ and } \psi(t_f) = U \tag{4.167g}$$

$$H(0) = 0. \tag{4.167h}$$

Equation (4.167e) is derived using *Pontryagin's minimum principle (PMP)*.

4.4.1 Singular Fuel Constrained Time-Optimal Control

The same proof which illustrated that flexible structures with rigid body modes cannot support singular fuel/time optimal control is valid for the fuel constrained time-optimal control.

4.4.2 Rigid Body

With the knowledge that the fuel constrained time-optimal control is bang-off-bang, the control profile can be represented as the output of a time-delay filter subject to a step input which is parameterized to generate a bang-off-bang control profile. For example, the fuel constrained time-optimal control profile for a double integrator,

$$\ddot{\theta} = \phi_0 u, \tag{4.168}$$

which in state-space form is

$$\left\{ \begin{matrix} \dot{x}_1 \\ \dot{x}_2 \end{matrix} \right\} = \begin{bmatrix} 0 & 1 \\ 0 & 0 \end{bmatrix} \left\{ \begin{matrix} x_1 \\ x_2 \end{matrix} \right\} + \begin{bmatrix} 0 \\ \phi_0 \end{bmatrix} u. \tag{4.169}$$

has the Hamiltonian:

$$H = 1 + \lambda_1 x_2 + \lambda_2 \phi_0 u + \mu |u|. \tag{4.170}$$

The resulting costate equations are

$$\dot{\lambda}_1 = 0 \tag{4.171}$$

$$\dot{\lambda}_2 = -\lambda_1 \tag{4.172}$$

$$\dot{\mu} = 0 \tag{4.173}$$

which can be solved in closed form resulting in the equations

$$\lambda_1(t) = E \tag{4.174}$$

$$\lambda_2(t) = -Et + F \tag{4.175}$$

$$\mu(t) = G. \tag{4.176}$$

Since the boundary conditions for the rest-to-rest maneuver of the system are

$$\theta(0) = 0, \dot{\theta}(0) = 0, \theta(t_f) = \theta_f, \dot{\theta}(t_f) = 0. \tag{4.177}$$

and since the Hamiltonian is equal to zero for all time, we have

$$H(0) = 1 + \lambda_2(0)\phi_0 + \mu = 0 \tag{4.178}$$

which results in the solution

$$\lambda_2(0) = \frac{-1 - \mu}{\phi_0} = F. \tag{4.179}$$

Since the switching curve $\phi_0 \lambda_2$ is a straight line, the fuel constrained time-optimal control profile is a two-switch bang-off-bang profile and can be parameterized as shown in Figure 4.24. The initial control pulse accelerates the mass to reach a cruising velocity which reduce time to the desired terminal location while consuming half of the available fuel, while the antisymmetric pulse is needed to decelerate the system to settle into the desired terminal state.

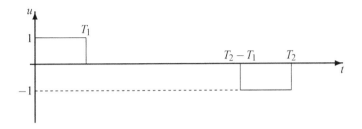

Figure 4.24: Fuel Constrained Time-Optimal Control Profile

$$u(t) = 1 - \mathscr{H}(t - T_1) - \mathscr{H}(t - (T_2 - T_1)) + \mathscr{H}(t - T_2) \tag{4.180a}$$

$$u(s) = \frac{1}{s}(1 - \exp(-sT_1) - \exp(-s(T_2 - T_1)) + \exp(-sT_2)) = \frac{1}{s}G_c(s) \tag{4.180b}$$

which is antisymmetric about the mid-maneuver time and satisfies the terminal velocity constraint. Integrating the second order model of the rigid body system for the parameterized control input, we have

$$\theta(t) = \frac{\phi_0}{2}(t^2 - (t - T_1)^2 \mathscr{H}(t - T_1) - (t - (T_2 - T_1))^2 \mathscr{H}(t - (T_2 - T_1)) + (t - T_2)^2 \mathscr{H}(t - T_2)) \tag{4.181}$$

which when evaluated at the final time T_2 should satisfy the displacement boundary condition

$$\theta(T_2) = \theta_f = (T_2^2 - (T_2 - T_1)^2 - T_1^2). \tag{4.182}$$

Since the time-optimal control will consume the maximum permitted fuel, we have the constraint

$$2T_1 = U \tag{4.183}$$

Solving Equations (4.182) and (4.183), we have

$$T_1 = \frac{U}{2}, \quad T_2 = \frac{2\theta_f}{\phi_0 U} + \frac{U}{2}. \tag{4.184}$$

Since, the switching function $\phi_0\lambda_2$ at the first and second switch time should be $-\mu$ and μ respectively, we have

$$\lambda_2(T_1) = -E\frac{U}{2} + \frac{-1-\mu}{\phi_0} = -\frac{\mu}{\phi_0} \tag{4.185}$$

$$\lambda_2(T_2 - T_1) = -E\frac{2\theta_f}{\phi_0 U} + \frac{-1-\mu}{\phi_0} = \frac{\mu}{\phi_0} \tag{4.186}$$

which can be solved for E and μ resulting in the solutions

$$\mu = \frac{2\theta_f}{\phi_0 U^2} - \frac{1}{2} = G \tag{4.187}$$

$$E = -\frac{2}{\phi_0 U}. \tag{4.188}$$

The closed form expression for the control is

$$u = -dez\left(\frac{2\phi_0 U^2}{4\theta_f - \phi_0 U^2}(\frac{2}{U}t - 1 - \mu)\right) \tag{4.189}$$

The variation of μ as a function of maximum permitted fuel is plotted using Equation (4.187), is shown in Figure 4.25. For a maneuver distance of $\theta_f = 10$, we can see from Equation (4.187), that μ equals zero when

$$U = \sqrt{\frac{4\theta_f}{\phi_0}} = \sqrt{80} = 8.9443 \tag{4.190}$$

which corresponds to the unconstrained time-optimal solution.

4.4.3   Fuel Constrained Time-Optimal Rest-to-Rest Maneuver

Since the structure of the fuel constrained time-optimal control is similar to the fuel/time optimal control when the fuel limits are smaller than the fuel consumed by the unconstrained time-optimal control, the same constraints are valid for this problem. Parameterizing the control profile as

$$u = \frac{1}{s}G_c(s) = \frac{1}{s}(1 + \sum_{i=1}^{n} \pm\exp(-sT_i)$$
$$+ \sum_{i=1}^{n} \pm\exp(-s(2T_{n+1} - T_i)) + \exp(-2sT_{n+1})). \tag{4.191}$$

The resulting parameter optimization problem for an undamped system is

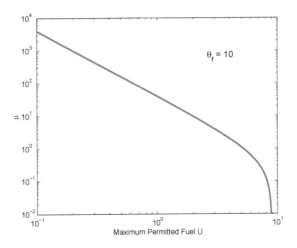

Figure 4.25: Variation of μ vs. U

$$\min \ J = T_{n+1} \tag{4.192a}$$

subject to $\qquad\qquad\qquad\qquad\qquad\qquad\qquad\qquad\qquad\qquad$ (4.192b)

$$\sum_{i=1}^{N} \pm\cos(\omega_j(T_{n+1} - T_i)) + \cos(\omega_j T_{n+1}) = 0, \quad \forall \omega_j, j = 1,2,..,m$$

$$\tag{4.192c}$$

$$2\left(T_1 + \sum_{i=2,4,6,...} (T_{i+1} - T_i)\right) = U \tag{4.192d}$$

$$y_f = \lim_{s \to 0} G(s)G_c(s) \tag{4.192e}$$

$$T_i - T_{i+1} < 0 \quad \forall i \tag{4.192f}$$

where $G(s)$ is the input-output transfer function of the system.

4.4.4 Sufficiency Conditions

Unlike the fuel/time optimal control, where the weighting parameter α is known, the corresponding costate μ for the fuel constrained time-optimal control problem, is unknown. To verify optimality of the solutions generated by the parameter optimization problem, the rest of the necessary conditions which deal with the costates need to be satisfied.

For optimality, the switching function

$$B^T\lambda = B^T\exp(-A^T t)\lambda(0) \tag{4.193}$$

should equal $\pm\mu$ at every switch time. We can solve for $\lambda(0)$ and μ from the equation:

$$
P\left\{\begin{array}{c} \lambda(0) \\ \mu \end{array}\right\} = \begin{bmatrix} B^T \exp(-A^T T_1) & \pm 1 \\ B^T \exp(-A^T T_2) & \pm 1 \\ B^T \exp(-A^T T_3) & \pm 1 \\ \cdots \\ B^T \exp(-A^T (2T_{n+1} - T_2)) & \pm 1 \\ B^T \exp(-A^T (2T_{n+1} - T_1)) & \pm 1 \end{bmatrix} \left\{\begin{array}{c} \lambda(0) \\ \mu \end{array}\right\} = \begin{bmatrix} 0 \\ 0 \\ 0 \\ \cdots \\ 0 \\ 0 \end{bmatrix} \quad (4.194)
$$

which requires calculating the null space of the matrix P. If the control profile determined from

$$
u = -dez\left(\frac{B^T \lambda(t)}{\mu}\right) \quad (4.195)
$$

switches at the times determined from the parameter optimization problem, then the solution is optimal.

4.4.5 Benchmark Problem

In this section, the fuel constrained time-optimal control of the benchmark problem is presented. With the knowledge that the fuel constrained time-optimal control profile has a bang-off-bang structure, the control profile for the undamped system can be parameterized to reflect the antisymmetry about the mid-maneuver time. Since, the control profile is three-switch bang-bang when an unlimited fuel supply is available, one can conjecture that the three-switch bang-bang control profile morphs to a six-switch bang-off-bang profile (Figure 4.14). Having parameterized the bang-off-bang control profile, the following parameter optimization problem can be solved to determine a candidate optimal solution:

$$
\min \ J = T_4 \quad (4.196a)
$$

subject to

$$
\cos(\omega(T_4 - T_3)) - \cos(\omega(T_4 - T_2)) - \cos(\omega(T_4 - T_1)) + \cos(\omega T_4) = 0 \quad (4.196b)
$$

$$
\frac{-2T_1^2 - 2T_2^2 + 2T_3^2 - 4T_3 T_4 + 4T_2 T_4 + 4T_1 T_4}{4} = x_f \quad (4.196c)
$$

$$
2T_1 + 2(T_3 - T_2) = U \quad (4.196d)
$$

$$
T_4 > T_3 > T_2 > T_1 > 0 \quad (4.196e)
$$

where $\omega = \sqrt{2}$ is the natural frequency of the system and $x_f = 1$, is the final displacement of the masses. With the knowledge that the time-optimal control profile for the benchmark problem for a unit displacement is

$$u(t) = 1 - 2\mathscr{H}(t - 1.0026) + 2\mathscr{H}(t - 2.1089) - 2\mathscr{H}(t - 3.2152) + \mathscr{H}(t - 4.2178) \quad (4.197)$$

the fuel consumed is $U = 4.2178$. Therefore, the effect of variation of the permitted fuel is studied for $U \leq 4.2178$.

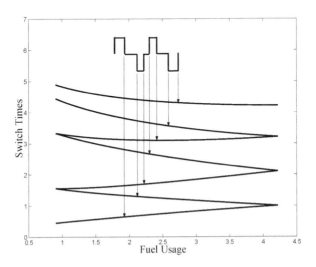

Figure 4.26: Variation of Switch Times vs. U

Figure 4.26 illustrates the variation of the switch times as a function of the fuel usage. It can be seen that the second and third switches coalesce as do the fourth and the fifth switches for a permitted fuel of $U = 0.9003$ at which point the control profile is a two-switch bang-off-bang profile. The two-switch control profile is parameterized as shown in Figure 4.27.

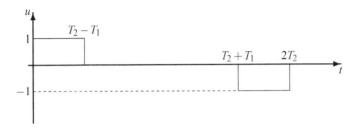

Figure 4.27: Two-Switch Fuel Constrained Time-Optimal Control Profile

$$u = \frac{1}{s}(1 - e^{-s(T_2-T_1)} - e^{-s(T_2+T_1)} + e^{-2sT_2}). \tag{4.198}$$

To cancel the mode with frequency ω, we require

$$- \cos(\omega T_1) + \cos(\omega T_2) = 0 \tag{4.199}$$

which results in the equation

$$\omega T_2 = \pm \omega T_1 + 2n\pi. \tag{4.200}$$

The next constraint is the satisfaction of the fuel constraint which results in the equation

$$2(T_2 - T_1) = U. \tag{4.201}$$

Solving for T_1 from Equations (4.200) and (4.201), we have

$$\omega \left(\frac{U}{2} + T_1 \right) = \pm \omega T_1 + 2n\pi \tag{4.202}$$

which results in the equation

$$T_1 = \frac{n\pi}{\omega} - \frac{U}{4}. \tag{4.203}$$

To satisfy the rigid-body boundary conditions, we require,

$$\theta_f = \phi_0 \left(T_2^2 - T_1^2 \right) \tag{4.204}$$

which leads to

$$x_f = \frac{n\pi U}{2\omega}. \tag{4.205}$$

for the benchmark problem.

For fuel usage smaller than $U = 0.9003$, one can conjecture that in the first half of the maneuver, no negative pulse will exist since these tend to decelerate the motion of the system. However, a two-switch control profile cannot be optimal since it cannot satisfy all the necessary constraints except for displacements which match that given by Equation (4.205).

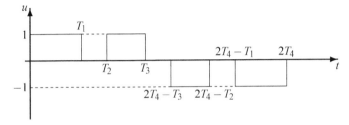

Figure 4.28: Type 2 Fuel Constrained Time-Optimal Control Profile

Figure 4.28 represents the six-switch parameterization of the bang-off-bang profile, which results in the parameter optimization problem:

$$\min\ J = T_4 \tag{4.206a}$$

subject to

$$-\cos(\omega(T_4 - T_3)) + \cos(\omega(T_4 - T_2)) - \cos(\omega(T_4 - T_1)) + \cos(\omega T_4) = 0 \tag{4.206b}$$

$$\frac{-2T_1^2 + 2T_2^2 - 2T_3^2 + 4T_3 T_4 - 4T_2 T_4 + 4T_1 T_4}{4} = x_f \tag{4.206c}$$

$$2T_1 + 2(T_3 - T_2) = U \tag{4.206d}$$

$$T_4 > T_3 > T_2 > T_1 > 0 \tag{4.206e}$$

Equation (4.205) states that a two-switch bang-off-bang control profile is optimal for fuel usage of

$$U = \frac{2\omega x_f}{n\pi}. \tag{4.207}$$

The variation of the switches for U satisfying the constraint $\frac{2\omega x_f}{2\pi} \le U \le \frac{2\omega x_f}{\pi}$ is illustrated in Figure 4.29.

It can be seen from Figure 4.29 that the second and third switches coalesce and the fourth and fifth switches coalesce, resulting in a two-switch control profile, as U is lowered.

Figure 4.30 illustrates typical control profile and the corresponding switching functions for the fuel constrained time-optimal control.

4.4.6   Effect of Damping

As was illustrated in the time-optimal and fuel/time optimal cases, the addition of damping into the system results in the loss of symmetry about the mid-maneuver time. To illustrate the variations of the structure of the fuel constrained time-optimal control profile, the benchmark problem with a damping coefficient $c = 0.1$ is studied. For a rest-to-rest maneuver with a final displacement of unity, the time-optimal

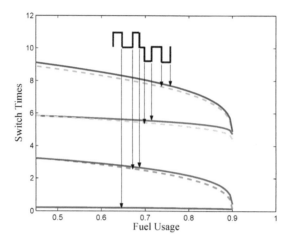

Figure 4.29: Variation of Switch Times vs. U

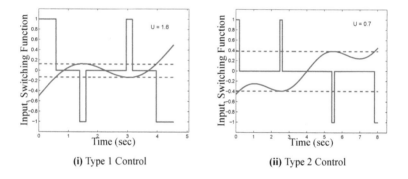

(i) Type 1 Control   (ii) Type 2 Control

Figure 4.30: Typical Fuel Limited Time-Optimal Control Profiles

controller's fuel consumption is:

$$\int_0^{t_f} |u| \, dt = 4.2211$$

where the control is constrained to lie in the range $-1 \leq u \leq 1$. For a fuel constraint smaller than 4.2211, it is clear that a six-switch bang-off-bang controller

$$u(s) = \frac{1}{s} \left(1 - e^{-sT_1} - e^{-sT_2} + e^{-sT_3} + e^{-sT_4} - e^{-sT_5} - e^{-sT_6} + e^{-sT_7} \right) \qquad (4.208)$$

is optimal. As the available fuel is reduced the switch time change as illustrated in Figure 4.31. For a permitted fuel of $U \leq 1.1458$, the fourth- and fifth-switch coalesce and the resulting optimal control profile is parameterized as:

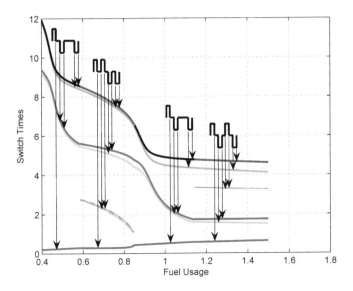

Figure 4.31: Variation of Switch Times vs. U

$$u(s) = \frac{1}{s}\left(1 - e^{-sT_1} - e^{-sT_2} + e^{-sT_3} - e^{-sT_6} + e^{-sT_7}\right) \qquad (4.209)$$

As the available fuel is reduced to $U = 0.85$, two new switches are introduced resulting in a six-switch bang-off-bang profile parameterized as:

$$u(s) = \frac{1}{s}\left(1 - e^{-sT_1} + e^{-sT_2} - e^{-sT_3} - e^{-sT_4} + e^{-sT_5} - e^{-sT_6} + e^{-sT_7}\right) \qquad (4.210)$$

When the permitted fuel is below $U = 0.587$, the structure of the fuel constrained time-optimal control profile for the damped benchmark profile reduces to a four-switch bang-off-bang profile parameterized by Equation (4.209). Figure 4.31 illustrates the spectrum of the structure of the optimal control profile over a permitted fuel ranging from 0.4 to 1.5.

One can see the change in the structure of the optimal control profile as a function of varying the damping constant. Figure 4.32 illustrates the variation of the switch times of a six- to a four-switch-control profile at $c = 0.156$, with increasing damping constant.

Biomimetics or Bio inspired design has captured the imagination of numerous researchers in the robotics and control communities. Emulate nature and you cannot go wrong is an axiom numerous researchers have taken to heart. Recently, Ben-Itzak and Karniel [18] illustrated that humans reaching motion is one which is jerk limited. This has prompted the focusing of research interest in the development of control strategies which satisfy constraints on jerk. Jerk which also goes by the moniker *jolt*,

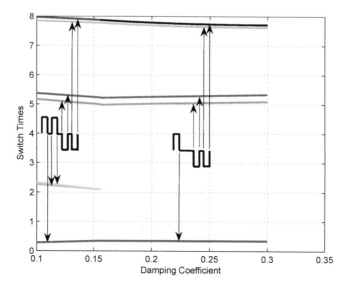

Figure 4.32: Variation of Switch Times vs. Damping Constant

surge, or *lurch* influences passenger comfort in people mover systems (the *Handbook of Railway Vehicle Dynamics* lists it at $1.5m/sec^3$ [19]), and is a factor in motion of mechanisms when considering its longevity. This motivates development of jerk limited controller which will be presented in the next section.

4.5   Jerk Limited Time-Optimal Control

Jerk is the time rate of change of the inertia forces, i.e., the derivative of the control input and is a measure of the impact level. Impact in mechanical systems is manifested in the form of noise and reduces life due to fatigue. For various flexible structures, higher level of jerk correlates to exciting the higher modes of the system which commonly are commonly referred to as *ringing*. For systems where the actuator is not an on-off device, but can be throttled, jerk limited time-optimal control profiles are desirable since they can reduce noise and the consequences of fatigue. Furthermore, the actuators are better able to track the optimal control profile since bandwidth limitations make it difficult to track bang-bang profiles.

This section addresses the problem of design of jerk limited time-optimal control profiles for rest-to-rest motion of systems. To include limits on the rate of change of the control, a new state is introduced into the model which represents the control input to the system. The rate of change of the control is now the input to be determined to satisfy the boundary conditions of the rest-to-rest maneuver. The state space system

$$\dot{x}' = A'x' + B'u \tag{4.211}$$

is augmented with the equation

$$\dot{u} = v \tag{4.212}$$

resulting in the new state space model

$$\dot{x} = Ax + Bv \tag{4.213a}$$

$$A = \begin{bmatrix} A' & B' \\ 0 & 0 \end{bmatrix}, \ B = \begin{bmatrix} 0 \\ 1 \end{bmatrix}, \ x = \begin{bmatrix} x' \\ u \end{bmatrix}. \tag{4.213b}$$

The optimal control problem is now stated as:

$$\min \ J = \int_0^{t_f} dt \tag{4.214a}$$

subject to

$$\dot{x} = Ax + Bv \tag{4.214b}$$

$$x'(0) = x'_0, \ u(0) = 0, x'(t_f) = x'_f, \text{ and } u(t_f) = 0 \tag{4.214c}$$

$$-1 \le u \le 1 \forall t \tag{4.214d}$$

$$-J \le v = \dot{u} \le J \ \forall t \tag{4.214e}$$

where J is the maximum permitted jerk [20, 21]. Defining the Hamiltonian as

$$H = 1 + \lambda^T (Ax + Bv) + \mu_1(u - 1) + \mu_2(-u - 1), \tag{4.215a}$$

$$H = 1 + \lambda'^T (A'x' + B'u) + \psi v + \mu_1(u - 1) + \mu_2(-u - 1) \tag{4.215b}$$

$$\text{where } \begin{cases} \mu_1 = 0 & \text{if } u - 1 < 0 \\ \mu_1 > 0 & \text{if } u - 1 = 0 \end{cases} \tag{4.215c}$$

$$\text{and } \begin{cases} \mu_2 = 0 & \text{if } -u - 1 < 0 \\ \mu_2 > 0 & \text{if } -u - 1 = 0 \end{cases} \tag{4.215d}$$

where $\lambda^T = [\lambda' \ \psi]^T$, the necessary conditions for optimality can be derived using calculus of variations, resulting in the equations

$$\dot{x} = \frac{\partial H}{\partial \lambda} = Ax + Bv \tag{4.216a}$$

$$\dot{\lambda}' = -\frac{\partial H}{\partial x'} = -A'^T \lambda' \tag{4.216b}$$

$$\psi = -\frac{\partial H}{\partial u} = \begin{cases} -B'^T \lambda' & \text{if } |u| - 1 < 0 \\ -B'^T \lambda' - \mu_1 & \text{if } u - 1 = 0 \\ -B'^T \lambda' + \mu_2 & \text{if } -u - 1 = 0 \end{cases} \tag{4.216c}$$

$$x'(0) = x'_0 \text{ and } x'(t_f) = x'_f \tag{4.216d}$$

$$u(0) = 0 \text{ and } u(t_f) = 0 \tag{4.216e}$$

$$H(0) = 0. \tag{4.216f}$$

The subarc of $v(t)$ when $u(t)$ is not ± 1, is called an interior arc and the control $v(t)$ is said to lie on a boundary arc, otherwise. The optimal jerk profile (v) when the inequality constraint $-1 < u < 1$ is satisfied, which corresponds to an interior arc is

$$v = -J sgn(\lambda^T B) = -J sgn(\psi). \tag{4.217}$$

For the situation when $u = \pm 1$, i.e., when $v(t)$ lies on a boundary arc, the control is determined from the equations

$$\frac{d}{dt}(u - 1) = \dot{u} = v = 0, \qquad\qquad \text{when } u - 1 = 0 \tag{4.218a}$$

$$\frac{d}{dt}(-u - 1) = -\dot{u} = -v = 0, \qquad \text{when } -u - 1 = 0 \tag{4.218b}$$

$$\tag{4.218c}$$

4.5.1   Rigid Body

The augmented state space model of the jerk limited time-optimal control for the rigid body

$$\ddot{\theta} = \phi_0 u \tag{4.219}$$

or

$$\dddot{\theta} = \phi_0 v, \text{ since } \dot{u} = v, \tag{4.220}$$

is

$$\left\{ \begin{matrix} \dot{x}_1 \\ \dot{x}_2 \\ \dot{u} \end{matrix} \right\} = \begin{bmatrix} 0 & 1 & 0 \\ 0 & 0 & \phi_0 \\ 0 & 0 & 0 \end{bmatrix} \left\{ \begin{matrix} x_1 \\ x_2 \\ u \end{matrix} \right\} + \begin{bmatrix} 0 \\ 0 \\ 1 \end{bmatrix} v, \tag{4.221}$$

and the corresponding Hamiltonian is:

$$H = 1 + \lambda_1 x_2 + \lambda_2 \phi_0 u + \psi v + \mu_1(u - 1) + \mu_2(-u - 1). \tag{4.222}$$

The resulting costate equations are

$$\dot{\lambda}_1 = 0 \tag{4.223a}$$

$$\dot{\lambda}_2 = -\lambda_1 \tag{4.223b}$$

$$\dot{\psi} = \begin{cases} -\phi_0\lambda_2 & \text{if } |u| - 1 < 0 \\ -\phi_0\lambda_2 - \mu_1 & \text{if } u - 1 = 0 \\ -\phi_0\lambda_2 + \mu_2 & \text{if } -u - 1 = 0 \end{cases} \tag{4.223c}$$

4.5.1.1 Case 1

For the situation where u is not at its constraint boundary, the costate equations can be solved in closed form resulting in the equations

$$\lambda_1(t) = E \tag{4.224}$$

$$\lambda_2(t) = -Et + F \tag{4.225}$$

$$\psi(t) = \frac{E}{2}t^2 - Ft + G. \tag{4.226}$$

since μ_1 and μ_2 are zero. Since the boundary conditions for the rest-to-rest maneuver of the system are

$$\theta(0) = 0, \dot{\theta}(0) = 0, \theta(t_f) = \theta_f, \dot{\theta}(t_f) = 0, u(0) = 0, u(t_f) = 0. \tag{4.227}$$

and since the Hamiltonian is equal to zero for all time, we have

$$H(0) = 1 + \psi(0)J = 0 \tag{4.228}$$

which results in the solution

$$\psi(0) = G = -\frac{1}{J} \tag{4.229}$$

Since the switching curve ψ is a quadratic curve, the jerk limited time-optimal control profile is at a maximum a three switch bang-bang profile when u is not at its constraint boundary and can be parameterized as (Figure 4.33(i)):

$$v(t) = J(1 - 2\mathcal{H}(t - T_1) + 2\mathcal{H}(t - (2T_2 - T_1)) - \mathcal{H}(t - 2T_2)) \tag{4.230a}$$

$$v(s) = \frac{J}{s}(1 - 2\exp(-sT_1) + 2\exp(-s(2T_2 - T_1)) - \exp(-2sT_2)) = \frac{J}{s}G_c(s) \tag{4.230b}$$

which is symmetric about the mid-maneuver time as shown in Figure 4.33(i). Since $G_c(s)$ should cancel the three poles at the origin of the augmented system we require

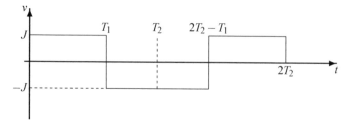

(i) Parameterization of the Jerk Profile

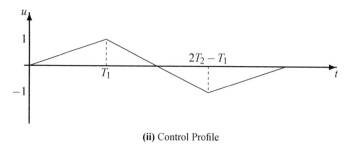

(ii) Control Profile

Figure 4.33: Case 1: Jerk Limited Time-Optimal Control

$$G_c(0) = (1 - 2\exp(-0T_1) + 2\exp(-0(2T_2 - T_1)) - \exp(-0(2T_2))) = 0 \quad (4.231\text{a})$$

$$\left.\frac{dG_c(s)}{ds}\right|_{s=0} = -2T_1 + 2(2T_2 - T_1) - 2T_2 = 2(T_2 - 2T_1) = 0 \quad (4.231\text{b})$$

$$\left.\frac{d^2G_c(s)}{ds^2}\right|_{s=0} = -2T_1^2 + 2(2T_2 - T_1)^2 - 2T_2^2 = 4T_2(T_2 - 2T_1) = 0 \quad (4.231\text{c})$$

Thus when $T_2 = 2T_1$, all the poles at the origin are canceled and the terminal velocity constraint is satisfied. Integrating the second-order model of the rigid body system for the parameterized control input, we have

$$\theta(t) = \frac{J\phi_0}{6}\left(t^3 - 2(t - T_1)^3 \mathscr{H}(t - T_1) + 2(t - (2T_2 - T_1))^3 \mathscr{H}(t - (2T_2 - T_1))\right)$$
$$-(t - 2T_2)^3 \mathscr{H}(t - 2T_2)) \quad (4.232)$$

which when evaluated at the final time $2T_2$ should satisfy the displacement boundary condition

$$\theta(2T_2) = \theta_f = 2J\phi_0 T_1^3 \quad (4.233)$$

since $T_2 = 2T_1$.

Since, the switching function $\psi(t)$ at the first and second switch times should be 0, we have

$$\psi(T_1) = \frac{ET_1^2}{2} - FT_1 - \frac{1}{J} = 0 \tag{4.234}$$

$$\psi(2T_2 - T_1) = \frac{E(2T_2 - T_1)^2}{2} - F(2T_2 - T_1) - \frac{1}{J} = 0 \tag{4.235}$$

which can be solved for E and F resulting in the solutions

$$\begin{bmatrix} E \\ F \end{bmatrix} = \begin{bmatrix} -\frac{2}{3JT_1^2} \\ -\frac{4}{3JT_1} \end{bmatrix} = \begin{bmatrix} \frac{-2\sqrt[3]{2J\phi_0}^2}{3J\sqrt[3]{\theta_f}^2} \\ -\frac{4\sqrt[3]{2J\phi_0}}{3J\sqrt[3]{\theta_f}} \end{bmatrix} \tag{4.236}$$

4.5.1.2 Case 2

There exists boundary conditions for the rest-to-rest maneuver for which the constraint is active. The optimal jerk profile (v) is a bang-off-bang profile as shown in Figure 4.34(i).

Since, the optimal control profile is a combination of a number of time-delayed ramp functions, the requirement of time-optimality can be satisfied by ensuring that $u(t)$ is in saturation for as much time as possible during the maneuver. This implies that

$$T_1 = \frac{1}{J}, \ (2T_3 - T_2) - T_2 = 2T_1 = \frac{2}{J} \tag{4.237}$$

assuming that $-1 \leq u(t) \leq 1$ and the maximum permitted jerk is J. The response of the system given by Equation (4.221), to a step input of magnitude J is

$$\begin{Bmatrix} x_1(t) \\ x_2(t) \\ u(t) \end{Bmatrix} = \begin{Bmatrix} J\phi_0 t^3/6 \\ J\phi_0 t^2/2 \\ Jt \end{Bmatrix} = \begin{Bmatrix} k_1 t^3 \\ k_2 t^2 \\ Jt \end{Bmatrix}. \tag{4.238}$$

Thus, the states of Equation (4.221) at the final time, to the jerk profile parameterized as

$$v(t) = J(1 - \mathcal{H}(t - T_1) - \mathcal{H}(t - T_2) + \mathcal{H}(t - (2T_3 - T_2))$$
$$+ \mathcal{H}(t - (2T_3 - T_1)) - \mathcal{H}(t - 2T_3)) \tag{4.239a}$$

$$v(s) = \frac{J}{s}(1 - \exp(-sT_1) - \exp(-sT_2) + \exp(-s(2T_3 - T_2))$$
$$+ \exp(-s(2T_3 - T_1)) - \exp(-2sT_3)) \tag{4.239b}$$

are

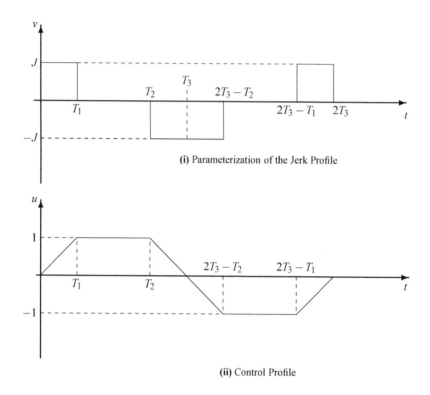

(i) Parameterization of the Jerk Profile

(ii) Control Profile

Figure 4.34: Case 2: Jerk Limited Time-Optimal Control

$$
\begin{Bmatrix} x_1(2T_3) \\ x_2(2T_3) \\ u(2T_3) \end{Bmatrix} = \begin{Bmatrix} k_1(2T_3)^3 \\ k_2(2T_3)^2 \\ J(2T_3) \end{Bmatrix} - \begin{Bmatrix} k_1(2T_3 - T_1)^3 \\ k_2(2T_3 - T_1)^2 \\ J(2T_3 - T_1) \end{Bmatrix} - \begin{Bmatrix} k_1(2T_3 - T_2)^3 \\ k_2(2T_3 - T_2)^2 \\ J(2T_3 - T_2) \end{Bmatrix}
$$
$$
+ \begin{Bmatrix} k_1(2T_3 - (2T_3 - T_2))^3 \\ k_2(2T_3 - (2T_3 - T_2))^2 \\ J(2T_3 - (2T_3 - T_2)) \end{Bmatrix} + \begin{Bmatrix} k_1(2T_3 - (2T_3 - T_1))^3 \\ k_2(2T_3 - (2T_3 - T_1))^2 \\ J(2T_3 - (2T_3 - T_1)) \end{Bmatrix} \tag{4.240}
$$

Substituting $T_1 = 1/J$ and $T_3 - T_2 = 1/J$, we have

$$
\begin{Bmatrix} x_1(2T_3) \\ x_2(2T_3) \\ u(2T_3) \end{Bmatrix} = \begin{Bmatrix} \phi T_3 (T_3 J - 1)/J \\ 0 \\ 0 \end{Bmatrix} \tag{4.241}
$$

which satisfies the constraint that the terminal velocity and control be zero. The maneuver time T_3 is

$$T_3 = \frac{\phi_0 \pm \sqrt{\phi_0^2 + 4J^2\phi_0\theta_f}}{2J\phi_0} \qquad (4.242)$$

Since, T_3 cannot be negative, the mid-maneuver time is given by the equation

$$T_3 = \frac{\phi_0 + \sqrt{\phi_0^2 + 4J^2\phi_0\theta_f}}{2J\phi_0} \qquad (4.243)$$

Knowledge of the switch time and the maneuver time can now be used to determine the costate equations and the switching function. The costate equations can be written as

$$\lambda_1(t) = E \qquad (4.244)$$

$$\lambda_2(t) = -Et + F \qquad (4.245)$$

$$\psi(t) = \begin{cases} \frac{\phi_0 E}{2}t^2 - \phi_0 Ft + G & \text{for } 0 < t \le T_1 \\ \frac{\phi_0 E}{2}t^2 - \phi_0 Ft + G - \eta_1(t) & \text{for } T_1 < t \le T_2 \\ \frac{\phi_0 E}{2}t^2 - \phi_0 Ft + G - \eta_1(T_2) & \text{for } T_2 < t \le (2T_3 - T_2) \\ \frac{\phi_0 E}{2}t^2 - \phi_0 Ft + G + \eta_2(t) & \text{for } (2T_3 - T_2) < t \le (2T_3 - T_1) \\ \frac{\phi_0 E}{2}t^2 - \phi_0 Ft + G + \eta_2(2T_3 - T_1) & \text{for } (2T_3 - T_1) < t \le 2T_3 \end{cases}$$

$$(4.246)$$

where

$$\eta_1(t) = \int \mu_1(t)dt = -\phi_0(-\frac{E}{2}t^2 + Ft + G) \text{ for } T_1 < t \le T_2 \qquad (4.247)$$

$$\eta_2(t) = \int \mu_2(t)dt = \phi_0(-\frac{E}{2}t^2 + Ft + G) \text{ for } (2T_3 - T_2) < t \le (2T_3 - T_1) \qquad (4.248)$$

Since the initial states for the rest-to-rest maneuver are zero and the initial control is 1, the initial value of the switching function, ψ can be calculated from the equation which requires the Hamiltonian to be zero at the initial time

$$H(0) = 1 + \psi(0)J = 0 \qquad (4.249)$$

which results in the solution

$$\psi(0) = G = -\frac{1}{J} \qquad (4.250)$$

Since the switching curve ψ is equal to zero at the switch times T_1 and $(2T_3 - T2)$, we can solve for E and F in closed form:

$$\begin{bmatrix} E \\ F \end{bmatrix} = \begin{bmatrix} -\dfrac{2J}{\phi_0(2T_3J-1)} \\[4mm] -\dfrac{2JT_3}{\phi_0(2T_3J-1)} \end{bmatrix} = \begin{bmatrix} -\dfrac{2J\phi_0}{\sqrt{\phi_0^2+4\phi_0J^2\theta_f}} \\[4mm] -\dfrac{\phi_0+\sqrt{\phi_0^2+4\phi_0J^2\theta_f}}{\sqrt{\phi_0^2+4\phi_0J^2\theta_f}} \end{bmatrix} \tag{4.251}$$

where T_3 is given by Equation (4.242).

To determine the transition from the Case 2 control profile to the Case 1 control profile, the two parameters T_2 and T_1 should be equal. Since,

$$T_2 = T_3 - \frac{1}{J} = T_1 = \frac{1}{J}, \tag{4.252}$$

we have

$$T_3 = \frac{2}{J} = \frac{\phi_0 + \sqrt{\phi_0^2 + 4J^2\phi_0\theta_f}}{2J\phi_0} \tag{4.253}$$

which results in the equation

$$\theta_f = \frac{2\phi_0}{J^2}. \tag{4.254}$$

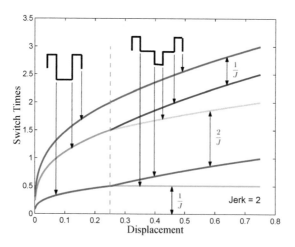

Figure 4.35: Variation of Switch Times vs. Displacement

Figure 4.35 illustrates the variation of the switch time as a function of displacement. It can be see that the structure of the control profile changes from a two-switch bang-bang to a four-switch bang-off-bang profile.

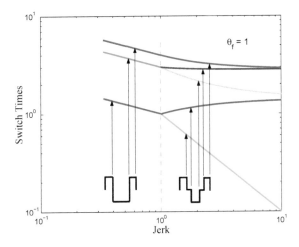

Figure 4.36: Variation of Switch Times vs. Jerk

Figure 4.36 illustrates the variation of the switch time as a function of permitted jerk and it can be seen that as the constraint on the jerk increases for a specified displacement, the bang-bang control profile transitions to a four-switch bang-off-bang profile.

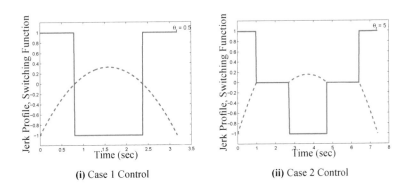

(i) Case 1 Control   **(ii)** Case 2 Control

Figure 4.37: Jerk Limited Time-Optimal Control Profiles

Figure 4.37 illustrates the jerk limited time-optimal profiles of the rate of change of the control and the corresponding switching function for final displacements of θ_f = 0.5, and 5. It can be seen that the displacement of 5 results in a singular solution.

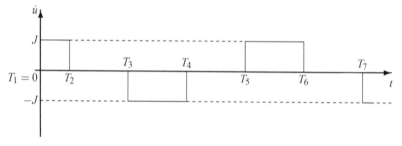

(i) Parameterization of the Jerk Profile

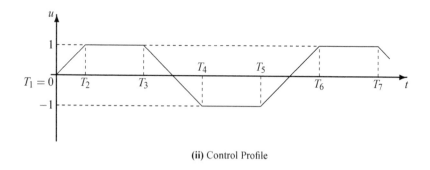

(ii) Control Profile

Figure 4.38: Typical Jerk and Control Profiles

4.5.2   Jerk Limited Time-Optimal Rest-to-Rest Maneuver

The optimal jerk-constrained control profile will be the combination of a number of time-delayed ramp functions with different slopes. The requirement of time-optimality can be satisfied by ensuring that $u(t)$ is in saturation for as much time as possible during the maneuver. This is justified by the bang-bang principle [22], which states that *if an optimal control exists, then there is always a bang-bang control profile that is optimal. Hence, if the optimal control is unique, it is bang-bang.* For the controller presented in this section, the jerk is the control variable. Thus, the jerk profile will be bang-bang, resulting in a ramping control input. For larger permissible amounts of jerk, the control input will eventually reach the saturation level. Then a bang-off-bang profile will be used. In the off-phases, the control input $u(t)$ remains saturated, exerting the maximum possible control authority.

It will be assumed that the actuator has symmetric limits in $u(t)$, therefore $u_{max} = -u_{min} = 1$. The time rate of change is limited by $|\dot{u}(t)| \leq J$ and is parameterized as shown in Figure 4.38(i). The resulting control profile is shown in Figure 4.38(ii). $\dot{u}(t)$ which is $v(t)$, will be represented as

$$v(t) = J\left(\sum_{i=1}^{n} A_i \mathcal{H}(t - T_i)\right), \tag{4.255}$$

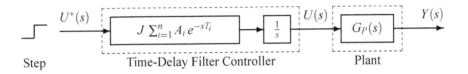

Step   Time-Delay Filter Controller   Plant

Figure 4.39: Time-Delay Filter Structure

where $\mathscr{H}$ denotes the Heaviside function and A_i is restricted to be

$$A_i \in \{-2, -1, 1, 2\} \; \forall \, i. \tag{4.256}$$

For the first switch, i.e., $i = 1$, $A_1 = 1$ and $T_1 = 0$. Similarly, for the last switch, i.e., $i = n$, $A_n \in \{-1, 1\}$ and $T_n = T_{final}$.

The time-delay filter structure drives an integrator, which ensures that the control input reaches its saturation limit over time. The input $u(t)$ will consist of a sum of delayed ramp signals,

$$u(t) = J \left(\sum_{i=1}^{n} A_i \langle t - T_i \rangle \right), \tag{4.257}$$

where $\langle x \rangle \equiv x \mathscr{H}(x)$. This control profile can be realized by means of a time-delay filter structure as shown in Figure 4.39. The symbol $U^*(s)$ denotes the control input used to drive the time-delay filter structure, which will be a step input.

Parameterizing the jerk profile to be point symmetric about the mid-maneuver time, the transfer function of the time-delay filter can be represented as

$$G_c(s) = \sum_{i=1}^{n} \left(A_i \, e^{-sT_i} - A_i \, e^{-s(2T_{mid} - T_i)} \right) \tag{4.258}$$

To cancel the three poles at the origin of the s-plane, Equation (4.258) and its first two derivatives with respect to s at $s = 0$ should equal zero, resulting in

$$G_c(s)|_{s=0} = \left(\sum_{i=1}^{n} \left(A_i \, e^{-sT_i} - A_i \, e^{-s(2T_{mid} - T_i)} \right) \right) \Bigg|_{s=0} = 0, \tag{4.259}$$

$$\begin{aligned}
\frac{dG_c(s)}{ds} \bigg|_{s=0} &= \frac{d}{ds} \left(\sum_{i=1}^{n} \left(A_i \, e^{-sT_i} - A_i \, e^{-s(2T_{mid} - T_i)} \right) \right) \bigg|_{s=0} \\
&= 2 \left(-\sum_{i=1}^{n} A_i \, T_i + \sum_{i=1}^{n} A_i \, T_{mid} \right) = 0.
\end{aligned} \tag{4.260}$$

and

$$\left.\frac{d^2 G_c(s)}{ds^2}\right|_{s=0} = \left.\frac{d^2}{ds^2}\left(\sum_{i=1}^{n}\left(A_i e^{-sT_i} - A_i e^{-s(2T_{mid}-T_i)}\right)\right)\right|_{s=0}$$

$$= -4T_{mid}\left(-\sum_{i=1}^{n}A_i T_i + \sum_{i=1}^{n}A_i T_{mid}\right) = 0. \qquad (4.261)$$

This shows that either both constraints (Equations 4.260 and 4.261) are satisfied at the same time or not at all. These constraints are identical to requiring that $u(t)$ is zero at the mid-maneuver time and point-symmetric to T_{mid}. $u(t)$ is given by

$$u(t) = J\left(\sum_{i=1}^{n}\left(A_i \langle t - T_i\rangle - A_i \langle t - (2T_{mid} - T_i)\rangle\right)\right). \qquad (4.262)$$

At the mid-maneuver time, it evaluates to

$$u(t)|_{t=T_{mid}} = J\sum_{i=1}^{n}A_i\left(T_{mid} - T_i\right) = -J\sum_{i=1}^{n}A_i T_i + J\sum_{i=1}^{n}A_i T_{mid} = 0. \qquad (4.263)$$

Thus, satisfying Equation (4.263), results in the cancelation of the triple pole at the origin.

In addition, it can be shown that for any $s = \pm j\omega$, the requirements $\Re\{G_c(j\omega)\} = 0$ and $\Im\{G_c(j\omega)\} = 0$ result in the same constraint

$$\sum_{i=1}^{n}A_i \sin(\omega\tilde{T}_i) = \sum_{i=1}^{n}A_i \sin\left(\omega(T_{mid} - T_i)\right) = 0 \qquad (4.264)$$

where $\tilde{T}_i$ is defined as

$$T_i = T_{mid} - \tilde{T}_i \ \forall \ i \in 1 \ldots n \qquad (4.265)$$

Finally, for the input-output transfer function

$$\frac{Y(s)}{U(s)} = G_p(s), \qquad (4.266)$$

the output displacement of y_f results in the constraint

$$y_f = \lim_{s \to 0} \frac{1}{s} G_p(s) G_c(s). \qquad (4.267)$$

4.5.3 Sufficiency Conditions

Unlike the three previous class of problems: time-optimal, fuel/time optimal, and fuel constrained time-optimal control, the jerk limited time-optimal control problem supports singular solutions. To confirm the optimality of the solution determined by the parameter optimization problem which only satisfies the state constraints, the costate constraints have to be studied. The switching function for the augmented system $\lambda^T B = \psi$ should equal zero when control reaches a boundary arc, i.e., when one of the state constraint becomes active. Figure 4.40 illustrates the evolution of the switching function (solid line) and the evolution of the switching function with no state constraints (dashed line). Equation (4.216c) states that the derivative of the costate equation ψ is discontinuous when the state constraint becomes active and the associated singular control is given by the equations

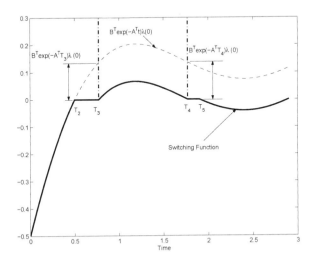

Figure 4.40: Switching Function

$$\dot{u} = v = 0 \text{ since } \begin{cases} \frac{d}{dt}(u-1) = 0 & \text{when } u = 1 \\ \frac{d}{dt}(-u-1) = 0 & \text{when } u = -1 \end{cases} \qquad (4.268)$$

Thus

$$\mu_1 = -B'^T \lambda' \qquad\qquad \text{when } u = 1 \qquad (4.269a)$$

$$\mu_2 = B'^T \lambda' \qquad\qquad \text{when } u = -1 \qquad (4.269b)$$

which results in $\dot{\psi}$ being zero when the state constraints are active.

Accounting for μ_1 and μ_2, the initial value of the costate can be solved from the unconstrained switching function by appropriately accounting for the effect of the discontinuity as shown in Figure 4.40. Since the switching function is equal to zero every time the state constraints become active, the initial costates can be determined from the equation

$$P\lambda(0) = \begin{bmatrix} B^T \exp(-A^T T_2) \\ B^T \exp(-A^T T_4) - B^T \exp(-A^T T_3) \\ B^T \exp(-A^T T_6) - B^T \exp(-A^T T_5) \\ \cdots \\ B^T \end{bmatrix} \lambda(0) = \begin{bmatrix} 0 \\ 0 \\ 0 \\ \cdots \\ -\frac{1}{J} \end{bmatrix} \tag{4.270}$$

where the last equations is derived from the requirement that the Hamiltonian be zero at the initial time. A and B are the system and input matrices of the augmented system. It is assumed that the jerk profile is one which results in the control input reaching its maximum and minimum limits. Solving for the costates using the equations

$$\dot{\lambda}' = -\frac{\partial H}{\partial x'} = -A'^T \lambda' \tag{4.271a}$$

$$\dot{\psi} = -\frac{\partial H}{\partial u} = \begin{cases} -B'^T \lambda' & \text{if } |u| - 1 < 0 \\ -B'^T \lambda' - \mu_1 & \text{if } u - 1 = 0 \\ -B'^T \lambda' + \mu_2 & \text{if } -u - 1 = 0 \end{cases} \tag{4.271b}$$

with the initial costates determined from Equation (4.270), and if the control determined from

$$u = -sgn\left(B^T \lambda(t)\right) \tag{4.272}$$

switches at the times determined from the parameter optimization problem, then the solution is optimal.

4.5.4   Benchmark Problem

The augmented model for the benchmark problem is

$$\begin{Bmatrix} \dot{y_1} \\ \dot{y_2} \\ \ddot{y_1} \\ \ddot{y_2} \\ \dot{u} \end{Bmatrix} = \dot{x} = \begin{bmatrix} 0 & 0 & 1 & 0 & 0 \\ 0 & 0 & 0 & 1 & 0 \\ -1 & 1 & 0 & 0 & 1 \\ 1 & -1 & 0 & 0 & 0 \\ 0 & 0 & 0 & 0 & 0 \end{bmatrix} x + \begin{bmatrix} 0 \\ 0 \\ 0 \\ 0 \\ 1 \end{bmatrix} v \tag{4.273}$$

and the corresponding Hamiltonian is:

$$H = 1 + \lambda_1 x_3 + \lambda_2 x_4 + \lambda_3(-x_1 + x_2 + u) + \lambda_4(x_1 - x_2) + \psi v + \mu_1(u - 1) + \mu_2(-u - 1). \tag{4.274}$$

The resulting costate equations are

$$\dot{\lambda}_1 = \lambda_3 - \lambda_4 \tag{4.275a}$$

$$\dot{\lambda}_2 = -\lambda_3 + \lambda_4 \tag{4.275b}$$

$$\dot{\lambda}_3 = -\lambda_1 \tag{4.275c}$$

$$\dot{\lambda}_4 = -\lambda_2 \tag{4.275d}$$

$$\dot{\psi} = \begin{cases} -\lambda_3 & \text{if } |u| - 1 < 0 \\ -\lambda_3 - \mu_1 & \text{if } u - 1 = 0 \\ -\lambda_3 + \mu_2 & \text{if } -u - 1 = 0 \end{cases} \tag{4.275e}$$

$$\tag{4.275f}$$

4.5.4.1 Case 1

With the knowledge that the time-optimal control with no constraints on the permitted jerk is

$$u(t) = 1 - 2\mathcal{H}(t - 1.0026) + 2\mathcal{H}(t - 2.1089) - 2\mathcal{H}(t - 3.2152) + \mathcal{H}(t - 4.2178) \tag{4.276}$$

the transition to jerk limited control profile results in a parameterization of an eight-switch bang-off-bang jerk profile. This results from the fact that each corner of the unconstrained time-optimal control profile corresponding to a switch of the jerk profile.

It can be seen from Figure 4.41 that the jerk profile can be described by two parameters, the time of the first zero crossing T_1 of the control input u, and the mid-maneuver time, T_{mid}. This is the reflection of the fact that for large jerk, the first switch is equal to $1/J$, i.e., the time required for the control input to reach its saturation limit of unity and the transition time from the upper and lower bound of the control takes $2/J$. The jerk profile can therefore be parameterized as

$$v = \frac{G_F(s)}{s} = \frac{J}{s}\left(1 - e^{-s\frac{1}{J}} - e^{-s\left(T_1 - \frac{1}{J}\right)} + e^{-s\left(T_1 + \frac{1}{J}\right)}\right.$$

$$+ e^{-s\left(T_{mid} - \frac{1}{J}\right)} - e^{-s\left(T_{mid} + \frac{1}{J}\right)} - e^{-s\left(2T_{mid} - T_1 - \frac{1}{J}\right)} \tag{4.277}$$

$$\left. + e^{-s\left(2T_{mid} - T_1 + \frac{1}{J}\right)} + e^{-s\left(2T_{mid} - \frac{1}{J}\right)} - e^{-s2T_{mid}}\right).$$

The transfer function $G_F(s)$ must include zeros which cancel the rigid-body poles as well as the poles corresponding to the oscillatory poles. The resulting parameter optimization problem is

$$\min \ f = T_{mid} \tag{4.278a}$$

subject to

$$\sin(\omega T_{mid}) - \sin(\omega(T_{mid} - \frac{1}{J})) - \sin(\omega(T_{mid} - T_1 + \frac{1}{J}))$$

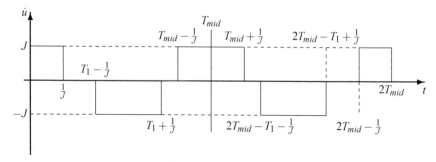

(i) Uncollapsed Jerk Profile

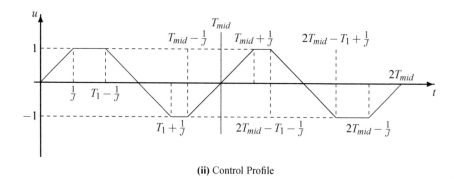

(ii) Control Profile

Figure 4.41: Large Jerk Profile

$$+ \sin(\omega(T_{mid} - T_1 - \frac{1}{J})) + \sin(\frac{\omega}{J}) = 0 \tag{4.278b}$$

$$\frac{J}{12}\left(-\frac{12T_1^2}{J} - \frac{6T_{mid}^2}{J} - \frac{6T_{mid}}{J^2} + \frac{24T_{mid}T_1}{J}\right) = y_f \tag{4.278c}$$

$$T_{mid} - \frac{2}{J} > T_1 > \frac{2}{J} \tag{4.278d}$$

where y_f refers to the final displacement of the rigid body mode.

The optimality of the results of the parameter optimization problem are confirmed by solving for the initial costates from the equation

$$
\underbrace{\begin{bmatrix} B^T \exp(-A^T \frac{1}{J}) \\ B^T \exp(-A^T (T_1 + \frac{1}{J})) - B^T \exp(-A^T (T_1 - \frac{1}{J})) \\ B^T \exp(-A^T (T_{mid} + \frac{1}{J})) - B^T \exp(-A^T (T_{mid} - \frac{1}{J})) \\ B^T \exp(-A^T (2T_{mid} - T_1 + \frac{1}{J})) - B^T \exp(-A^T (2T_{mid} - T_1 - \frac{1}{J})) \\ B^T \end{bmatrix}}_{P} \lambda(0) = \begin{bmatrix} 0 \\ 0 \\ 0 \\ \cdots \\ -\frac{1}{J} \end{bmatrix}
$$

$$(4.279)$$

and determining if the switch time of the control profile determined from the switching curve coincide with those determined from the parameter optimization problem.

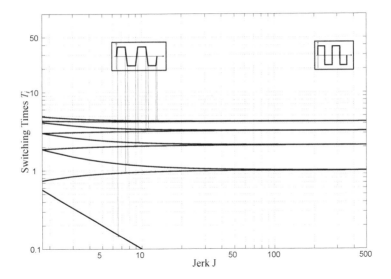

Figure 4.42: Case 1: Variation of Switch Times vs. Jerk

Figure 4.42 illustrates the variation of the switch time as a function of permitted jerk. It can be see that the structure of the jerk profile changes from an eight-switch bang-off-bang to a six-switch profile. This is due to the fact that the third and fourth and the fifth and sixth switches collapse into each other. For a displacement of unity, this occurs for a jerk of $J = 1.7539$. At this transition, i.e., when the third and fourth and the fifth and sixth collapse into each other, the jerk profile shown in Figure 4.43 can be parameterized as

$$v = \frac{G_F(s)}{s} = \frac{J}{s}\left(1 - e^{-s\frac{1}{J}} - e^{-s\left(T_{mid}-\frac{3}{J}\right)} + 2e^{-s\left(T_{mid}-\frac{1}{J}\right)}\right.$$

$$\left. - 2e^{-s\left(T_{mid}+\frac{1}{J}\right)} + e^{-s\left(T_{mid}+\frac{3}{J}\right)} + e^{-s\left(2T_{mid}-\frac{1}{J}\right)} - e^{-s2T_{mid}}\right). \tag{4.280}$$

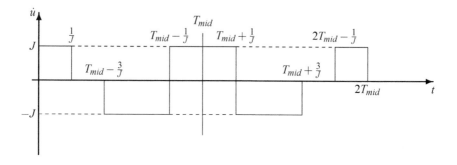

Figure 4.43: Partially Collapsed Jerk Profile

It can be shown from the pole cancelation constraint that

$$T_{mid} = \frac{\arcsin \dfrac{\sin\left(\frac{3\omega}{J}\right)-2\sin\left(\frac{\omega}{J}\right)}{\sqrt{\left(1-\cos\left(\frac{\omega}{J}\right)\right)^2+\left(\sin\left(\frac{\omega}{J}\right)\right)^2}} - \arctan \dfrac{\sin\left(\frac{\omega}{J}\right)}{1-\cos\left(\frac{\omega}{J}\right)}}{\omega}. \tag{4.281}$$

The constraint governing the displacement of the *center of mass* leads to the equation

$$J_{1,2} = \frac{1}{2}\frac{T_{mid} \pm \sqrt{33\,T_{mid}^2 - 32y_f}}{T_{mid}^2 - y_f}. \tag{4.282}$$

Using this result, an equation relating the mid-maneuver time to the frequency of the oscillatory mode ω and the final displacement y_f can be written as

$$\sin(\omega T_{mid})\left(1 - \cos\left(\frac{\omega}{\frac{1}{2}\frac{-T_{mid}\pm\sqrt{33\,T_{mid}^2-32y_f}}{-T_{mid}^2+y_f}}\right)\right)$$

$$+ \cos(\omega T_{mid})\sin\left(\frac{\omega}{\frac{1}{2}\frac{-T_{mid}\pm\sqrt{33\,T_{mid}^2-32y_f}}{-T_{mid}^2+y_f}}\right) \qquad (4.283)$$

$$= \sin\left(\frac{3\omega}{\frac{1}{2}\frac{-T_{mid}\pm\sqrt{33\,T_{mid}^2-32y_f}}{-T_{mid}^2+y_f}}\right) - 2\sin\left(\frac{\omega}{\frac{1}{2}\frac{-T_{mid}\pm\sqrt{33\,T_{mid}^2-32y_f}}{-T_{mid}^2+y_f}}\right).$$

Equation (4.283) is a transcendental function which can be used to solve for T_{mid} given a desired final displacement of y_f.

It is conceivable that instead of the third and fourth and the fifth and sixth switches coalescing, the first and second and the seventh and eighth could coalesce for a different set of model parameters. The transition equations for this scenario can be derived as shown above.

4.5.4.2 Case 2

For jerk values below that corresponding to the transition given by Equation (4.283), the control magnitude does not saturate in the region where the switches have collapsed. The maximum magnitude of the control input in this region becomes a variable to be optimized for.

Figure 4.44 illustrates the jerk profile parameterized in terms of the mid-maneuver time T_{mid}, and u_2 which corresponds to the maximum magnitude of the control in the region where the switches have collapsed. The jerk profile can be described by the equation

$$v = \frac{G_F(s)}{s} = \frac{J}{s}\left(1 - e^{-s\frac{1}{J}} - e^{-s\left(T_{mid}-\frac{1+2u_2}{J}\right)} + 2e^{-s\left(T_{mid}-\frac{u_2}{J}\right)}\right.$$

$$\left. - 2e^{-s\left(T_{mid}+\frac{u_2}{J}\right)} + e^{-s\left(T_{mid}+\frac{1+2u_2}{J}\right)} \right. \qquad (4.284)$$

$$\left. + e^{-s\left(2T_{mid}-\frac{1}{J}\right)} - e^{-s2T_{mid}}\right).$$

The parameter optimization problem for the Case 2 jerk profile is

$$\min \; f = T_{mid} \qquad (4.285a)$$

subject to

$$\sin(\omega T_{mid}) - \sin(\omega(T_{mid}-\frac{1}{J})) - \sin(\omega\frac{1+2u_2}{J}) + 2\sin(\omega\frac{u_2}{J}) = 0$$
$$\qquad (4.285b)$$

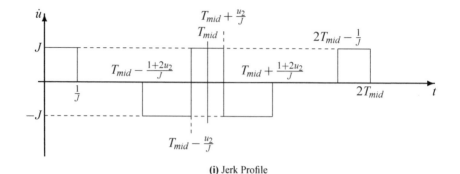

(i) Jerk Profile

(ii) Control Profile

Figure 4.44: Case 2: Control Profile

$$\frac{J}{12}\left(-\frac{12u_2}{J^3}-\frac{6T_{mid}}{J^2}-\frac{24u_2^2}{J^3}-\frac{12u_2^3}{J^3}+\frac{6T_{mid}^2}{J}\right)=y_f \qquad (4.285c)$$

$$T_{mid}-\frac{1+2u_2}{J}>\frac{1}{J}>\frac{u_2}{J} \qquad (4.285d)$$

The initial costates for the optimal solution are determined from the equation

$$P\lambda(0)=\begin{bmatrix}B^T\exp(-A^T\frac{1}{J})\\B^T\exp(-A^T(T_{mid}-\frac{u_2}{J}))-B^T\exp(-A^T(T_{mid}-\frac{1+2u_2}{J}))\\B^T\exp(-A^T(T_{mid}+\frac{u_2}{J}))-B^T\exp(-A^T(T_{mid}-\frac{1+2u_2}{J}))\\B^T\exp(-A^T(T_{mid}+\frac{1+2u_2}{J}))-B^T\exp(-A^T(T_{mid}-\frac{1+2u_2}{J}))\\B^T\end{bmatrix}\lambda(0)=\begin{bmatrix}0\\0\\0\\0\\-\frac{1}{J}\end{bmatrix}$$

$$(4.286)$$

which are subsequently used to confirm optimality of the solution of the parameter optimization problem.

Figure 4.45 illustrates the variation of the switch time as a function of permitted jerk. It can be see that the structure of the jerk profile changes from a six-switch bang-off-bang to a four-switch profile. This is due to the fact that the first and second

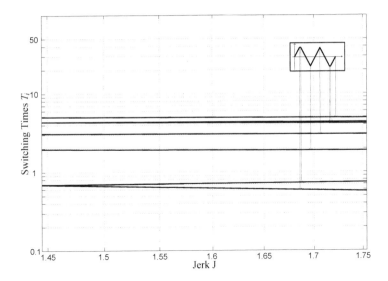

Figure 4.45: Case 2: Variation of Switch Times vs. Jerk

and the fifth and sixth collapse into each other. For a displacement of unity, this occurs for a jerk of $J = 1.445065$. At this transition, i.e., when the first and second and the fifth and sixth collapse into each other, the jerk profile parameterized by Equation (4.284) requires the first switch to coincide with the second resulting in the equation

$$\frac{1}{J} = T_{mid} - \frac{1 + 2u_2}{J} \Rightarrow u_2 = \frac{JT_{mid}}{2} - 1. \tag{4.287}$$

The final displacement constraint can be used to solve for the jerk

$$J_{1,2} = \frac{1}{2T_{mid}^3} \left(6T_{mid}^2 - 4y_f \pm 2\sqrt{5T_{mid}^4 - 12T_{mid}^2 y_f + 4y_f^2} \right) \tag{4.288}$$

and the constraint to cancel the undamped flexible model reduces to the equation

$$2\sin(\omega \frac{u_2}{2}) - 2\sin(\omega(T_{mid} - \frac{1}{J})) + \sin(\omega T_{mid}) = 0 \tag{4.289}$$

which can be used to solve for u_2.

4.5.4.3 Case 3

For jerk values lower than that given by Equation (4.288), the control never saturates and the design problem degenerates into a state unconstrained time-optimal design. Here the control input u is considered the state and the jerk v, the input. Figure 4.46 illustrates the parameterization of the jerk profile in terms of the maximum magnitude of the control input u at the end of every ramp. The resulting jerk profile is

$$G_F(s) = J\left(1 - 2e^{-s\frac{u_1}{J}} + 2e^{-s\frac{2u_1+u_2}{J}} - 2e^{-s\frac{2u_1+3u_2}{J}}\right.$$
$$\left. + 2e^{-s\frac{3u_1+4u_2}{J}} - e^{-s\frac{4u_1+4u_2}{J}}\right). \qquad (4.290)$$

The parameter optimization problem for the Case 3 jerk profile is

$$\min\ f = \frac{2u_1 + 2u_2}{J} \qquad (4.291\text{a})$$

subject to

$$2\sin(\omega\frac{u_2}{J}) - 2\sin(\omega\frac{u_1+2u_2}{J}) + 2\sin(\omega\frac{2u_1+2u_2}{J}) = 0 \qquad (4.291\text{b})$$

$$\frac{J}{12}\left(\frac{12u_1^3}{J^3} + \frac{24u_1^2 u_2}{J^3} - \frac{12u_2^3}{J^3}\right) = y_f \qquad (4.291\text{c})$$

$$u_1 > 0,\ u_2 > 0 \qquad (4.291\text{d})$$

The corroboration of optimality for this case is identical to that of the time-optimal control problem since the state constraints are not active.

Figure 4.47 illustrates the variation of the switch time as a function of permitted jerk. It can be see that the structure of the jerk profile changes from a four-switch bang-bang to a two-switch profile. This is referred to as the triangular control profile. This transition profile can be parameterized as

$$G_F(s) = J\left(1 - 2e^{-s\frac{T_{mid}}{2}} + 2e^{-s\frac{3T_{mid}}{2}} - e^{-s2T_{mid}}\right). \qquad (4.292)$$

By applying the pole-cancelation technique, an equation for T_{mid} can be derived as

$$\sin\left(\omega T_{mid}\right) - 2\sin\left(\frac{1}{2}\omega T_{mid}\right) = 0$$

$$\leftrightarrow \sin\left(\frac{\omega T_{mid}}{2}\right)\left(\cos\left(\frac{\omega T_{mid}}{2}\right) - 1\right) = 0$$

$$\leftrightarrow T_{mid} = \frac{2n\pi}{\omega}, n \in \mathcal{N}, \qquad (4.293)$$

where $\mathcal{N}$ denotes the set of natural numbers. Once T_{mid} is known, the final displacement constraint is given as

$$\frac{J}{6}\left(T_{mid}^3 - 2\left(\frac{T_{mid}}{2}\right)^3\right) = \frac{y_f}{2}. \qquad (4.294)$$

This equation can be solved for the jerk J as

$$J = \frac{4y_f}{T_{mid}^3} = \frac{4y_f \omega^3}{(2n\pi)^3}, n \in \mathcal{N}. \qquad (4.295)$$

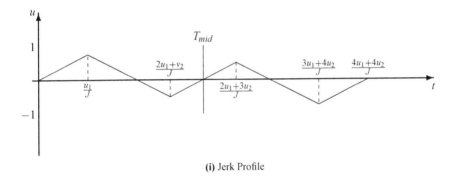

(i) Jerk Profile

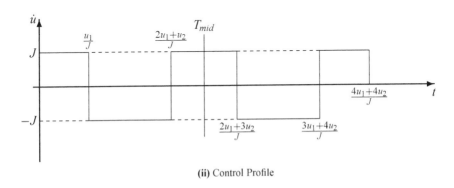

(ii) Control Profile

Figure 4.46: Case 3: Jerk and Control Profiles

For the first transition, $n = 1$. Since n can be any natural number, this triangular profile will also be the optimal solution for some other, smaller values of jerk. The very small jerk profile, which is shown in the next section will collapse to the triangular profile for some discrete values of jerk, which corresponds to $n = 2$.

4.5.4.4 Case 4

For permitted jerk values smaller than those which result in the triangular control profile, four more switches are introduced into the optimal jerk profile as shown in Figure 4.49. This jerk profile can be parameterized as

$$G_F(s) = J\left(1 - 2e^{-sT_1} + 2e^{-sT_2} - 2e^{-sT_3} + 2e^{-s(2T_{mid}-T_3)} \right.$$
$$\left. -2e^{-s(2T_{mid}-T_2)} + 2e^{-s(2T_{mid}-T_1)} - e^{-2sT_{mid}}\right) \tag{4.296}$$

The parameter optimization problem for the Case 4 jerk profile is

$$\min \ f = T_{mid} \tag{4.297a}$$

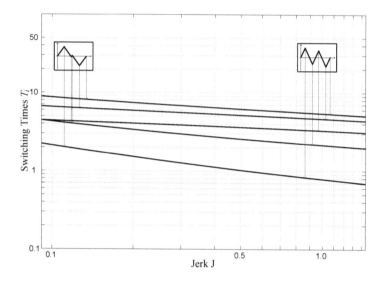

Figure 4.47: Case 3: Variation of Switch Times vs. Jerk

subject to

$$2T_1 - 2T_2 + 2T_3 - T_{mid} = 0 \tag{4.297b}$$

$$-2\sin(\omega(T_{mid} - T_3)) + 2\sin(\omega(T_{mid} - T_2))$$

$$-2\sin(\omega(T_{mid} - T_1)) + \sin(\omega T_{mid}) = 0 \tag{4.297c}$$

$$\frac{J}{12}\left(4(T_1^3 - T_2^3 + T_3^3 - 2T_{mid}^3) + 24T_{mid}^2(T_3 - T_2 + T_1)\right.$$

$$\left. +12T_{mid}(-T_3^2 + T_2^2 - T_1^2)\right) = y_f \tag{4.297d}$$

$$T_{mid} > T_3 > T_2 > T_1 > 0 \tag{4.297e}$$

Since the state constraints are not active for this case as well, the corroboration of optimality for this case is identical to that of the time-optimal control problem.

Figure 4.50 illustrates that the switching curves start from a triangular profile and as the permitted jerk is decreased, ends up with a triangular control profile again. It can be seen that two switches are introduced earlier to the first switch of the triangular control profile and as the jerk is decreases, the second switch drifts towards the third and coalesces with it reducing the jerk profile to one which generates a triangular control input.

To illustrate the entire spectrum of control profiles, the entire suite of profiles and the variations of switch times as a function of jerk are presented in Figure 4.51. It can be seen from Figure 4.51 that reducing the jerk from $J \to \infty$ to $J = 2$, the final time increases from 4.2179 to 4.8017, a small increase in maneuver time while the fuel consumed drops from 4.2179 to 2.8017, a significant percentage. To study the

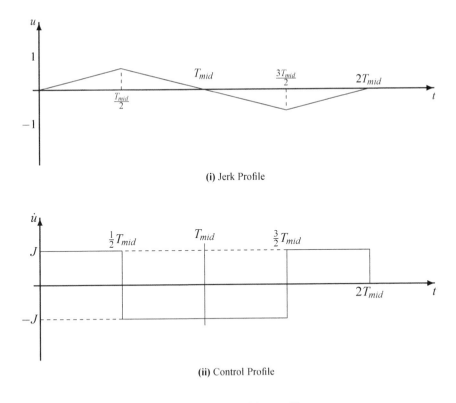

(i) Jerk Profile

(ii) Control Profile

Figure 4.48: Third Transition Profile

frequency spectrum of the energy, the Fourier transform of the input signal $u(t)$ is calculated and its magnitude is plotted in Figure 4.52 which clearly demonstrates that the energy injected into the higher frequency modes is reduced.

Figure 4.53 illustrates the optimal jerk profiles and the corresponding switching function for the four distinct cases described in this section. It can be seen that for Cases 1 and 2, singular intervals exist. For Cases 3 and 4, the control input never saturates and the switching function does not include any singular intervals.

For the Hubble telescope, the rate of change of jerk (snap) was considered in the design process. One can easily extend the techniques presented in this chapter to design controllers that are subject to constraints on snap, or the rate of change of snap (crackle), or the rate of change of crackle (pop), to successively develop control profiles whose frequency spectrum has an increased roll off at higher frequencies resulting in smoother control profiles.

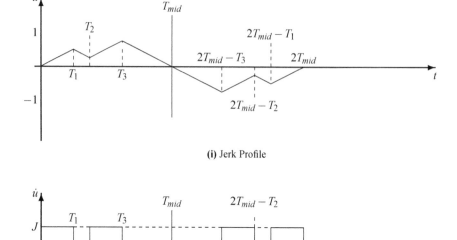

(i) Jerk Profile

(ii) Control Profile

Figure 4.49: Case 4: Jerk and Control Profiles

4.5.5   Summary

The objective of this chapter was to present design of controllers with control constraints. The cost function studied are:

- Time-Optimal Control

- Fuel/Time Optimal Control

- Fuel Limited Time-Optimal Control

- Jerk Limited Time-Optimal Control

All the cost functions selected for design resulted in controllers which can be characterized as bang-bang or bang-off-bang controllers. Such control profiles can be represented as the sum of time-delayed step inputs of different amplitudes. The amplitudes define whether the resulting control profile is bang-bang or bang-off-bang. The combined time-delayed step inputs can be generated by a time-delay filter subject to a unit step input which corresponds to the reference input for a rest-to-rest

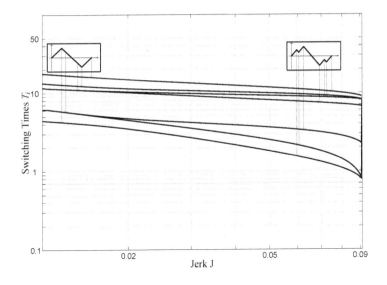

Figure 4.50: Case 4: Variation of Switch Times vs. Jerk

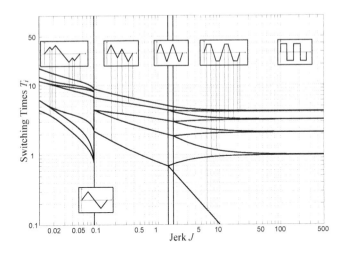

Figure 4.51: Variation of Switch Times vs. Jerk

maneuver. The design of the controllers where the control switches between discrete values, can then be posed as the design of the time-delays of a time-delay filter so as to cancel all the poles of the system and satisfy the rigid body boundary conditions.

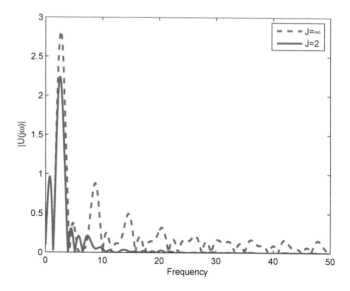

Figure 4.52: Frequency Spectra of Control Profile

The Hamiltonian formulation is used to derive the necessary conditions for optimality. Since only part of the necessary conditions for optimality are used to arrive at the parameter optimization problem, the satisfaction of the remaining necessary conditions need to be corroborated to guarantee optimality. A simple technique is used to determine the initial costates which can then be used to determine the switching function. If the control profile resulting from the switching function is coincident with that determined from the parameter optimization problem, optimality is guaranteed.

For all the cost functions considered in this chapter, closed form solutions for the optimal control for a rigid body system are derived. This is followed by a study of the undamped floating oscillator benchmark problem. Parametric studies related to variation of the final displacement, damping ratio, weighting factor of the cost function, and so forth, are studied for their effect on the structure of the optimal control profiles. Except for the jerk limited time-optimal control problem, none of the other cost functions resulted in singular control profiles for rest-to-rest maneuvers.

Exercises

4.1   Design a time-optimal controller for the system:

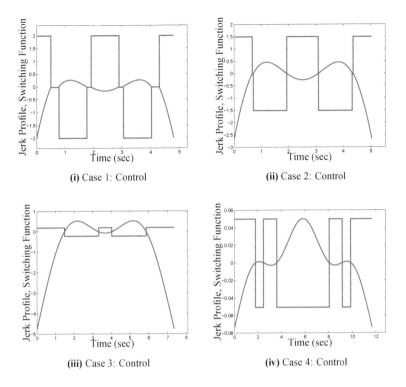

(i) Case 1: Control

(ii) Case 2: Control

(iii) Case 3: Control

(iv) Case 4: Control

Figure 4.53: Spectrum of Jerk Limited Time-Optimal Control Profiles

$$m_1\ddot{x}_1 + c_1(\dot{x}_1 - \dot{x}_2) + k_1(x_1 - x_2) = u$$
$$m_2\ddot{x}_2 - c_1(\dot{x}_1 - \dot{x}_2) - k_1(x_1 - x_2) + c_2(\dot{x}_2 - \dot{x}_3) + k_2(x_2 - x_3) = 0$$
$$m_3\ddot{x}_3 - c_2(\dot{x}_2 - \dot{x}_3) - k_2(x_2 - x_3) = 0$$

for the boundary conditions

$$x_1(0) = x_2(0) = x_3(0) = \dot{x}_1(0) = \dot{x}_3(0) = \dot{x}_3(0) = 0$$

$$x_1(t_f) = x_2(t_f) = x_3(t_f) = 1, \dot{x}_1(t_f) = \dot{x}_3(t_f) = \dot{x}_3(t_f) = 0$$

Assume $m_1 = m_2 = m_3 = 1$, $k_1 = k_2 = 1$, $c_1 = c_2 = 0.1$

4.2   Design a controller which minimizes the cost function:

$$J = \int_0^{t_f} (1 + |u|)\, dt$$

for the system:

$$L\frac{di}{dt} + Ri = V$$

$$J\ddot{\theta} = Ki$$

Assume $R = 1$, $L = 1$, $J = 1$, $K = 1$, with the boundary conditions

$$\theta(0) = \dot{\theta}(0) = i(0) = 0, \text{ and } \theta(t_f) = 1, \dot{\theta}(t_f) = 0, i(t_f) = 0.$$

V is the control input and satisfies the constraint

$$-1 \le V \le 1$$

4.3   For the two-mass-spring system (Figure 4.54), design a controller which minimizes the cost function

$$J = \frac{1}{2} \int_0^{t_f} (1 + |u|) \, dt$$

subject to the constraints

$$-1 \le u(t) \le 1 \text{ and } -5 \le \dot{u}(t) \le 5$$

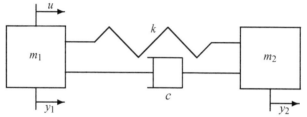

Figure 4.54: Floating Oscillator

for a rest-to-rest motion with the initial and final conditions

$$\begin{Bmatrix} y_1 \\ y_2 \\ \dot{y}_1 \\ \dot{y}_2 \end{Bmatrix} (0) = \begin{Bmatrix} 0 \\ 0 \\ 0 \\ 0 \end{Bmatrix} \text{ and } \begin{Bmatrix} y_1 \\ y_2 \\ \dot{y}_1 \\ \dot{y}_2 \end{Bmatrix} (\pi) = \begin{Bmatrix} 1 \\ 1 \\ 0 \\ 0 \end{Bmatrix}$$

Assume the nominal value of the parameters are

$$\begin{Bmatrix} m_1^{nom} \\ m_2^{nom} \\ k^{nom} \\ c^{nom} \end{Bmatrix} = \begin{Bmatrix} 1.0 \\ 1.0 \\ 1.0 \\ 0 \end{Bmatrix}$$

4.4   Design a closed form expression for the time-optimal controller for the system with a transfer function:

$$\frac{Y(s)}{U(s)} = G_p(s) = \frac{s+2}{s^2}$$

subject to the boundary conditions

$$y(0) = 0 \text{ and } y(t_f) = 1$$

and the control constraint:

$$-1 \le u(t) \le 1$$

This class of problem is referred to as output-transition control as opposed to state-transition control.

4.5   Design a time-optimal controller for the system with a transfer function:

$$\frac{Y(s)}{U(s)} = G_p(s) = \frac{s+K}{s^2(s+2)}$$

subject to the boundary conditions

$$y(0) = 0 \text{ and } y(t_f) = 1$$

and the control constraint:

$$-1 \le u(t) \le 1$$

Assume $K = 1$.

4.6   Design a fuel limited time-optimal control for the slosh problem [23]. The dynamics are given as:

$$\dot{x} = \begin{bmatrix} -0.42 & -21 & 0 & 0 \\ 21 & 0 & 0 & 0 \\ 0 & 0 & 0 & 0 \\ 0 & 0 & 1 & 0 \end{bmatrix} x + \begin{bmatrix} 0.0749 \\ 0 \\ 1 \\ 0 \end{bmatrix} u$$

and are subject to the constraints:

$$-1 \le u(t) \le 1 \text{ and } \int_0^{t_f} |u|\,dt \le 10$$

The initial and boundary conditions are

$$\begin{Bmatrix} x_1 \\ x_2 \\ x_3 \\ x_4 \end{Bmatrix}(0) = \begin{Bmatrix} 0 \\ 0 \\ 0 \\ 0 \end{Bmatrix} \text{ and } \begin{Bmatrix} x_1 \\ x_2 \\ x_3 \\ x_4 \end{Bmatrix}(t_f) = \begin{Bmatrix} 0 \\ 0 \\ 0 \\ 1 \end{Bmatrix}$$

4.7   Harmonic drives which are often used in robots are characterized by high torque capacity, but suffer from high flexibility. Figure 4.55 illustrate a single link robot driven by a harmonic drive where the flexibility is represented by a torsional spring. The equations of motion are:

$$J_m \ddot{\psi} + k(\psi - \phi) = \tau$$
$$J \ddot{\phi} - k(\psi - \phi) = 0.$$

Design a jerk limited fuel/time optimal controller assuming the jerk limit is 5 and the model parameters are: $J_m = 1$, $J = 5$, $k = 10$, and the control input has to satisfy the constraint:

$$-2 \leq \tau \leq 2$$

The initial and final conditions are:

$$\psi(0) = \dot{\psi}(0) = \phi(0) = \dot{\phi}(0) = 0$$

and

$$\psi(t_f) = \phi(t_f) = \frac{\pi}{2}, \quad \dot{\psi}(t_f) = \dot{\phi}(t_f) = 0$$

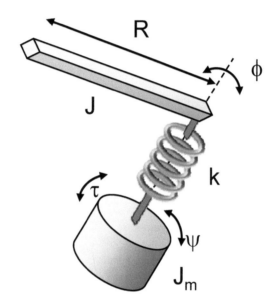

Figure 4.55: Harmonic Drive Robot

4.8 Design a time-optimal controller for the pitch control of an aircraft (Figure 4.56). The transfer function relating the pitch angle θ to the elevator deflection δ_e is: (Taken from: http://www.engin.umich.edu/group/ctm/examples/pitch/Mpitch.html.)

$$\frac{\theta(s)}{\delta_e(s)} = \frac{1.151s + 0.1774}{s^3 + 0.739s^2 + 0.921s}$$

The elevator deflection is constrained by:

$$-\frac{\pi}{6} \leq \delta_e \leq \frac{\pi}{6}$$

Design a time-optimal control to change the pitch angle from an initial position of rest to a final pitch angle of 0.2 rad.

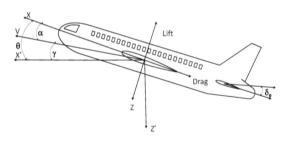

Figure 4.56: Airplane Pitch Dynamics Variables

4.9 Rapid human movement is referred to as Ballistic movement since feedback is not used in the maneuver [18]. It has been shown that that human motion can be represented as the solution of a jerk limited time-optimal control problem. Figure 4.57 illustrates a point-to-point motion. A simplified model for the motion of the arm is:

$$\left(m_1 \frac{L_1^2}{4} + m_2 L_1^2 + m_2 \frac{L_2^2}{4} + m_2 L_1 L_2 \right) \ddot{\theta}_1 + \left(m_2 \frac{L_2^2}{4} + m_2 \frac{1}{2} L_1 L_2 \right) \ddot{\theta}_2 = u_1$$

$$\left(m_2 \frac{L_2^2}{4} \right) \ddot{\theta}_2 + \left(m_2 \frac{L_2^2}{4} + m_2 \frac{1}{2} L_1 L_2 \right) \ddot{\theta}_1 = u_2$$

Design a jerk limited time-optimal control where the initial and boundary conditions are:

$$\begin{bmatrix} \theta_1 \\ \theta_2 \\ \dot{\theta}_1 \\ \dot{\theta}_2 \end{bmatrix}(0) = \begin{bmatrix} 40^o \\ 42^o \\ 0 \\ 0 \end{bmatrix}, \quad \begin{bmatrix} \theta_1 \\ \theta_2 \\ \dot{\theta}_1 \\ \dot{\theta}_2 \end{bmatrix}(t_f) = \begin{bmatrix} 32^o \\ 35^o \\ 0 \\ 0 \end{bmatrix}$$

where

$$L_1 = 0.3m, \quad L_2 = 0.4m, \quad m_1 = 1.89kg, \quad m_2 = 1.5kg$$

and the jerk is limited to the range

$$-60\frac{m}{s^3} \le \dddot{\psi} \le 60\frac{m}{s^3}$$

where

$$\psi = L_1 \theta_1 + L_2 \theta_2$$

is an approximate displacement of the end of the arm. The acceleration of the end of the links are limited to the range

$$-5\frac{m}{s^2} \le L_1 \ddot{\theta}_1 \le 5\frac{m}{s^2}$$

$$-5\frac{m}{s^2} \le L_2 \ddot{\theta}_2 \le 5\frac{m}{s^2}$$

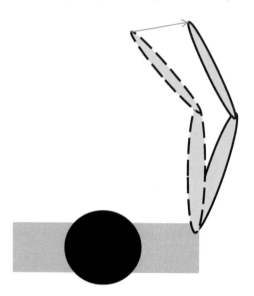

Figure 4.57: Human Ballistic Motion

References

[1] B. Wie and D. S. Bernstein. Benchmark problmes for robust control design. *J. of Guid., Cont. and Dyn.*, 15(5):1057–1059, 1992.

[2] M. Athans and P. L. Falb. *Optimal Control.* McGraw Hill, 1966.

[3] Arthur E. Bryson and Yu-Chi Ho. *Applied Optimal Control: Optimization, Estimation and Control.* Hemisphere Publishing, 1975.

[4] S. Scrivner and R. C. Thompson. Survey of time-optimal attitude maneuvers. *J. of Guid., Cont. and Dyn.*, 17(2):225–233, 1992.

[5] G. Singh, P. T. Kabamba, and N. H. McClamroch. Planar time-optimal control, rest-to-rest slewing of flexible spacecraft. *J. of Guid., Cont. and Dyn.*, 12(1):71–81, 1989.

[6] J. Ben-Asher, J. A. Burns, and E. M. Cliff. Time-optimal slewing of flexible spacecraft. *J. of Guid., Cont. and Dyn.*, 15(2):360–367, 1992.

[7] Q. Liu and B. Wie. Robust time-optimal control of uncertain flexible spacecraft. *J. of Guid., Cont. and Dyn.*, 15(3):597–604, 1992.

[8] B. H. Hablani. Zero-residual-energy, single-axis slew of flexible spacecraft

using thrusters. dynamics approach. *J. of Guid., Cont. and Dyn.*, 15(1):104–113, 1992.

[9] L. Y. Pao. Minimum-time control characteristics of flexible structures. *J. of Guid., Cont. and Dyn.*, 19(1):123–129, 1996.

[10] T. Singh. Effect of damping on the structure of time-optimal controllers. *Journal of Guidance, Control and Dynamics*, 19(5):1182–1184, September 1996.

[11] W. E. Vander Velde and J. He. Design of space structure control systems using on-off thrusters. *Journal of Guidance, Control and Dynamics*, 6(1):53–60, January 1983.

[12] M. Lopes de Oliveria e Souza. Exactly solving the weighted time/fuel optimal control of an undamped harmonic oscillator. *Journal of Guidance, Control and Dynamics*, 11(6):488–494, 1983.

[13] R. Hartmann and T. Singh. Fuel/time optimal control of flexible structures: A frequency domain approach. *Journal of Vibration and Control*, 5(5):795–817, 1999.

[14] B. Wie, R. Sinha, J. Sunkel, and K. Cox. Robust fuel- and time-optimal control of uncertain flexible space strcutures. In *AIAA Guidance, Navigation and Control Conference*, Monterey, CA, 1993.

[15] T. Singh. Fuel/time optimal control of the benchmark problem. *J. of Guid., Cont. and Dyn.*, 18(6):1225–1231, 1995.

[16] T. Singh and S. R. Vadali. Robust time-optimal control: Frequency domain approach. *J. of Guid., Cont. and Dyn.*, 17(2):346–353, 1994.

[17] W. Singhose, T. Singh, and W. P. Seering. On-off control with specified fuel usage. *ASME J. of Dyn. Systems, Measurements, and Cont.*, 121(2):206–212, 1999.

[18] S. Ben-Itzak and A. Karniel. Minimum acceleration criterion with constraints implies bang-bang control as an underlying principle for optimal trajectories of arm reaching movements. *Neural Computation*, 20:779–812, 2008.

[19] S. Iwnicki. *Handbook of Railway Vehicle Dynamics*. CRC/Taylor & Francis Group, 2006.

[20] M. Muenchhof and T. Singh. Jerk limited time optimal control of flexible structures. *ASME J. of Dyn. Systems, Measurements, and Cont.*, 125(1):139–142, 2003.

[21] M. Muenchhof and T. Singh. Desensitized jerk limited-time optimal control of multi-input systems. *J. of Guid., Cont. and Dyn.*, 25(3):474–481, 2002.

[22] H. Hermes and J. P. Lasalle. *Functional Analysis and Time Optimal Control*. Academic Press, 1969.

[23] M. Grundelius and B. Bernhardsson. Control of liquid slosh in an industrial packaging machine. In *IEEE International Conference on Control Applications*, pages 1654–1659, Hawaii, HI, August 1999.

5

Minimax Control

If people do not believe that mathematics is simple, it is only because they do not realize how complicated life is.

John von Neumann (1903-1957); Hungarian–American mathematician

CHAPTERS 2, 3, and 4 described various techniques for the control of systems with vibratory modes. These techniques are model–based and are often sensitive to errors in model parameters. In Chapter 2, the sensitivity of the pole-zero cancelation constraints with respect to the uncertain model parameters was used to design filters/controllers. The resulting control profiles are robust around the nominal model of the system. Thus, for shaping the reference input to a system with modeling errors, it was shown that by cascading multiple instances of the time-delay filter designed to cancel the poles of the system, resulted in a filter that was insensitive to error in modeled natural frequency and damping ratio. The idea of locating multiple zeros of a time-delay filter at the estimated location of the poles of the system has been exploited to design robust time-optimal control [1,2], robust fuel/time optimal control [3], fuel constrained time-optimal control [4], and so forth, as shown in Chapter 4. The concept of using terminal state sensitivities with respect to the uncertain model parameters can be extended to nonlinear systems undergoing rest-to-rest maneuver, and has been shown to result in robust control by Liu and Singh [5].

If knowledge of the range of uncertainty or the probability distribution function (pdf) of the uncertain parameter is available, developing control algorithms that can exploit this knowledge in the design of the pre-filters or feedback controllers seems patent. One can consider a worst case design over the domain of uncertainty. This would entail design of a controller which minimizes the maximum magnitude of the cost function over the domain of uncertainty. If the knowledge is in the form of a pdf then minimizing the expected performance index over that pdf is an alternative to minimizing the maximum over domain which includes regions of extreme unlikelihood. Particularly in the case where the region of support is unbounded (e.g., Gaussian), one must minimax over the entire space, which is computationally challenging. For a system with one uncertain parameter, Figure 5.1 illustrates the variation of the cost function over the range of uncertainty. The objective of the control design is to minimize the maximum magnitude of the cost over the range p_{lb} and p_{ub}, the lower and upper bound of the uncertain parameter p. In this chapter, a minimax optimization problem is formulated to design controllers which are robust over a specified range of uncertainties. This approach can accommodate distribution of the uncertainties. First, a technique for the design of time-delay pre-filters which shape the reference input to the system is presented. The objective of the design is to arrive at an optimal pre-filter which minimizes the maximum magnitude of the residual vibration of a system undergoing rest-to-rest maneuvers. This is followed by the development of feedback controllers which minimize a performance index which is an infinite time horizon, integral of a quadratic function of the system states.

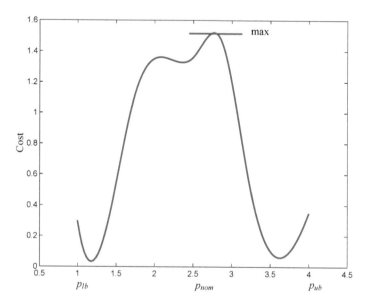

Figure 5.1: Cost Function Variation over Uncertain Domain

5.1 Minimax Time-Delay Filters

Minimax is a word used in decision theory, game theory and statistics when one is interested in minmizing the worst outcome. In a control design setting, one can consider the uncertain model parameter such as the spring stiffness to take a value which maximizes the residual energy over the domain of uncertainty, for a rest-to-rest maneuver. The minimax control objective is to determine the parameters of the controller to minimize the worst (i.e., largest) residual energy. This worst case design obviously trades off performance in the nominal case for robustness. This section will formulate the optimization problem for the design of time-delay filters for uncertain mechanical systems undergoing rest-to-rest maneuvers.

5.1.1 Cost Function

Consider the linear mechanical system

$$M\ddot{x} + C(p)\dot{x} + K(p)x = Du \tag{5.1}$$

where x represents the generalized coordinates and M the mass matrix is positive definite, C, and K the damping and stiffness matrices are positive definite or positive semidefinite. The stiffness matrix is positive semidefinite when the system includes rigid body modes. p is a vector of uncertain parameters which satisfies the constraints:

$$p_i^{lb} \le p_i \le p_i^{ub} \tag{5.2}$$

where p_i^{lb} and p_i^{ub} represent the lower and upper bound on the p_i^{th} parameter. Since the objective of a rest-to-rest maneuver is to move the system from its initial state of rest to the final state of rest, it is natural to consider the residual energy of the system at the end of the maneuver as the cost function. For systems without rigid body modes, the sum of the kinetic and potential energy at the end of the maneuver

$$F(t_f) = \frac{1}{2}\dot{x}^T M\dot{x} + \frac{1}{2}(x - x_f)^T K(p)(x - x_f) \tag{5.3}$$

evaluated at time t_f captures the total residual energy in the system. x_f corresponds to the final position of the displacement states. For systems with rigid body modes, to force $F(t_f)$ to be positive definite, a pseudo-potential energy term needs to be added to the kinetic and potential energy [6]. The pseudo-potential energy term is equal to zero when the rigid body mode has zero energy. The augmented cost function is

$$F(t_f) = \frac{1}{2}\dot{x}^T M\dot{x} + \frac{1}{2}x^T K(p)x + \frac{1}{2}(x_r - x_{rf})^2 \tag{5.4}$$

where x_r refers to the rigid body displacement and x_{rf} is the final desired position of the rigid body. In the rest of this chapter, the term residual energy will be used to refer to both functions given by Equations (5.3) or (5.4).

For the design of robust controllers where the domain of uncertainty of the vector of uncertain parameter p is specified, a minimax optimization problem can be formulated to minimize the maximum magnitude of the residual energy over the domain of uncertainty. Assuming the distribution of the uncertain parameters is uniform, the minimax cost is

$$\min_{q} \max_{p} \left(\frac{1}{2}\dot{x}^T M\dot{x} + \frac{1}{2}x^T Kx + \frac{1}{2}(x_r - x_{rf})^2 \right) \tag{5.5}$$

where q is the vector of parameters which defines the controller.

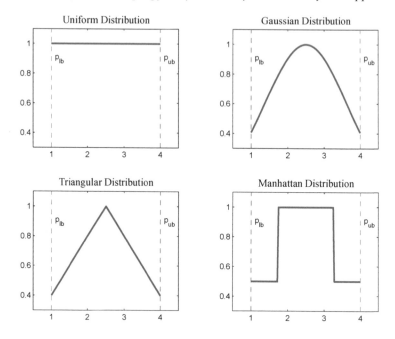

Figure 5.2: Typical Uncertainty Distributions

Provided knowledge of the distribution of the uncertainties exist, they can be easily included into the cost function. If the distribution is Gaussian, the cost function is

$$\min_{q} \max_{p} \; e^{-(p-p_{nom})^T \Gamma^{-1}(p-p_{nom})} \left(\frac{1}{2}\dot{x}^T M\dot{x} + \frac{1}{2}x^T Kx + \frac{1}{2}(x_r - x_{rf})^2 \right) \quad (5.6)$$

where Γ is the covariance matrix of the Gaussian distribution and p_{nom} is the vector of nominal value of the uncertain parameters. The Γ matrix is diagonal if all the uncertain parameters are independent. For distributions whose support is unbounded, one can limit the support to include 95–99% of the most likely values. Figure 5.2 illustrate some of the distributions for a single uncertain parameter case, that one can exploit in the design of the minimax controller. These distributions can be normalized such that the area under the curve is equal to one. However, this is not necessary since the optimization algorithm uses these distributions to differentially weight various parameter sets over the uncertain domain.

5.1.2   Van Loan Identity

First, a technique for the design of pre-filters for the vibratory systems will be presented where the pre-filter is parameterized as a time-delay filter

$$u = \sum_{i=0}^{N} A_i \mathcal{H}(t - T_i) \quad (5.7)$$

where $\mathcal{H}(t - T_i)$ is the Heaviside function, T_0 is zero and N is the number of delays in the filter. From Equation (5.7), it is clear that the input u can be represented by a sum of delayed step inputs and since the system we are considering is linear, the magnitude of the states at any time can be determined by the sum of the states at any time due to each of the delayed step inputs. The van Loan identity is an efficient technique for the calculation of the states of a system subject to a step input, at any time.

Consider the state-space model of a system

$$\dot{z} = Az + Bu \text{ where } z = \begin{bmatrix} x \\ \dot{x} \end{bmatrix} \in \mathcal{R}^n, u \in \mathcal{R}^1. \tag{5.8}$$

To determine the response of the system (Equation 5.8) to a unit step input, construct the matrix

$$P = \begin{bmatrix} A & B \\ 0 & 0 \end{bmatrix} \tag{5.9}$$

which is a $\mathcal{R}^{n+1 \times n+1}$ matrix. Using the van Loan identity [7], one can show that

$$Z = e^{PT} = \begin{bmatrix} e^{AT} & \int_0^T e^{A(T-\tau)}B\, d\tau \\ 0 & I \end{bmatrix} \tag{5.10}$$

It can be seen that the upper right hand term of the matrix Z is the convolution integral of the system given by Equation (5.8) subject to a unit step input. Thus, the value of the states at time T for a unit step input are given by the first n rows of the last column of Z. This permits us to calculate the final states for a step input accurately, without numerical integration. This is very attractive for numerical optimization where a significant cost of optimizing dynamical system is contributed by the numerical simulation of the response of the system. For instance, the response of the system represented by Equation (5.8) to the input represented by Equation (5.7) is given by the first n rows of the last column of the matrix

$$\Phi = \sum_{i=0}^{n} A_i e^{P(T_n - T_i)}. \tag{5.11}$$

The optimization algorithms which are used to solve minimax problems are generally gradient based. Thus, the accuracy and the speed of the optimization can be increased by providing analytical gradients to the optimization algorithm. Fortunately, for the time-delay control, closed form equations representing the gradients of the cost and constraints can be easily derived as shown below.

For the optimization algorithm, we require the value of the gradient of the cost F and the constraints, with respect to the controller parameters. For the control given by Equation (5.7), the gradients of F (Equation 5.3) with respect to A_i and T_i are given by

$$\frac{dF}{dA_i} = \dot{x}^T M \frac{d\dot{x}}{dA_i} + (x - x_f)^T K \frac{dx}{dA_i} \tag{5.12}$$

and

$$\frac{dF}{dT_i} = \dot{x}^T M \frac{d\dot{x}}{dT_i} + (x - x_f)^T K \frac{dx}{dT_i} \tag{5.13}$$

evaluated at the final time. To determine $\frac{dx}{dA_i}$, $\frac{d\dot{x}}{dA_i}$ and $\frac{dx}{dT_i}$, $\frac{d\dot{x}}{dT_i}$, we require the derivative of the state Equation (5.8). The solution of the equation

$$\frac{d\dot{z}}{dA_i} = A \frac{dz}{dA_i} + B\mathcal{H}(t - T_i) \text{ i} = 0, 1, 2, 3, \dots \tag{5.14}$$

can be derived using the van Loan identity as described earlier. Similarly the solution of equation

$$\frac{d\dot{z}}{dT_i} = A \frac{dz}{dT_i} - B(A_i \delta(t - T_i)) \text{ i} = 1, 2, 3, \dots \tag{5.15}$$

where $\delta(.)$ is the dirac delta function, can be shown to be

$$\frac{dz}{dT_i}(T_f) = -A_i \exp(A(T_f - T_i))B. \tag{5.16}$$

With the analytical gradients, we can expedite the convergence of the optimization algorithm.

5.1.3 Pre-Filter Design

This section focuses on the design of time-delay filters which modify the reference input to the underdamped system so as to minimize the maximum magnitude of the residual vibration over the range of uncertainty, for a rest-to-rest maneuver. The benchmark spring-mass-dashpot system will be used to illustrate the minimax controller.

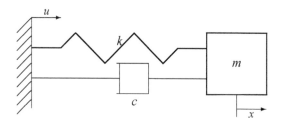

Figure 5.3: Spring-Mass-Dashpot System

The equation of motion of the spring-mass-dashpot system illustrated in Figure 5.3 is

$$m\ddot{x} + c\dot{x} + kx = ku \tag{5.17}$$

where it is assumed that the coefficient of stiffness is uncertain and lies in the range

$$0.7 \leq k \leq 1.3. \tag{5.18}$$

In Chapter 2, it was shown that a single time-delay filter that is designed to locate a pair of zeros of the time-delay filter at the nominal location of the underdamped poles of the system results in a finite-time interval for a rest-to-rest maneuver. This solution corresponds to the *Posicast* solution [8]. It has also been shown that by cascading this single time-delay filter, i.e., locating multiple pairs of zeros of the time-delay filter at the nominal location of the underdamped poles, results in robustness around the nominal poles of the system [9]. In this design, the only information necessary is the location of the nominal poles of the system. Often, the range of uncertainty and the distribution of uncertainty is known in addition to the nominal parameters of the system. The technique presented in this chapter endeavors to exploit this knowledge in the design of a robust pre-filter. Since the basic robust time-delay filter consists of two time-delays, we start by studying the minimax problem by considering the two-time-delay filter. The transfer function of the time-delay filter is:

$$\frac{u(s)}{r(s)} = A_0 + A_1 \exp(-sT_1) + A_2 \exp(-sT_2). \tag{5.19}$$

To ensure that the steady-state value of $u(s)$ is the same as the reference input $r(s)$, we require

$$A_0 + A_1 + A_2 = 1. \tag{5.20}$$

The optimization problem can be stated as the determination of A_0, A_1, A_2, T_1, and T_2 of the time-delay filter so as to

$$\min_{T_i, A_i} \max_{k} \sqrt{\frac{1}{2}m\dot{y}^2 + \frac{1}{2}k(y-1)^2} \tag{5.21}$$

evaluated at T_2.

The optimization toolbox of MATLAB is used to solve the minimax optimization algorithm. The optimal minimax time-delay filter for a system with a nominal mass of $m = 1$, nominal damping coefficient of $c = 0.2$, and with the uncertain stiffness range being specified by Equation (5.18), is given by the transfer function

$$\frac{u(s)}{r(s)} = 0.3452 + 0.4730e^{-3.1703s} + 0.1818e^{-6.3405s}. \tag{5.22}$$

Figure 5.4 (dotted line) illustrates the variation of the residual energy of the system as a function of the uncertain parameter k. It can be seen that the maximum magnitude of the residual energy in the range of the uncertain parameters occurs at the bounding limits ($k = 0.7$, $k = 1.3$) and at a value of k which lies between the limits. It is also clear that the maximum magnitude of the residual energy is significantly smaller than that resulting from the robust time-delay filter defined by Equation (2.40) over the entire range of k. However, at the nominal value of $k = 1$, the minimax solution has a large magnitude of residual vibration. The minimax solution is similar to the *extra-insensitive input shaper* proposed by Singhose et al. [10] where an optimization problem is formulated by defining the magnitude of residual vibration permitted at the nominal value of the uncertain parameter and solving for the magnitudes of a sequence of impulses. The impulse sequence is required to satisfy the constraints that the magnitude of the residual vibration is zero at two frequencies which flank the nominal value and the slope of the residual energy distribution is zero at the nominal value of the uncertain parameter.

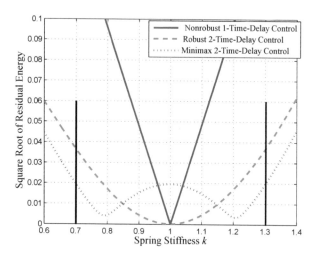

Figure 5.4: Residual Vibration Distribution

The trade-off resulting from the minimax design is evident from Figure 5.4. Minimizing the maximum magnitude of the residual vibration over the range of uncertainty has required an increase in the magnitude of residual vibration at and near the nominal value of the uncertain parameter. To address this disadvantage, an additional constraint can be included in the optimization problem which requires the magnitude of the residual vibration at the nominal value of k, be forced to zero. This can only be achieved by adding a time-delay to the two time-delay filter, which implies that the maneuver will take longer. The transfer function of this pre-filter is

$$\frac{u(s)}{r(s)} = A_0 + A_1 \exp(-sT_1) + A_2 \exp(-sT_2) + A_3 \exp(-sT_3). \qquad (5.23)$$

The unknown parameters of Equation (5.23) are solved for using the solution of the parameters of three nonrobust time-delay filters in cascade, as the initial guess.

The transfer function of the minimax time-delay controller with the constraint to force the residual vibration to be zero at the nominal value of k can be shown to be

$$\frac{u(s)}{r(s)} = 0.2052 + 0.4141e^{-3.1652s} + 0.3015e^{-6.3304s} + 0.07924e^{-9.4956s}. \quad (5.24)$$

Figure 5.5 illustrates the distribution of the residual energy of the time-delay filter designed by cascading three nonrobust time-delay filters (solid line) and the minimax time-delay filter (dashed line). It is clear from Figure 5.5 that the maximum magnitude of the residual energy of the minimax controller over the uncertain range ($0.7 \leq k \leq 1.3$), is significantly smaller than the robust three-time-delay controller, which is a metric to gauge the robustness of the controllers. One should note that the improved performance at and near the nominal value of the uncertain parameter is acquired at a cost of an increase in the maneuver time compared to the two-time-delay filter.

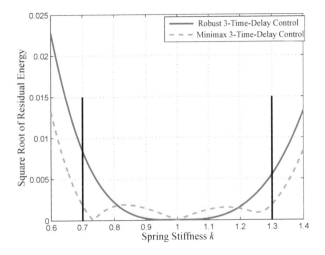

Figure 5.5: Residual Vibration Distribution

The proposed technique for the design of minimax controller can be extended to systems with multiple uncertain parameters. For instance, assuming that the coefficient of stiffness and damping for the single mass-spring-dashpot system, lie in the uncertain region:

$$0.7 \leq k \leq 1.3, \text{ and } 0.1 \leq c \leq 0.2 \quad (5.25)$$

the resulting two-time-delay minimax pre-filter has the transfer function

$$\frac{u(s)}{r(s)} = 0.3380 + 0.4732e^{-3.1541s} + 0.1888e^{-6.3082s} \quad (5.26)$$

which is slightly different from the filter designed assuming that only the coefficient of stiffness is uncertain. Figure 5.6 illustrates the variation of the residual energy as a function of the two uncertain parameters.

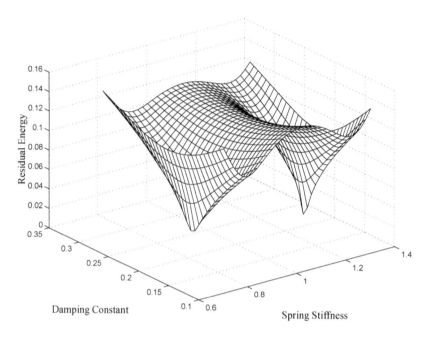

Figure 5.6: Residual Vibration Distribution

Including the constraint that the residual energy be zero at the nominal parameters of the system results in a three-time-delay filter with a transfer function:

$$\frac{u(s)}{r(s)} = 0.1961 + 0.4123e^{-3.1574s} + 0.3092e^{-6.3148s} + 0.0823e^{-9.4722s}. \qquad (5.27)$$

Figure 5.7 illustrates the variation of the residual energy as a function of the two uncertain parameters where it is clear that the maximum magnitude of the residual energy is significantly smaller than that of the two-time-delay pre-filter case.

5.1.4 Minimax Filter Design for Multi-Input Systems

The technique for the design of minimax prefilters can be extended to multi-input systems [11]. Figure 5.8 illustrates a two-mass-spring system with two inputs. The equations of motion of the system are

$$\begin{bmatrix} m_1 & 0 \\ 0 & m_2 \end{bmatrix} \begin{Bmatrix} \ddot{y}_1 \\ \ddot{y}_2 \end{Bmatrix} + \begin{bmatrix} k_1 + k_2 & -k_2 \\ -k_2 & k_2 \end{bmatrix} \begin{Bmatrix} y_1 \\ y_2 \end{Bmatrix} =$$

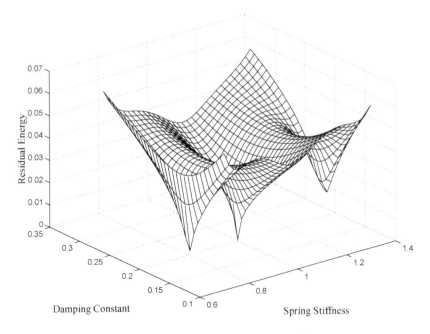

Figure 5.7: Residual Vibration Distribution

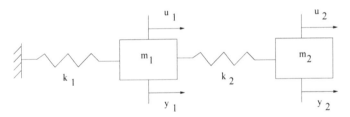

Figure 5.8: Two-Input Oscillator

$$\begin{bmatrix} 1 & 0 \\ 0 & 1 \end{bmatrix} \begin{Bmatrix} u_1 \\ u_2 \end{Bmatrix}. \tag{5.28}$$

The goal is to design a time-delay filter with the objective of completing a rest-to-rest maneuver with the boundary conditions:

$$y_1(0) = y_2(0) = \dot{y}_1(0) = \dot{y}_2(0) = 0 \tag{5.29}$$

$$y_1(t_f) = 2, y_2(t_f) = 1, \dot{y}_1(t_f) = \dot{y}_2(t_f) = 0, \tag{5.30}$$

and which it is robust to uncertainties in the spring stiffness k_1 and k_2. It is assumed that the uncertain parameters lie in the range:

$$0.7 \le k_1 \le 1.3 \text{ and } 0.7 \le k_2 \le 1.3. \tag{5.31}$$

To permit design of time-delay filters whose final values when subject to a unit step input are the same as the desired final displacement of the two outputs, the control input is rewritten as

$$\left\{ \begin{matrix} u_1 \\ u_2 \end{matrix} \right\} = \left[\begin{matrix} k_1 + k_2 & -k_2 \\ -k_2 & k_2 \end{matrix} \right]_{\text{nominal}} \left\{ \begin{matrix} v_1 \\ v_2 \end{matrix} \right\} \tag{5.32}$$

The final values of the new control inputs to satisfy the boundary conditions are now given as

$$\left\{ \begin{matrix} v_1 \\ v_2 \end{matrix} \right\} = \left\{ \begin{matrix} 2 \\ 1 \end{matrix} \right\} \tag{5.33}$$

since the final displacements of the two masses are $y_1(t_f) = 2, y_2(t_f) = 1$. Assuming the nominal values of the spring stiffness are $k_1 = 1$, $k_2 = 1$, the final values of the control inputs can be calculated to be $u_1 = 3$, $u_2 = -1$. A minimax problem is formulated to minimize the maximum magnitude of the residual energy of the system over the domain of uncertainty defined by Equation (5.31). The transfer functions of the time-delay filters are parameterized as

$$G_1(s) = \sum_{i=0}^{N} A_{1i} \exp(-sT_{1i}) \text{ where } T_{10} = 0 \tag{5.34}$$

$$G_2(s) = \sum_{i=0}^{N} A_{2i} \exp(-sT_{2i}) \text{ where } T_{20} = 0 \tag{5.35}$$

where N is the number of delays in the time-delay filter. The minimax optimization problem is used to solve for A_{1i}, T_{1i}, A_{2i}, and T_{2i}. The following constraints are imposed on the optimization problem. To ensure that the systems states are quiescent at the final time, and since the system modes cannot be decoupled, we require

$$T_{1N} = T_{2N}. \tag{5.36}$$

Further, we require

$$\left\{ \begin{matrix} \sum_{i=0}^{N} A_{1i} \\ \sum_{i=0}^{N} A_{2i} \end{matrix} \right\} = \left\{ \begin{matrix} y_1(t_f) \\ y_2(t_f) \end{matrix} \right\} \tag{5.37}$$

where t_f is the final maneuver time.

To compare the performance of the minimax time-delay filter to conventional filters, two classes of filters are considered. The first is designed to cancel the undamped poles of the system which is given by the time-delay filters

$$G_1(s) = 2(\frac{1}{2} + \frac{1}{2}\exp(-s\frac{\pi}{\omega_1}))$$

$$(\frac{1}{2} + \frac{1}{2}\exp(-s\frac{\pi}{\omega_2})) \tag{5.38}$$

$$G_2(s) = 1(\frac{1}{2} + \frac{1}{2}\exp(-s\frac{\pi}{\omega_1}))$$

$$(\frac{1}{2} + \frac{1}{2}\exp(-s\frac{\pi}{\omega_2})) \tag{5.39}$$

where ω_1 and ω_2 are the nominal frequencies of the controlled system. Next, a minimax problem is solved for the same number of delays as in Equations (5.38) and (5.39). The optimal parameters of the minimax time-delay filter are given in Table 5.1.

A_{1i}	0.5200	0.5954	0.5344	0.3503
T_{1i}	0.0	3.3688	6.3719	12.1535
A_{2i}	0.3630	−0.5139	0.7001	0.4508
T_{2i}	0	1.6853	6.8284	12.1535

Table 5.1: Minimax Time-Delay Filter Parameters

Figures 5.9 and 5.10 illustrate the variation of the residual energy of the nonrobust and the corresponding minimax prefilters. It is clear that for the nominal model, the nonrobust filter results in zero residual energy. However, over the entire uncertain region, the minimax filter outperforms the nonrobust filter.

The second is the robust time-delay filter designed by locating multiple zeros of the time-delay filter at the estimated location of the poles of the system, which is given by the transfer functions:

$$G_1(s) = 2(\frac{1}{4} + \frac{1}{2}\exp(-s\frac{\pi}{\omega_1}) + \frac{1}{4}\exp(-2s\frac{\pi}{\omega_1}))$$

$$(\frac{1}{4} + \frac{1}{2}\exp(-s\frac{\pi}{\omega_2}) + \frac{1}{4}\exp(-2s\frac{\pi}{\omega_2})) \tag{5.40}$$

$$G_2(s) = 1(\frac{1}{4} + \frac{1}{2}\exp(-s\frac{\pi}{\omega_1}) + \frac{1}{4}\exp(-2s\frac{\pi}{\omega_1}))$$

$$(\frac{1}{4} + \frac{1}{2}\exp(-s\frac{\pi}{\omega_2}) + \frac{1}{4}\exp(-2s\frac{\pi}{\omega_2})) \tag{5.41}$$

Next, a minimax time-delay filter is designed to minimize the maximum magnitude of the residual energy of the uncertain domain and the optimal parameters of the time-delay filter are given in Table 5.2.

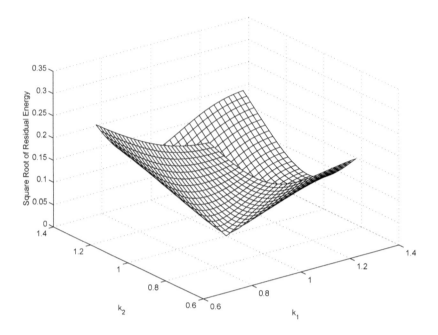

Figure 5.9: Residual Vibration Distribution (Nonrobust Filter)

From Figures 5.11 and 5.12, it can be seen that the maximum magnitude of the square root of the residual energy over the uncertain region reduces from 0.0762 to 0.0116, an 85% reduction.

The minimax approach presented in this section for the design of pre-filters for systems which are open-loop stable, or which have been stabilized using feedback controller provide a simple approach for desensitizing the performance variation of uncertain linear systems, provided the domain and distribution of uncertainty is

A_{1i}	T_{1i}	A_{2i}	T_{2i}
0.1014	0	0.1003	0
0.1568	2.2897	−0.0867	5.0907
0.3376	4.7851	0.1551	5.0907
0.2147	7.0169	0.0678	6.6232
0.2609	8.0249	0.1571	8.2987
0.2843	10.2515	−0.1312	11.6788
0.3379	12.2812	0.3658	11.8754
0.2127	14.8886	0.2104	15.1427
0.0938	18.6908	0.1613	18.6908

Table 5.2: Minimax Time-Delay Filter Parameters

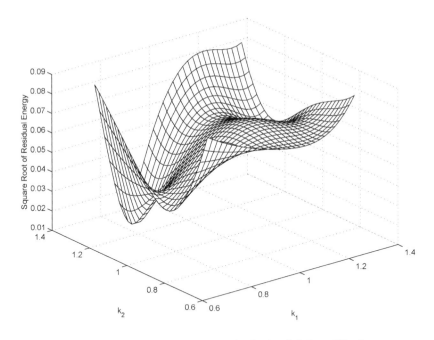

Figure 5.10: Residual Vibration Distribution (Minimax Filter)

known. The applicability of the minimax optimization algorithm for the design of full state feedback controller is an obvious question to pose. The next section will expound techniques that have been proposed by Bryson et al. [12, 13] for the design of controllers which minimize the maximum quadratic function of the systems states and control.

5.2   Minimax Feedback Controllers

We will first consider the design of feedback controller for uncertain systems where knowledge of the range and distribution of uncertainties are known. A minimax optimization problem is formulated to arrive at feedback gains to minimize the worst performance of a plant over the uncertain space. The cost function which will be minimized is the integral of a quadratic function of the states and control:

$$J = \int_0^\infty \left(x^T Q x + u^T R u \right) \, dt \tag{5.42}$$

where we assume that the initial conditions of the system are modeled as a Gaussian random vector with zero mean and a covariance of Σ

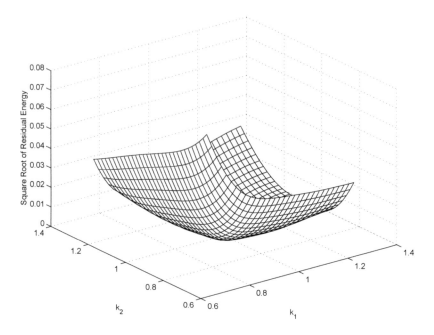

Figure 5.11: Residual Vibration Distribution (Robust Control)

$$\Sigma = E\left[x(0)x(0)^T\right].$$ (5.43)

Consider the set of control laws specified in a full state feedback form,

$$u = -Kx$$ (5.44)

the cost function can be represented as

$$J = \int_0^\infty x^T (Q + K^T RK)x \, dt$$ (5.45)

where the closed-loop dynamics are:

$$\dot{x} = Ax + Bu = (A - BK)x,$$ (5.46)

$$\Rightarrow x(t) = \exp((A - BK)t)x(0).$$ (5.47)

The resulting cost function is

$$J = \int_0^\infty x(0)^T \exp((A - BK)t)^T (Q + K^T RK) \exp((A - BK)t)x(0) \, dt$$ (5.48)

which is a function of the uncertain initial conditions as well as K. Define

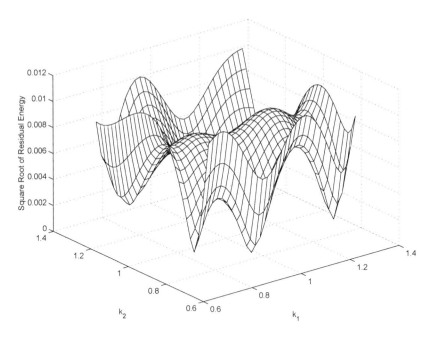

Figure 5.12: Residual Vibration Distribution (Minimax Control)

$$P = \int_0^\infty \exp((A - BK)t)^T (Q + K^T RK) \exp((A - BK)t) \, dt. \qquad (5.49)$$

which permits representing the cost function as:

$$J = x(0)^T Px(0). \qquad (5.50)$$

From the definition of derivatives, it can be shown that

$$\frac{d}{dt} \int_{t_0}^t \phi(t, \tau) \, d\tau = \phi(t, t) + \int_{t_0}^t \frac{\partial \phi}{\partial t}(t, \tau) \, d\tau. \qquad (5.51)$$

Using this result, the derivative of P with respect to time results in

$$\frac{dP}{dt} = \exp((A - BK)0)^T (Q + K^T RK) \exp((A - BK)0) +$$

$$\int_0^\infty \frac{\partial}{\partial t} \left(\exp((A - BK)t)^T \right) (Q + K^T RK) \exp((A - BK)t) \, dt + \qquad (5.52)$$

$$\int_0^\infty \exp((A - BK)t)^T (Q + K^T RK) \frac{\partial}{\partial t} \left(\exp((A - BK)t) \right) \, dt$$

which can be simplified to

$$\frac{dP}{dt} = (Q + K^T R K) +$$

$$(A - BK)^T \int_0^\infty \left(\exp((A - BK)t)^T \right) (Q + K^T R K) \exp((A - BK)t) \, dt + \quad (5.53)$$

$$\int_0^\infty \exp((A - BK)t)^T (Q + K^T R K) \left(\exp((A - BK)t) \right) \, dt \, (A - BK)$$

or

$$\frac{dP}{dt} = (Q + K^T R K) + (A - BK)^T P + P(A - BK). \quad (5.54)$$

Since, for stable systems, P tends to a constant matrix as time tends to ∞, P at steady state can be solved from Equation (5.54) by forcing $\frac{dP}{dt}$ to zero, which satisfies the Algebraic Lyapunov Equation (ALE)

$$P(A - BK) + (A - BK)^T P = -(Q + K^T R K). \quad (5.55)$$

The cost function can now be rewritten as

$$J = \int_0^\infty x^T (Q + K^T R K) x \, dt$$

$$= - \int_0^\infty x^T (P(A - BK) + (A - BK)^T P) x \, dt \quad (5.56)$$

$$= - \int_0^\infty x^T P \dot{x} + \dot{x}^T P x \, dt = - \int_0^\infty \frac{d}{dt} x^T P x \, dt$$

which can be simplified to

$$J = - \left(x^T P x \big|_{t_f} - x^T P x \big|_{t_0} \right) = x(0)^T P x(0) \quad (5.57)$$

where $x(0)$ corresponds to the random initial conditions. The cost can be rewritten as

$$J = E \left[x(0)^T P x(0) \right] = tr \left(E \left[x(0) x(0)^T P \right] \right) = tr \left(\Sigma P \right) \quad (5.58)$$

where $tr()$ is the trace operator. Thus, solving the ALE Equation (5.55), one can calculate the integral quadratic cost for any feedback gain matrix. This simple technique can be exploited to optimize the feedback gain matrix to minimize the integral cost function. A simple example is used to illustrate the determination of the integral quadratic cost as a function of an uncertain variable.

Example 5.1: Consider a unity mass, spring-mass-dashpot system

$$\ddot{y} + c\dot{y} + ky = 0 \quad (5.59)$$

where the damping factor c is a variable. The state space model of the system is

$$\begin{Bmatrix} \dot{x}_1 \\ \dot{x}_2 \end{Bmatrix} = \begin{bmatrix} 0 & 1 \\ -k & -c \end{bmatrix} \begin{Bmatrix} x_1 \\ x_2 \end{Bmatrix} \tag{5.60}$$

where $x_1 = y$. The quadratic cost function

$$J = \int_0^\infty \begin{Bmatrix} x_1 \\ x_2 \end{Bmatrix}^T \begin{bmatrix} 1 & 0 \\ 0 & 1 \end{bmatrix} \begin{Bmatrix} x_1 \\ x_2 \end{Bmatrix} dt, \tag{5.61}$$

for random initial conditions with a variance of Σ is

$$J = tr \left(\Sigma \begin{bmatrix} \frac{c^2+k^2+k}{2ck} & \frac{1}{2k} \\ \frac{1}{2k} & \frac{k+1}{2kc} \end{bmatrix} \right) \tag{5.62}$$

Figure 5.13 illustrates the variation of J as a function of c assuming that the covariance matrix Σ is the identity matrix. The solid line corresponds to the variable J. Since the system is stable only for c greater than 0, we are only interested in the graph which corresponds to $c > 0$. It can be seen that for $k = 1$, $c = 2$ is the optimal solution which corresponds to the critically damped case. However, it can be seen that the graph has a discontinuity around the origin which corresponds to the transition from the stable to the unstable region. To prevent this, a scaled version of the negative of the reciprocal of J is plotted and is shown by the dashed line. It is clear that the minimum of this function coincides with that of J, but it does not include the discontinuity around the origin which is a useful property for gradient based optimization algorithms.

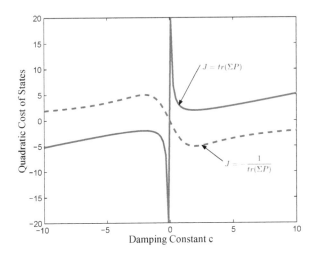

Figure 5.13: Quadratic Cost

The previous example provides the motivation for the use of the negative of the reciprocal of the integral quadratic cost to preclude pathological conditions in the use of gradient based optimizers.

The cost function which will be used for the design of minimax controllers is

$$\min_{K} \max_{p} J(K,p)$$

$$J(K,p) = -1/tr\,(\Sigma P). \tag{5.63}$$

The design problem is to determine the feedback gain which minimizes the maximum magnitude of the cost (Equation 5.63) over the entire range of uncertainty. For one or two uncertain parameters, it is reasonable to discretize the uncertain region and determine the cost function at each of the grid points. However, for systems with a large number of uncertainty, this does not result in a feasible approach for the determination of the minimax solution. Bryson and Mills [12] propose to determine the cost at the corners of the uncertain hypercube and minimize the maximum magnitude of the cost over all the vertices (Figure 5.14). The small cube at the center of the uncertain cube represents the location of the nominal plant parameter set and the vertices of the cube are functions of the lower or upper bounds of the uncertain parameters.

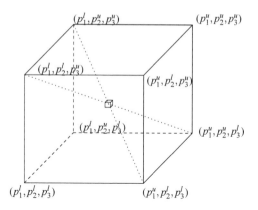

Figure 5.14: Uncertain Cube

Gradient-based optimization algorithms are used for solving the minimax problem. The cost function at any point in the uncertain hypercube is

$$J = -1/tr\,(\Sigma P) \tag{5.64}$$

where P is the solution of the Lyapunov equation

$$P(A - BK) + (A - BK)^T P = -(Q + K^T RK). \tag{5.65}$$

The derivative of Equation (5.65) with respect to the feedback gain k_i is

$$\frac{dP}{dk_i}(A - BK) + (A - BK)^T \frac{dP}{dk_i} = PB\frac{dK}{dk_i} + \frac{dK}{dk_i}^T B^T P - K^T R\frac{dK}{dk_i} - \frac{dK}{dk_i}^T RK \quad (5.66)$$

which is also a Lyapunov equation. The gradient of the cost J with respect to k_i is

$$\frac{dJ}{dk_i} = \frac{1}{(tr(\Sigma P))^2} tr\left(\Sigma \frac{dP}{dk_i}\right) \quad (5.67)$$

which can be exploited in the optimization algorithm. This technique is illustrated next, on the floating oscillator benchmark problem.

Example 5.2: Consider the uncertain floating oscillator problem shown in Figure 5.15 where the masses, spring stiffness and damping constant are all uncertain and the uncertain region is

$$0.7 \left\{ \begin{array}{c} m_1^{nom} \\ m_2^{nom} \\ k^{nom} \\ c^{nom} \end{array} \right\} \leq \left\{ \begin{array}{c} m_1 \\ m_2 \\ k \\ c \end{array} \right\} \leq 1.3 \left\{ \begin{array}{c} m_1^{nom} \\ m_2^{nom} \\ k^{nom} \\ c^{nom} \end{array} \right\} \quad (5.68)$$

where $(.)^{nom}$ refers to the nominal values of the uncertain parameters. The 4-dimensional uncertain space results in 16 vertices of the uncertain hypercube.

The equations of motion of the benchmark problem are

$$\begin{bmatrix} m_1 & 0 \\ 0 & m_2 \end{bmatrix} \left\{ \begin{array}{c} \ddot{y}_1 \\ \ddot{y}_2 \end{array} \right\} + \begin{bmatrix} c & -c \\ -c & c \end{bmatrix} \left\{ \begin{array}{c} \dot{y}_1 \\ \dot{y}_2 \end{array} \right\} + \begin{bmatrix} k & -k \\ -k & k \end{bmatrix} \left\{ \begin{array}{c} y_1 \\ y_2 \end{array} \right\} = \begin{bmatrix} 1 \\ 0 \end{bmatrix} u \quad (5.69)$$

and the associated random initial conditions have the property

$$E\left[x(0)x(0)^T\right] = \Sigma = \begin{bmatrix} 1 & 0 & 0 & 0 \\ 0 & 1 & 0 & 0 \\ 0 & 0 & 1 & 0 \\ 0 & 0 & 0 & 1 \end{bmatrix}. \quad (5.70)$$

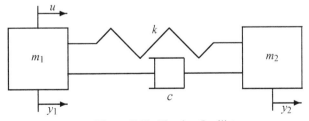

Figure 5.15: Floating Oscillator

First, an optimization problem is solved to determine the feedback gains to minimize the cost function given by Equation (5.63) for the nominal plant. The optimal gains for a nominal plant with parameters

$$\left\{ \begin{array}{c} m_1^{nom} \\ m_2^{nom} \\ k^{nom} \\ c^{nom} \end{array} \right\} = \left\{ \begin{array}{c} 1.0 \\ 1.0 \\ 1.0 \\ 0.1 \end{array} \right\} \tag{5.71}$$

are

$$K_{nom} = \begin{bmatrix} 0.2447 & 0.4597 & 0.8920 & 0.2556 \end{bmatrix} \tag{5.72}$$

and the cost $J = -0.0570$.

Next, the minimax problem is solved to minimize the maximum magnitude of the cost at the vertices of the uncertain hypercube. The resulting feedback gain is

$$K_{minimax} = \begin{bmatrix} 0.4067 & 0.2803 & 0.8498 & 0.2917 \end{bmatrix} \tag{5.73}$$

Figure 5.16 illustrates the variations of $tr(\Sigma P)$ calculated at the vertices for the gains K_{nom} and $K_{minimax}$ for the 16 corners. Circles and squares at the top of the stems correspond to K_{nom} and $K_{minimax}$ respectively. The final stem corresponds to the cost of the nominal plant. The dashed lines correspond to the gain K_{nom} and it is clear that the maximum magnitude of the cost for the gain $K_{minimax}$ had been reduced. Concurrently the cost of the nominal plant has increased, as a trade-off.

5.2.1   Exponentially Weighted LQR Cost

The Q and the R matrices of the standard LQR cost function define the relative penalty of the states and the control input. The penalty is not a function of time and all state and control variations from zero are equally penalized for all time. For applications such as the docking or the intercept problem, it is desirable to design controllers where the penalty on the state and control excursions from zero are penalized with an exponentially growing function. The following development presents controller design for an exponentially weighted quadratic-cost function.

Consider the cost which penalizes the integral-quadratic cost of states and control, exponentially with time

$$J = \int_0^\infty e^{2\lambda t} \left(x^T Q x + u^T R u \right) dt \tag{5.74}$$

where the initial conditions are again assumed to be random with a covariance of Σ. Assuming a control of the form

$$u = -Kx \tag{5.75}$$

the cost function can be represented as

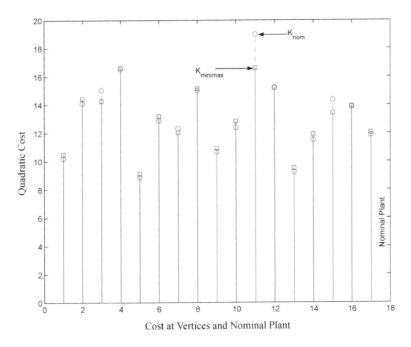

Figure 5.16: Quadratic Cost

$$J = \int_0^\infty e^{2\lambda t} x^T (Q + K^T RK) x \, dt \tag{5.76}$$

where the closed-loop dynamics is given by the equation

$$x(t) = \exp((A - BK)t) x(0) \tag{5.77}$$

resulting in the cost function

$$J = \int_0^\infty e^{2\lambda t} x(0)^T \exp((A - BK)t)^T (Q + K^T RK) \exp((A - BK)t).x(0) \, dt \tag{5.78}$$

Defining

$$P = \int_0^\infty \exp((A - BK + \lambda I)t)^T (Q + K^T RK) \exp((A - BK + \lambda I)t) \, dt, \tag{5.79}$$

it can be easily shown that the steady state value of P satisfies the Lyapunov equation

$$P(A - BK + \lambda I) + (A - BK + \lambda I)^T P = -(Q + K^T RK). \tag{5.80}$$

and the cost function can be written in terms of the initial conditions as

$$J = x(0)^T P x(0) \tag{5.81}$$

where $x(0)$ corresponds to the random initial conditions. The cost can be rewritten as

$$J = E\left[x(0)^T P x(0)\right] = tr\left(E\left[x(0)x(0)^T P\right]\right) = tr\left(\Sigma P\right) \tag{5.82}$$

Since, J has a discontinuity for parameters corresponding to the transition from the stable to the unstable region, the cost function used for the minimax optimization problem is

$$J = -1/tr\left(\Sigma P\right). \tag{5.83}$$

The gradient of J with respect to the controller parameters requires the derivative of the Lyapunov Equation (5.80),

$$\frac{dP}{dk_i}(A - BK + \lambda I) + (A - BK + \lambda I)^T \frac{dP}{dk_i} = PB\frac{dK}{dk_i} + \frac{dK}{dk_i}^T B^T P - K^T R \frac{dK}{dk_i} - \frac{dK}{dk_i}^T RK \tag{5.84}$$

which is also a Lyapunov equation. The gradient of the cost J with respect to k_i is then

$$\frac{dJ}{dk_i} = \frac{1}{(tr\left(\Sigma P\right))^2} tr\left(\Sigma\frac{dP}{dk_i}\right) \tag{5.85}$$

Example 5.3:

The exponentially weighted minimax control design is illustrated on the floating oscillator benchmark problem with the same uncertainty as in Example 5.2. The coefficient of exponential penalty λ is selected to be 0.1. First, the exponentially weighted cost function is used to solve for the feedback gains which is optimal for the nominal plant. The resulting feedback gains are

$$K_{nom} = \begin{bmatrix} 0.4816 & 0.2644 & 1.2703 & 0.2959 \end{bmatrix} \tag{5.86}$$

and the cost $J = -0.0168$.

Minimizing the maximum cost over the four-dimensional uncertain hypercube results in the feedback gains

$$K_{minimax} = \begin{bmatrix} 0.7874 & -0.1061 & 1.7298 & 0.3586 \end{bmatrix} \tag{5.87}$$

and the cost $J = -0.0128$ at the nominal plant. Figure 5.17 illustrates the variations of $tr(\Sigma P)$ calculated at the vertices for the gains K_{nom} and $K_{minimax}$ for the 16 corners. Circles and squares at the top of the stems correspond to K_{nom} and $K_{minimax}$ respectively. The final stem corresponds to the cost of the nominal plant. The dashed lines correspond to the gain K_{nom} and it is clear that the maximum magnitude of the

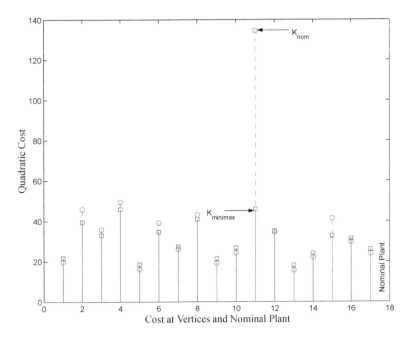

Figure 5.17: Quadratic Cost

cost for the gain $K_{minimax}$ had been reduced. Concurrently the cost of the nominal plant has increased, as a trade-off.

Figure 5.18 illustrates the variation of the reciprocal of the integral-quadratic function of the states and control input as a function of % variation of the model parameters from the nominal parameter set. A σ of zero corresponds to the cost function for the nominal plant and for any σ greater than 0, the largest cost evaluated over the set of plants corresponding to the vertices of the uncertain hypercube is plotted. When the graph goes to zero, that corresponds to the transition from a stable to an unstable region. These graphs are generated for the standard LQR problem, the proposed cost function evaluated for the nominal plant and a minimax problem solved for a σ of 0.3. Two other curves are generated where the integral-quadratic cost is penalized by an exponentially increasing function of time for a λ equal to 0.1.

The larger the cost for σ equal to 0, the better the integral-quadratic cost. The dashed line corresponds to the LQR problem, and the solid lines corresponds to the solution of a optimization problem using the cost $J = -1/tr(\Sigma P)$. Here the maximum magnitude of $J = -1/tr(\Sigma P)$ is minimized with a σ of 0.3. It can be seen that the solution of the minimax optimization results in increased gain margin compared to the curve corresponding the solution based on the nominal plant only. It is also clear that the performance of the nominal plant deteriorates. This is an expected trade-off. It can also be seen that as the exponential penalty increases, the gain-margin increases with a corresponding decrease in the performance for the

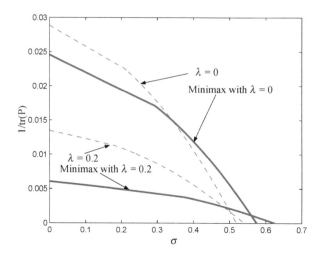

Figure 5.18: % Variation of Model Parameters

nominal plant. Thus, the effect of including an exponential weight has the same effect as designing a minimax controller with a larger uncertain hypercube.

5.2.2   Minimax Output Feedback Controller

This section considers the problem of design of an output feedback controller to minimize the integral-quadratic cost

$$J = \int_0^\infty \left(y^T Q y + u^T R u \right) \, dt \tag{5.88}$$

for the linear system

$$\dot{x}_p = A_p x_p + B_p u \tag{5.89}$$
$$y = C_p x_p \tag{5.90}$$

where the subscript p refers to the plant. x_p represents a $\mathscr{R}^n$ vector, u is a $\mathscr{R}^m$ dimensional input vector and y is a $\mathscr{R}^l$ dimensional output vector. Parameterizing the controller via a controllable canonical state-space model

$$\dot{x}_c = A_c x_c + B_c y \tag{5.91}$$
$$u = C_c x_c \tag{5.92}$$

where the input to the controller is the output of the plant and the output of the controller is the input to the plant. x_c represents a $\mathscr{R}^q$ vector and the matrix A_c is parameterized by q variables. The controller can be parameterized by $q + qm + ml$ variables. Equations (5.89) through (5.92) can be combined resulting in the augmented state-space model

$$\begin{Bmatrix} \dot{x}_p \\ \dot{x}_c \end{Bmatrix} = \begin{bmatrix} A_p & B_p C_c \\ B_c C_p & A_c \end{bmatrix} \begin{Bmatrix} x_p \\ x_c \end{Bmatrix} \tag{5.93}$$

and the cost function can be rewritten as

$$J = \int_0^\infty \begin{Bmatrix} x_p \\ x_c \end{Bmatrix}^T \begin{bmatrix} C_p^T Q C_p & 0 \\ 0 & C_c^T R C_c \end{bmatrix} \begin{Bmatrix} x_p \\ x_c \end{Bmatrix} dt \tag{5.94}$$

It can be easily shown that the cost J can be reduced to the form

$$J = x(0)^T P x(0) \tag{5.95}$$

where $x(0)$ corresponds to the initial states of the augmented system, i.e., $x(0)^T = \{x_p(0)\ x_c(0)\}$ and P is the solution of the Lyapunov equation

$$\begin{bmatrix} A_p & B_p C_c \\ B_c C_p & A_c \end{bmatrix}^T P + P \begin{bmatrix} A_p & B_p C_c \\ B_c C_p & A_c \end{bmatrix} = - \begin{bmatrix} C_p^T Q C_p & 0 \\ 0 & C_c^T R C_c \end{bmatrix}. \tag{5.96}$$

Example 5.4:   Consider the benchmark floating oscillator problem shown in Figure 5.19.

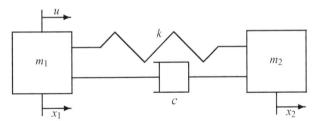

Figure 5.19: Floating Oscillator

The state-space model of the system is given as

$$\begin{Bmatrix} \dot{x}_1 \\ \dot{x}_2 \\ \ddot{x}_1 \\ \ddot{x}_2 \end{Bmatrix} = \begin{bmatrix} 0 & 0 & 1 & 0 \\ 0 & 0 & 0 & 1 \\ -\frac{k}{m_1} & \frac{k}{m_1} & -\frac{c}{m_1} & \frac{c}{m_1} \\ \frac{k}{m_2} & -\frac{k}{m_2} & \frac{c}{m_2} & -\frac{c}{m_2} \end{bmatrix} \begin{Bmatrix} x_1 \\ x_2 \\ \dot{x}_1 \\ \dot{x}_2 \end{Bmatrix} + \begin{Bmatrix} 0 \\ 0 \\ \frac{1}{m_1} \\ 0 \end{Bmatrix} u \tag{5.97}$$

$$y = \{1\ 0\ 0\ 0\} \begin{Bmatrix} x_1 \\ x_2 \\ \dot{x}_1 \\ \dot{x}_2 \end{Bmatrix} \tag{5.98}$$

where the uncertain parameters lie in the range

$$0.7 \begin{Bmatrix} m_1^{nom} \\ m_2^{nom} \\ k^{nom} \\ c^{nom} \end{Bmatrix} \leq \begin{Bmatrix} m_1 \\ m_2 \\ k \\ c \end{Bmatrix} \leq 1.3 \begin{Bmatrix} m_1^{nom} \\ m_2^{nom} \\ k^{nom} \\ c^{nom} \end{Bmatrix} \tag{5.99}$$

where $(.)^{nom}$ refers to the nominal values of the uncertain parameters.

Parameterizing the controller in state-space form as

$$\begin{Bmatrix} \dot{w}_1 \\ \dot{w}_2 \\ \dot{w}_3 \\ \dot{w}_4 \end{Bmatrix} = \begin{bmatrix} 0 & 1 & 0 & 0 \\ 0 & 0 & 1 & 0 \\ 0 & 0 & 0 & 1 \\ a_1 & a_2 & a_3 & a_4 \end{bmatrix} \begin{Bmatrix} w_1 \\ w_2 \\ w_3 \\ w_4 \end{Bmatrix} + \begin{Bmatrix} 0 \\ 0 \\ 0 \\ 1 \end{Bmatrix} y \tag{5.100}$$

$$u = \{c_1 \; c_2 \; c_3 \; c_4\} \begin{Bmatrix} w_1 \\ w_2 \\ w_3 \\ w_4 \end{Bmatrix} \tag{5.101}$$

where eight variables of the controller need to be determined so as to minimize the integral quadratic cost function.

First, an optimization problem is solved to determine the variables of the state-space controller to minimize the cost function

$$J = \int_0^\infty (y_1^2 + u^2) \; dt \tag{5.102}$$

for the nominal plant

$$\begin{Bmatrix} m_1^{nom} \\ m_2^{nom} \\ k^{nom} \\ c^{nom} \end{Bmatrix} = \begin{Bmatrix} 1 \\ 1 \\ 1 \\ 0.1 \end{Bmatrix}. \tag{5.103}$$

The optimal controller is given by the state-space equations

$$\begin{Bmatrix} \dot{w}_1 \\ \dot{w}_2 \\ \dot{w}_3 \\ \dot{w}_4 \end{Bmatrix} = \begin{bmatrix} 0 & 1 & 0 & 0 \\ 0 & 0 & 1 & 0 \\ 0 & 0 & 0 & 1 \\ -0.1389 & -0.6571 & -1.5979 & -2.7144 \end{bmatrix} \begin{Bmatrix} w_1 \\ w_2 \\ w_3 \\ w_4 \end{Bmatrix} + \begin{Bmatrix} 0 \\ 0 \\ 0 \\ 1 \end{Bmatrix} y \tag{5.104}$$

$$u = \{-0.0275 \; -0.2504 \; -0.4906 \; -1.3703\} \begin{Bmatrix} w_1 \\ w_2 \\ w_3 \\ w_4 \end{Bmatrix} \tag{5.105}$$

and the cost $J = 23.7517$. The controller written in a transfer function form is

$$\frac{u(s)}{y(s)} = \frac{-1.3703s^3 - 0.4906s^2 - 0.2504s - 0.0275}{s^4 + 2.7144s^3 + 1.5979s^2 + 0.6571s + 0.1389}. \tag{5.106}$$

A full state feedback controller with the same cost function results in the state feedback gain

$$K_{nom} = \begin{bmatrix} 0.6391 & 0.3609 & 1.0962 & 0.7143 \end{bmatrix} \tag{5.107}$$

and the cost $J = 3.9873$. Figure 5.20 illustrates the robustness of the state feedback and output feedback controllers optimized for the same cost function. It is clear that the gain margin of the state feedback controller is significantly greater than the output feedback controller. Furthermore, the performance of the state feedback controller for the nominal plant is better than that of the output feedback controller. This is due to the fact that the state at time t contains complete information of how the future will evolve when given the input, while the output does not.

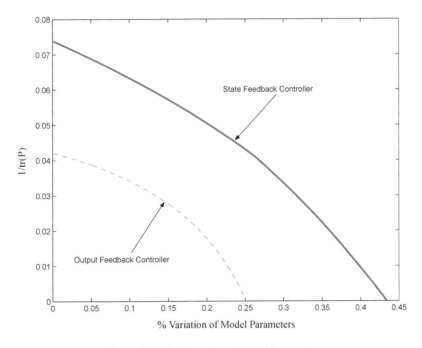

Figure 5.20: % Variation of Model Parameters

Next, the minimax problem which corresponds to the worst plant design, is completed. The state-space model of the controller is

$$
\begin{Bmatrix} \dot{w}_1 \\ \dot{w}_2 \\ \dot{w}_3 \\ \dot{w}_4 \end{Bmatrix} = \begin{bmatrix} 0 & 1 & 0 & 0 \\ 0 & 0 & 1 & 0 \\ 0 & 0 & 0 & 1 \\ -0.4678 & -1.3583 & -2.6722 & -4.6049 \end{bmatrix} \begin{Bmatrix} w_1 \\ w_2 \\ w_3 \\ w_4 \end{Bmatrix} + \begin{Bmatrix} 0 \\ 0 \\ 0 \\ 1 \end{Bmatrix} y \qquad (5.108)
$$

$$
u = \{-0.1004 \;\; -0.6255 \;\; -0.7456 \;\; -2.2246\} \begin{Bmatrix} w_1 \\ w_2 \\ w_3 \\ w_4 \end{Bmatrix} \qquad (5.109)
$$

and the cost for the nominal plant is $J = 27.5617$ which is slightly greater than that for the controller designed specifically for the nominal plant. The cost for the worst plant is $J = 36.8985$.

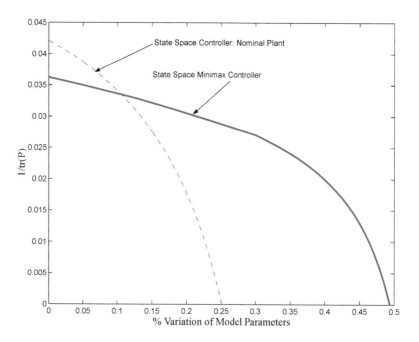

Figure 5.21: % Variation of Model Parameters

Figure 5.21 illustrates the variation of the worst plant cost as a function of the % variation of the model uncertainty, for the controller designed for the nominal plant and the worst plant. As is expected there is a trade-off of performance of the nominal plant for robustness. Figure 5.22 presents the cost of the vertices of an uncertain hypercube with a σ of 0.3. It is clear that the controller designed for the nominal plant whose cost is represented by circles results in instability for the plant corresponding to vertex 2, while the minimax controller results in reasonable performance over the entire domain of uncertainty.

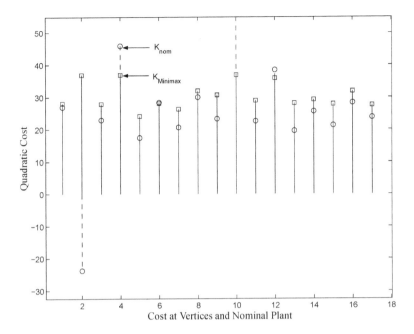

Figure 5.22: % Variation of Model Parameters

5.3 Summary

High-fidelity representation of dynamical models ensures experimental reproduction of the performance expected based on numerical simulations. However, experience informs us that model truncation, errors in estimated model parameters and variation of model parameters as a function of the environment preclude an accurate representation of the system model. In such scenarios one would like to ensure some baseline performance of the controlled system over the domain of model errors. This problem can be posed as a "worst case design," i.e, the system models which generates the worst performance are the ones which most influence the controller design. Under perfect nominal conditions, this minimax pre-filter or feedback controller cannot reproduce the performance of a controller designed for the nominal model only.

It can guarantee that no other controller will have a better performance over the entire uncertain domain. This chapter dealt with minimax design of time-delay filters and full state feedback controllers which minimize a quadratic cost. The trade-off between performance and robustness is illustrated for both the pre-filters and feedback controllers. The technique presented in this chapter can easily be extended to discrete-time systems.

Exercises

5.1   For a gantry crane, which is driven by a proportional-derivative controller, design a time-delay filter which minimizes the maximum magnitude of residual vibration for a rest-to-rest maneuver. The model of the crane is:

$$(m_1 + m_2)\ddot{\theta} + m_2 L\ddot{\phi} = u$$
$$m_2\ddot{\theta} + m_2 L\ddot{\phi} + m_2 g\phi = 0$$

m_1	m_2	L	g
1000 kg	8000 kg	1–25 m	9.81 m/s^2

Table 5.3: Gantry Crane Parameters

The initial and terminal conditions are:

$$\theta(0) = 0, \dot{\theta}(0) = 0, \phi(0) = 0, \dot{\phi}(0) = 0, \ \theta(t_f) = 1, \dot{\theta}(t_f) = 0, \phi(t_f) = 0, \dot{\phi}(t_f) = 0$$

where the length of the cable lies in the range:

$$10 < L < 20$$

The feedback controller for the system is:

$$u = -5000(\theta - \theta(t_f)) - 2000\dot{\theta}$$

Assume that the time-delay prefilter is parameterized as:

$$G(s) = A_0 + A_1 e^{-sT_1} + A_2 e^{-sT_2} + A_2 e^{-sT_3} + A_4 e^{-15s}$$

which requires the maneuver time to be 15 seconds, and

$$A_0 + A_1 + A_2 + A_3 + A_4 = 1.$$

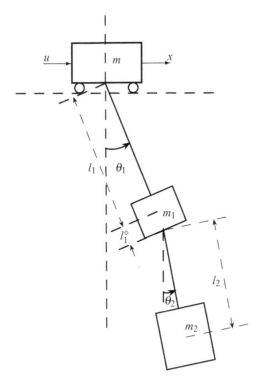

Figure 5.23: Double Pendulum Crane

5.2 Design a minimax jerk-limited time-optimal reference profile for a rest-to-rest maneuver of a spring-mass system given by the equation of motion:

$$\ddot{x} + \omega^2 x = \omega^2 u$$

where

$$0.7 \leq \omega \leq 1.3$$

and

$$0 \leq \dot{u} \leq 0.2$$

The initial and final states are

$$x(0) = \dot{x}(0) = 0, \quad x(t_f) = 2, \dot{x}(t_f) = 0.$$

5.3 Design a robust LQR controller for a double pendulum shown in Figure 5.23. The system model is

$$m_{11}\ddot{x} + m_{12}\ddot{\theta}_1 + m_{13}\ddot{\theta}_2 + c\dot{x} = u \tag{5.110}$$

$$m_{21}\ddot{x} + m_{22}\ddot{\theta}_1 + m_{23}\ddot{\theta}_2 + k_{12}\theta_1 = 0 \tag{5.111}$$

$$m_{31}\ddot{x} + m_{32}\ddot{\theta}_1 + m_{33}\ddot{\theta}_2 + k_{34}\theta_2 = 0 \tag{5.112}$$

where

$$
\begin{aligned}
m_{11} &= m + m_1 + m_2 & m_{12} &= m_1 l_1 + m_2 L^* & m_{13} &= m_2 l_2 \\
m_{21} &= m_1 l_1 + m_2 L^* & m_{22} &= m_1 l_1^2 + m_2 L^* L^* + I_1 & m_{23} &= m_2 l_2 L^* \\
m_{31} &= m_2 l_2 & m_{32} &= m_2 l_2 (l_1 + l_1^\circ) & m_{33} &= m_2 l_2^2 + I_2 \\
k_{33} &= m_2 g l_2 & k_{22} &= m_1 g l_1 + m_2 g L^* & L^* &= (l_1 + l_1^\circ)
\end{aligned}
$$

The nominal parameters are:

$$
\begin{array}{lll}
m_1 = 0.5 & m_2 = 1 & m_3 = 8 \\
l_1 = 1.5 & l_1^\circ = 0.1 & l_2 = 0.3 \\
I_1 = 0.0008 & I_2 = 0.06 &
\end{array}
$$

Design a robust minimax controller which minimizes the quadratic cost function where all the states are equally weighted, i.e., the Q matrix is identity and the control weight term is 0.1. The uncertain parameter are:

$$0.5 \leq m_2 \leq 1.5 \tag{5.113}$$

$$4 \leq m_3 \leq 12 \tag{5.114}$$

$$0.75 \leq l_1 \leq 2.25 \tag{5.115}$$

$$0.15 \leq l_2 \leq 0.45 \tag{5.116}$$

which correspond to a $\pm 50\%$ variation about the nominal value of the parameters. Plot the gain margin for the LQR and the minimax controllers.

5.4 Design a minimax time-delay filter for the flexible transmission system shown in Figure 5.24 for a rest-to-rest maneuver. The maximum magnitude of the cost function is minimized over the range of the load J_3 [14]. The transfer function of the system is

$$
\frac{\phi(s)}{\tau(s)} = \frac{(2Kr^2)^2}{(J_3 s^2 + fs + 2Kr^2)(J_2 s^2 + fs + 4Kr^2) - (2Kr^2)^2}
$$

where

$$J_2 = 8.5e-3, K_1 = K_2 = K = 170, f = 7.25e-3, r = 0.1$$

and the load parameter lies in the range

$$0.0424 - 0.017 \leq J_3 \leq 0.0424 + 0.017$$

The initial and desired final states are:

$$\phi(0) = 0, \dot{\phi}(0) = 0, \phi(t_f) = 1 \text{ rad}, \dot{\phi}(t_f) = 0$$

Note that the input to the system is the position of the first pulley.

5.5 Design a minimax controller for the pitch control of an aircraft (Figure 5.25). The transfer function relating the pitch angle θ to the elevator deflection δ_e is: (Source: www.engin.umich.edu/group/ctm/examples/pitch/Mpitch.html.)

$$
\frac{\theta(s)}{\delta_e(s)} = \frac{1.151s + 0.1774}{s^3 + 0.739s^2 + 0.921s}
$$

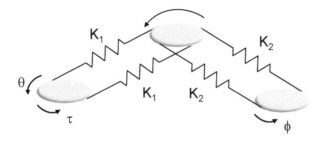

Figure 5.24: Flexible Transmission Problem

The cost function to be minimized is:

$$J = \int_0^\infty \left(\theta(t)^2 + \delta_e(t)^2 \right) dt$$

where all the coefficients of the transfer function lie in the range $0.5 p_i^n \leq p_i^n \leq 1.5 p_i^n$, where p_i^n are the nominal values of the ith coefficients.

$$p_i^n = \begin{bmatrix} 1.151 & 0.1774 & 0.739 & 0.921 \end{bmatrix}$$

Design an output feedback controller of the form:

$$\dot{x}_c = A_c x_c + B_c \theta \tag{5.117}$$
$$\delta_e = C_c x_c + D_c \theta \tag{5.118}$$

to minimize the maximum magnitude of J evaluated at the vertices of the uncertain hypercube. x_c is a third-order vector which represents the control states.

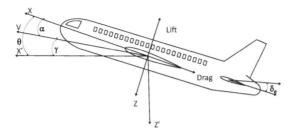

Figure 5.25: Airplane Pitch Dynamics Variables

References

[1] T. Singh and S. R. Vadali. Robust time-optimal control: Frequency domain approach. *J. of Guid., Cont. and Dyn.*, 17(2):346–353, 1994.

[2] Q. Liu and B. Wie. Robust time-optimal control of uncertain flexible space-craft. *J. of Guid., Cont. and Dyn.*, 15(3):597–604, 1992.

[3] T. Singh. Effect of damping on the structure of time-optimal controllers. *J. of Guid., Cont. and Dyn.*, 19(5):1225–1231, 1995.

[4] W. Singhose, T. Singh, and W. P. Seering. On-off control with specified fuel usage. *ASME J. of Dyn. Systems, Measurements, and Cont.*, 121(2):206–212, 1999.

[5] S.W. Liu and T. Singh. Robust time-optimal control of nonlinear structures with parameter uncertainties. *ASME Journal of Dynamic Systems, Measurement and Control*, 119(4):743–748, December 1997.

[6] T. Singh. Minimax design of robust controllers for flexible systems. *J. of Guid., Cont. and Dyn.*, 25(5):868–875, 2002.

[7] C. F. van Loan. Computing integrals involving the matrix exponential. *IEEE Trans. on Auto. Cont.*, (3):395–404, 1978.

[8] O. J. M. Smith. Posicast control of damped oscillatory systems. 45:1249–1255, September 1957.

[9] T. Singh and S. R. Vadali. Robust time delay control. *ASME J. of Dyn. Systems, Measurements, and Cont.*, 115(2):303–306, 1993.

[10] W. Singhose, S. Derezinski, and N. Singer. Extra-insensitive input shapers for controlling flexible spacecraft. *J. of Guid., Cont. and Dyn.*, 19(2):385–391, 1996.

[11] T. Singh and Y-L Kuo. Minimax design of prefilters for maneuvering flexible structures. In *AIAA Guidance, Navigation and Control Conference*, Monterey, California, 2002.

[12] A. E. Bryson and R. A. Mills. Linear-quadratic-Gaussian controllers with specified parameter robustness. *J. of Guid., Cont. and Dyn.*, 21(1):11–18, February 1998.

[13] L. El Ghaoui, A. Carrier, and A. E. Bryson. Linear quadratic minimax controllers. *IEEE Trans. on Cont. Systems Tech.*, 15(4):953–961, 1992.

[14] R. Ramirifar and N. Sadati. h_∞ control of a flexible transmission system. In *Proceedings of the Thirty-Sixth Southeastern Symposium on System Theory*, pages 536–540, Atlanta, GA, 2004.

6

Friction Control

In science the credit goes to the man who convinces the world, not to the man to whom the idea first occurred.

Sir William Osler (1849–1919); Canadian physician

FRICTION is a nonlinearity which is ubiquitous. In applications where very high precision is desired and where the system dynamics are characterized by flexible modes, the presence of friction poses a challenge to the design of controllers for rest-to-rest maneuvers. For systems where the coefficients of friction and the damping and natural frequencies of the flexible modes are uncertain, there is a need to design desensitized controllers.

Numerous applications such as hard disk drives, flexible arm robots, high-precision machine tools, servo-motors, and so forth, are applications where the presence of friction is a significant factor in the deterioration of the performance of these devices. Thus, there is a motivation to study controllers which can account for the presence of friction and track a desired output accurately. This chapter presents techniques for the design of controllers for systems undergoing rest-to-rest maneuvers in the presence of friction. First, the time-optimal control problem will be addressed. This will be followed by a detailed development of pulse-width pulse-amplitude modulated control techniques for rigid body system. The chapter will conclude with the extension of the pulse-width pulse-amplitude technique to the motion control of the flexible benchmark problem.

6.1 Time-Optimal Rest-to-Rest Maneuvers

Maneuvering structures from quiescent initial conditions to quiescent final conditions in minimum time have applications in hard disk drives [1, 2], digital versatile discs [3], industrial robots [4, 5], and container cranes [6], which are used to load and unload ships, and so forth. All these applications have joints that are subject to friction which significantly influences the positioning accuracy of controllers. This section will consider the simplest model of friction, Coulomb friction, and develop minimum-time control profiles for rest-to-rest maneuvers. As in earlier chapters, a rigid body is used as the first example to illustrate the design technique. This is followed by the development of the optimal controller for flexible structures.

6.1.1   Rigid Body

The rigid-body system subject to Coulomb friction is illustrated in Figure 6.1. The

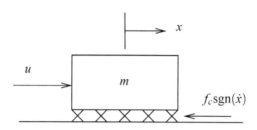

Figure 6.1: Rigid-Body System Subject to Coulomb Friction

time-optimal control problem of the rigid body system can be written as:

$$\min \quad \int_0^{T_f} dt$$

subject to

$$\dot{\underline{x}}(t) = A\underline{x}(t) + B\big(u(t) - f_c\mathrm{sgn}(\dot{x}(t))\big), \tag{6.1}$$

$$\underline{x}(0) = [x_1(0)\, x_2(0)]^T, \quad \underline{x}(T_f) = [x_1(T_f)\, x_2(T_f)]^T,$$
$$-U \le u(t) \le U,$$

where A and B are defined as:

$$A = \begin{bmatrix} 0 & 1 \\ 0 & 0 \end{bmatrix}, \quad B = \begin{bmatrix} 0 \\ \frac{1}{m} \end{bmatrix} \tag{6.2}$$

Formulating a Hamiltonian as:

$$\mathbf{H}(\underline{x},\underline{\lambda},u,t) = 1 + \underline{\lambda}^T\big(A\underline{x}(t) + B(u(t) - f_c\mathrm{sgn}(\dot{x}))\big), \tag{6.3}$$

the necessary conditions for optimality are given as

$$\dot{\underline{x}}(t) = \frac{\partial \mathbf{H}}{\partial \underline{\lambda}(t)} \tag{6.4}$$

$$\dot{\underline{\lambda}}(t) = -\frac{\partial \mathbf{H}}{\partial \underline{x}(t)} \tag{6.5}$$

From Pontryagin's principle, the optimal control input is the one that minimizes the Hamiltonian, which is

$$u = -U\text{sgn}\left(B^T \lambda(t)\right). \tag{6.6}$$

The resulting control profile becomes bang-bang, where the input switch is determined by the switching function, $B^T \lambda$.

6.1.1.1 Parameterization of Linearizing Input

In Equation (6.1), the equation of motion is nonlinear because of the signum function in the Coulomb friction model. To eliminate the nonlinearity attributed to the signum function, a simple technique is presented in this section which permits parameterization of the new input to the system which includes the external input and the friction force. This *net* input now acts on a linear system, permitting the use of all the tools available for the design of controllers for linear systems.

Assume that the time-optimal control profile for the problem in Equation (6.1) is a one-switch bang-bang profile. By defining a linearizing net input such that [7]

$$u_{net} = u - f_c\text{sgn}(\dot{x}), \tag{6.7}$$

the system equation of motion becomes linear. However, the new linearizing net input is not a bang-bang profile anymore. Figure 6.2 illustrates how the linearizing net input can be determined. First, the velocity profile of the optimal control profile is assumed as shown in the top plot of Figure 6.2. T_1 and T_3 correspond to the time when the velocity of the frictional body becomes zero, and T_2 is the time when the actual control input switches. The friction force can be determined from this velocity profile which is shown in the third plot of Figure 6.2. Then the linearizing net input can be determined as the sum of actual input and friction force as shown in the last plot of Figure 6.2. Now the linearizing net input can be parameterized by the unknown velocity and control switches, and final time $[T_1 \; T_2 \; T_3 \; T_f]$. Without losing generality, for a rigid body system, we can assume that the final position of the mass is always greater than the initial position of the mass, i.e., $x_1(T_f) > x_1(0)$. Then the spectrum of velocity profiles of the general time-optimal problem consist of four cases shown in Figure 6.3, depending on the initial and final states of the system. Based on these cases, the linearizing net control input for the corresponding velocity profile can be written as:

$$u_{net(i)} = \begin{cases} U - f_c & 0 \leq t < T_1 \\ -U - fc & T_1 \leq t < T_f \\ 0 & T_f \leq t \end{cases} \qquad u_{net(ii)} = \begin{cases} U + f_c & 0 \leq t < T_1 \\ U - f_c & T_1 \leq t < T_2 \\ -U - f_c & T_2 \leq t < T_f \\ 0 & T_f \leq t \end{cases} \tag{6.8}$$

$$u_{net(iii)} = \begin{cases} U - f_c & 0 \leq t < T_1 \\ -U - f_c & T_1 \leq t < T_2 \\ -U + f_c & T_2 \leq t < T_f \\ 0 & T_f \leq t \end{cases} \qquad u_{net(iv)} = \begin{cases} U + f_c & 0 \leq t < T_1 \\ U - f_c & T_1 \leq t < T_2 \\ -U - f_c & T_2 \leq t < t_3 \\ -U + f_c & T_3 \leq t < T_f \\ 0 & T_f \leq t \end{cases}$$

$$\tag{6.9}$$

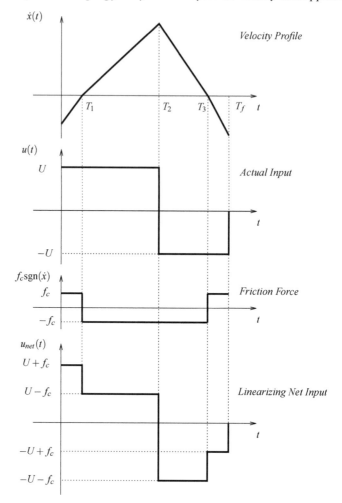

Figure 6.2: Linearizing Net Input Parameterization

Since the new control input is parameterized in terms of the unknown switch and final times, the system equation of motion can be integrated forward in time with the given initial states. With the first linearized net input $u_{net(i)}$ (Case (i)), states at the switch and final time become

$$\underline{x}(T_1) = e^{AT_1}\underline{x}_0 + \int_0^{T_1} (U - f_c)e^{A(t-\tau)}Bd\tau$$

$$\underline{x}(T_f) = e^{A(T_f-T_1)}\underline{x}(T_1) + \int_{T_1}^{T_f} (-U - f_c)e^{A(t-\tau)}Bd\tau. \tag{6.10}$$

Solving for the matrix exponential and integration analytically and exploiting the constraint $\underline{x}(T_f) = [x_1(T_f) \quad x_2(T_f)]^T$, the switch time and final time can be solved,

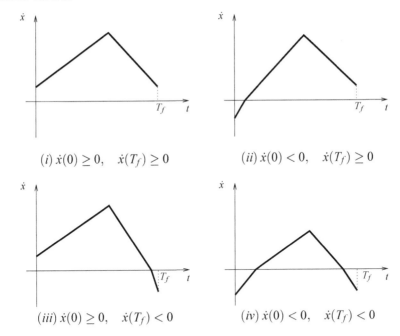

$(i) \ \dot{x}(0) \geq 0, \quad \dot{x}(T_f) \geq 0$

$(ii) \ \dot{x}(0) < 0, \quad \dot{x}(T_f) \geq 0$

$(iii) \ \dot{x}(0) \geq 0, \quad \dot{x}(T_f) < 0$

$(iv) \ \dot{x}(0) < 0, \quad \dot{x}(T_f) < 0$

Figure 6.3: Four Different Velocity Profiles of a Rigid Body

resulting in the following closed-form equations:

$$T_1 = \frac{(U + f_c)T_f + m(x_2(T_f) - x_2(0))}{2U}$$

$$T_f = -\frac{m(U + f_c)x_2(0) + m(U - f_c)x_2(T_f)}{U^2 - f_c^2}$$
$$+ \frac{\sqrt{2Um^2(U + f_c)x_2(0)^2 + m(U - f_c)x_2(T_f)^2 + 2(U^2 - f_c^2)(x_1(T_f) - x_1(0))}}{U^2 - f_c^2}.$$
$$(6.11)$$

The actual control input is found by subtracting the friction force from the resulting linearizing control input. Similarly, the states at the switch and final times with the input $u_{net(ii)}$ (Case (ii)), become

$$\underline{x}(T_1) = e^{AT_1}\underline{x}_0 + \int_0^{T_1}(U + f_c)e^{A(t-\tau)}Bd\tau$$
$$\underline{x}(T_2) = e^{A(T_2-T_1)}\underline{x}(T_1) + \int_{T_1}^{T_2}(U + f_c)e^{A(t-\tau)}Bd\tau \qquad (6.12)$$
$$\underline{x}(T_f) = e^{A(t_3-T_2)}\underline{x}(T_2) + \int_{T_2}^{t_3}-(U + f_c)e^{A(t-\tau)}Bd\tau.$$

In this net control profile, T_1 is the velocity switch and T_2 is the actual control input switch. Resulting states should satisfy the boundary state conditions at the final time as well as the zero velocity constraint at T_1 such that

$$\underline{x}(T_f) = [x_1(T_f) \quad x_2(T_f)]^T, \quad x_2(T_1) = 0. \tag{6.13}$$

The three unknown switch and final times, $[T_1 \ T_2 \ T_f]$ can now be solved from Equation (6.13). It is also possible to formulate a problem using time-delay filters. The linearizing net input, $u_{net(ii)}$, can be written in the form:

$$G(s) = (U + f_c) - 2f_c e^{-sT_1} - 2Ue^{-sT_2} + (U + f_c)e^{-sT_f}. \tag{6.14}$$

Then, zeros of $G(s)$ are designed to cancel the poles of the rigid body while satisfying the displacement constraint. Similar procedures are used to find the linearizing control inputs, $u_{net(iii)}$ and $u_{net(iv)}$.

6.1.1.2   Optimality Condition

For a rigid body with Coulomb friction, the control profile is a one-switch bang-bang with the switching curve $B^T \lambda$. To show that the resulting control profile is optimal, the switching curve is computed to show that it passes through zero at the actual switch time.

Assume that the control input has a switch at ℓ time instants $(T_i, i = 1, 2, \ldots, \ell)$ and the velocity of the rigid body changes its sign m times at the k_j^{th} time instant T_{k_j} $(j = 1 \ldots m)$. For example, in Figure 6.2, T_2 is the actual control switch and T_1 and T_3 are the velocity switches. Therefore, $\ell = 2$, $k_1 = 1$, $k_2 = 3$, and $m = 2$. In the general case, the optimal control problem in Equation (6.1) can be rewritten with the linearizing net input such that

$$\min \int_0^{T_f} dt$$

subject to

$$\dot{\underline{x}} = A\underline{x} + B(u + (-1)^j f_c) \quad T_{j-1} \leq t < T_j \tag{6.15}$$
$$x_2(T_{k_j}) = 0$$
$$\underline{x}(0) = [x_1(0) \ x_2(0)]^T, \quad \underline{x}(T_f) = [x_1(T_f) \ x_2(T_f)]^T$$
$$-U \leq u \leq U$$

for $j = 1 \cdots m + 1$. The Lagrangian of the new problem can be written as [8]

$$L = \underline{\psi}^T \underline{N} + \int_{t_0}^{T_f} (\mathbf{H} - \lambda^T \dot{\underline{x}}) dt \tag{6.16}$$

where,

Hamiltonian $\qquad \mathbf{H} = 1 + \lambda^T [A\underline{x} + B(u + (-1)^j f_c)], \quad T_{j-1} \leq t < T_j$
Interior point constraint $\underline{N} = [\dot{\underline{x}}(T_{k_1}) \ldots \dot{\underline{x}}(T_{k_m})] = 0$

$$\tag{6.17}$$

for $j = 1 \cdots m + 1$. The $(m + 1)^{th}$ index corresponds to the final time T_f, which is a free variable. The Lagrangian multipliers are defined as

$$\psi_j \begin{cases} \geq 0 \ t = T_{k_j} \\ = 0 \text{ elsewhere} \end{cases} \quad j = 1 \ldots m \tag{6.18}$$

The optimality condition is found from the variation of L such that

$$\underline{\dot{x}} = \frac{\partial \mathbf{H}}{\partial \underline{\lambda}}$$ (6.19)

$$\underline{\dot{\lambda}} = -\frac{\partial \mathbf{H}}{\partial \underline{x}}$$ (6.20)

$$u = -U\mathrm{sgn}\left(B^T \underline{\lambda}\right)$$ (6.21)

and the interior point condition,

$$\underline{\lambda}(T_{k_j}^-) = \underline{\lambda}(T_{k_j}^+) + \psi \frac{\partial N_{k_j}}{\partial \underline{x}(T_{k_j})} \quad \text{and} \quad \mathbf{H}(T_{k_j}^-) = \mathbf{H}(T_{k_j}^+) - \psi \frac{\partial N_{k_j}}{\partial T_{k_j}}$$ (6.22)

The Hamiltonian should be continuous at these interior points because the interior point constraint N_{k_j} in Equation (6.22) is not an explicit function of time. This results the the following equation:

$$\mathbf{H}(T_{k_j}^-) = \mathbf{H}(T_{k_j}^+).$$ (6.23)

Therefore, the jump discontinuity in the costates has to meet continuous Hamiltonian requirements. The costates are computed from differential equation of the costates in Equation (6.22) with the initial costates. Since the initial costate vector is unknown, the fact that the switching function is zero at the control switch is used. First, costates are integrated in terms of initial costates $\underline{\lambda}(0)$ forward in time. Since costates are discontinuous at the interior point where the velocity of the mass is zero, the discontinuity in the function is handled by defining

$$\frac{\partial N_{k_j}}{\partial \underline{x}(T_{k_j})} \equiv \gamma_j \lambda_2(T_{k_j}^-)$$ (6.24)

where γ can be found from the continuous Hamiltonian requirement, $\mathbf{H}(T_{k_j}^-) = \mathbf{H}(T_{k_j}^+)$. Then the costates can be integrated forward with $\underline{\lambda}(0)$ using Equation (6.20) and Equation (6.24). The resulting costates should satisfy the optimality condition to ensure optimality of the resulting control input.

6.1.1.3 Numerical Example

Consider the following time-optimal control problem.

$$\min \quad \int_0^{T_f} dt$$

subject to

$$\underline{\dot{x}}(t) = \begin{bmatrix} 0 & 1 \\ 0 & 0 \end{bmatrix} \underline{x}(t) + \begin{bmatrix} 0 \\ 1 \end{bmatrix} (u - f_c \mathrm{sgn}(\dot{x}))$$ (6.25)

$$\underline{x}(0) = [0 \ -1]^T, \quad \underline{x}(T_f) = [1 \ 0]^T$$

$$-1 \le u \le 1$$

where $f_c = 0.4$. The optimal control profile is a one-switch bang-bang with the switch time at $t = T_2$ and the velocity of the mass crosses zero once at $t = T_1$ (Figure 6.3b). The linearizing input is parameterized as:

$$u_{net} = \begin{cases} 1.4 & 0 \le t < T_1 \\ 0.6 & T_1 \le t < T_2 \\ -1.4 & T_2 \le t < T_f \\ 0 & T_f \le t. \end{cases} \tag{6.26}$$

The switch and final times can be found as:

$$\begin{aligned} T_1 &= 0.7143 \\ T_2 &= 2.4938 \\ T_f &= 3.2564. \end{aligned} \tag{6.27}$$

From the resulting switch time, the optimality condition can be verified. The optimal control problem in Equation (6.25) can be re-written as:

$$\begin{aligned} \min \quad & T_f \\ \text{subject to} \\ & \dot{x} = Ax + B(u + f_c) \quad 0 \le t < T_1 \\ & \dot{x} = Ax + B(u - f_c) \quad T_1 < t \le T_f \\ & x_2(T_1) = 0 \\ & \underline{x}(0) = [0 \ -1]^T, \quad \underline{x}(T_f) = [1 \ 0]^T \\ & -1 \le u \le 1. \end{aligned} \tag{6.28}$$

Formulate a Lagrangian such that

$$L = \psi x_2(T_1) + \int_{t_0}^{T_f} (H - \lambda^T \dot{x}) dt \tag{6.29}$$

where Hamiltonian of the problem is written as

$$H = \begin{cases} 1 + \lambda^T [Ax + B(u + f_c)], & 0 \le t < T_1 \\ 1 + \lambda^T [Ax + B(u - f_c)], & T_1 < t \le T_f. \end{cases} \tag{6.30}$$

The optimality condition is found from the variation of L such that

$$\dot{x} = \frac{\partial H}{\partial \underline{\lambda}} \tag{6.31}$$

$$\dot{\underline{\lambda}} = -\frac{\partial H}{\partial \underline{x}} \tag{6.32}$$

$$u = -\text{sgn}(B^T \underline{\lambda}) \tag{6.33}$$

and the interior-point constraint yields

$$\underline{\lambda}(T_1^-) = \underline{\lambda}(T_1^+) + \psi \frac{\partial N}{\partial \underline{x}(T_1)} \quad \text{and} \quad H(T_1^-) = H(T_1^+) \tag{6.34}$$

where $\underline{N}$ is a interior point constraint such that $N = x_2(T_1)$. The continuous Hamiltonian requirement at T_1 yields

$$1+[\lambda_1(T_1^-)\ \lambda_2(T_1^-)]\big(A\underline{x}(T_1^-)+B(1+f_c)\big) = 1+[\lambda_1(T_1^+)\ \lambda_2(T_1^+)]\big(A\underline{x}(T_1^+)+B(1-f_c)\big).$$
(6.35)

Since only λ_2 is discontinuous at $t = T_1$, assuming $\lambda_2(T_1^+) = \lambda_2(T_1^-) + \gamma\lambda_2(T_1^-)$, γ is found from Equation (6.35) to be

$$\gamma = \frac{2f_c}{1-f_c}.$$
(6.36)

Now the costates at the switch times are found in terms of $\underline{\lambda}(0)$ from Equation (6.32)

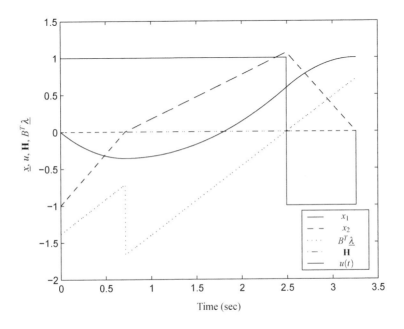

Figure 6.4: Time-Optimal Evolution of States

such that

$$\underline{\lambda}(T_1^-) = e^{-A^T T_1}\underline{\lambda}(0)$$
$$\underline{\lambda}(T_1^+) = (I+J)\underline{\lambda}(T_1^-)$$
$$\underline{\lambda}(T_2) = e^{-A^T(T_2-T_1)}\underline{\lambda}(T_1^+).$$
(6.37)

where J is defined as:

$$J \equiv \begin{bmatrix} 0 & 0 \\ 0 & \gamma \end{bmatrix}.$$
(6.38)

The switching curve has to be zero at the control switch $t = T_2$ such that

$$B^T e^{-A^T(T_2-T_1)}(I+J)e^{-A^T T_1}\underline{\lambda}(0) = 0.$$
(6.39)

The initial costates are found to satisfy Equation (6.39). Since Equation (6.39) can be satisfied with $\underline{\lambda}(0)=0$, which corresponds to the trivial solution, $\underline{\lambda}(0)$ has to lie in the null space of $B^T e^{-A^T (T_2 - T_1)}(I + J)e^{-A^T T_1}$ for the non-trivial solution. Since the Hamiltonian has to satisfy the constraint $\mathbf{H}(t = 0) = 0$, we have

$$\mathbf{H}(0) = 1 + \lambda^T(0)[A\underline{x}(0) + B(1 + f_c)] = 1 + \lambda^T(0)\begin{bmatrix} -1 \\ (1 + f_c) \end{bmatrix} = 0 \qquad (6.40)$$

which requires

$$\lambda^T(0)\begin{bmatrix} -1 \\ (1 + f_c) \end{bmatrix} = -1. \qquad (6.41)$$

Satisfying Equations (6.39) and (6.41) leads to:

$$\underline{\lambda}(0) = -\frac{\text{Null}(B^T e^{-A^T (T_2 - T_1)}(I + J)e^{-A^T T_1})}{[\text{Null}(B^T e^{-A^T (T_2 - T_1)}(I + J)e^{-A^T T_1}]^T[-1 \quad 1 + f_c]^T} \qquad (6.42)$$

where $\text{Null}(\cdot)$ represents the null space. The resulting control input, switching curve, Hamiltonian, and the evolution of the states are plotted in Figure 6.4. It is shown that the switching curve passes through the abscissa when there is a sign change in the control input.

6.1.2 Flexible Structure

In contrast to rigid-body systems which are only subject to the control and friction forces, flexible structures include additional spring and damping forces which complicate the analysis and synthesis of minimum-time control profiles. This section presents a detailed development of the design of time-optimal control profiles for rest-to-rest maneuvers of flexible structures which are represented as a chain of spring-mass-dampers. The assumption in this development is that the control and friction forces act on the same mass.

6.1.2.1 Problem Formulation

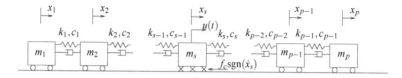

Figure 6.5: *p*-Mass-Spring-Damper System Subject to Coulomb Friction

Consider the minimum time problem for a general *p*-mass-spring-damper system shown in Figure 6.5. A single control input acts on the sth mass, which is also subject to the Coulomb friction f_c. The time-optimal problem statement for a rest-to-rest maneuver is given as:

$$\min \quad \int_0^{T_f} dt$$

subject to
$$\dot{\underline{x}}(t) = A\underline{x}(t) + B\big(u(t) - f_c\mathrm{sgn}(\dot{x}_s)\big) \qquad (6.43)$$
$$\underline{x}(0) = \underline{x}_0, \quad \underline{x}(T_f) = \underline{x}_f$$
$$-U \le u(t) \le U$$

where A and B are defined as

$$A = \begin{bmatrix} 0_{p\times p} & I_{p\times p} \\ -M^{-1}K & -M^{-1}C \end{bmatrix}, \quad B = \begin{bmatrix} 0_{p\times 1} \\ M^{-1}\underline{d} \end{bmatrix} \qquad (6.44)$$

M, K, and C are the mass matrix, stiffness matrix, and damping matrix, respectively, and $\underline{d}$ is a control influence vector of zeros with a nonzero value in the sth entry. f_c is the Coulomb friction coefficient which is assumed to be constant. The Hamiltonian of the problem is formulated as:

$$\mathbf{H}(\underline{x}, \underline{\lambda}, u, t) = 1 + \underline{\lambda}^T\big(A\underline{x}(t) + B\big(u(t) - f_c\mathrm{sgn}(\dot{x}_s)\big)\big) \qquad (6.45)$$

The necessary conditions for optimality are given as

$$\dot{\underline{x}}(t) = \frac{\partial \mathbf{H}}{\partial \underline{\lambda}(t)} \qquad (6.46)$$

$$\dot{\underline{\lambda}}(t) = -\frac{\partial \mathbf{H}}{\partial \underline{x}(t)}. \qquad (6.47)$$

From Pontryagin's minimum principle, the controller that minimizes the Hamiltonian is:

$$u = -U\mathrm{sgn}\big(B^T\underline{\lambda}(t)\big) \qquad (6.48)$$

where $B^T\underline{\lambda}(t)$ is the switching function. Therefore, the time-optimal solution for this system is a bang-bang profile. However, this may not be always true for the frictional flexible systems because of the additional state constraints imposed on the optimal control problem. This is due to the fact that the friction force varies to match the imposed forces when the velocity of the frictional body is zero. For a time-optimal control profile for the rigid body, it is guaranteed that the control input is always larger than the force needed to initiate the movement when there is a velocity reversal. For a flexible system, however, external forces acting on the frictional body include the input and spring forces, and the available friction force becomes a function of the external forces acting on the frictional mass, when the net force to the frictional body is not enough to initiate the motion (i.e., when stiction occurs). Therefore, the bang-bang control profile in Equation (6.48) is only valid when the external forces to the frictional body at the velocity reversal are greater than the resisting friction force. The bang-bang control input is parameterized in the next section and the transitions of the control input as a function of displacement are presented in later sections.

6.1.2.2 Linearizing Net Input and Parameterization of Control

It was shown previously that the time-delay filter can be used to cancel the dynamics of a flexible systems undergoing rest-to-rest maneuvers. The time-delay filter can be used for the flexible systems with friction if the linearizing net input formulation is used. The system equation shown in Equation (6.43) can be rewritten in a linear form as:

$$\dot{\underline{x}}(t) = A\underline{x}(t) + Bu_{net}(t) \tag{6.49}$$

The linearizing net input, $u_{net}(t)$, is defined in a manner similar to that for the rigid body, resulting in the equation

$$u_{net}(t) = u(t) - f_c \mathrm{sgn}(\dot{x}_s) \tag{6.50}$$

In order to solve the time-optimal problem using the time-delay filter technique, the zeros of the time-delay filter should cancel out the rigid body and flexible poles in addition to satisfying the displacement boundary conditions of the problem. With the net control input, the velocity of the frictional body must be zero when the sign of the frictional force changes. Figure 6.6 shows a schematic of the thought process in developing this technique for a two-mass-spring system where the control and frictional force are acting on the first mass. The first plot in Figure 6.6 shows the assumed velocity profile of the first mass. The second plot in Figure 6.6 is the actual input to the system, which is described by Equation (6.48), assuming no singular interval, to be bang-bang. The linearizing net input is parameterized and is shown in the final plot, which is the sum of actual input and friction force. For the general system shown in Figure 6.5, assuming that the control has n switch times at the ℓ_i^{th} time instant ($i = 1 \ldots n$). The velocity of the sth mass is assumed to change its sign m times at the k_j^{th} time instant ($j = 1 \ldots m$). For example, $\ell = [1, 3, 5]$ ($n = 3$) and $k = [2, 4]$ ($m = 2$) for the problem in Figure 6.6. The time-delay filter which generates the control profile when driven by a unit step input, can be parameterized as

$$G(s) = (U - f_c) + 2U \sum_{i=1}^{n} (-1)^i e^{-sT_{\ell_i}} + 2f_c \sum_{j=1}^{m} (-1)^{j+1} e^{-sT_{k_j}} + (U + f_c)e^{-sT_f} \tag{6.51}$$

$G(s)$ must cancel out all of the poles of the system while the control input satisfies the final boundary conditions shown in Equation (6.43). One of the rigid-body poles is canceled due to Equation (6.51) having a zero at $s = 0$. To cancel the other rigid body pole, the derivative of Equation (6.51) with respect to s must also have a zero at $s = 0$. This satisfies the velocity boundary condition for a rest-to-rest maneuver, which yields the following constraint:

$$\left.\frac{dG(s)}{ds}\right|_{s=0} = -2U \sum_{i=1}^{n} (-1)^i T_{\ell_i} - 2f_c \sum_{j=1}^{m} (-1)^{j+1} T_{k_j} - (U + f_c)T_f = 0 \tag{6.52}$$

The time-delay filter must also cancel out the flexible-mode poles, which will lead to the next $(2p - 2)$ constraints. Substituting $s_\mu = \sigma_\mu + i\omega_\mu$ into Equation (6.51),

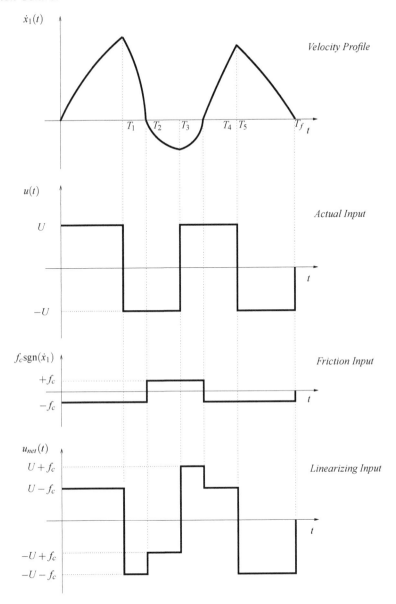

Figure 6.6: General Schematic of Linearizing Net Input Parameterization

yields Equation (6.53) and Equation (6.54).

$$0 = (U - f_c) + (U + f_c)e^{-T_f \sigma_\mu} \cos\left(T_f \omega_\mu\right) + 2U \sum_{i=1}^{n} (-1)^i e^{-T_{\ell_i} \sigma_\mu} \cos\left(T_{\ell_i} \omega_\mu\right)$$

$$+2f_c \sum_{j=1}^{m}(-1)^{j+1}e^{-T_{k_j}\sigma\mu}\cos\left(T_{k_j}\omega_\mu\right) \tag{6.53}$$

$$0 = (U+f_c)e^{-T_f\sigma\mu}\sin\left(T_f\omega_\mu\right) + 2U\sum_{i=1}^{n}(-1)^i e^{-T_{\ell_i}\sigma\mu}\sin\left(T_{\ell_i}\omega_\mu\right)$$

$$+2f_c \sum_{j=1}^{m}(-1)^{j+1}e^{-T_{k_j}\sigma\mu}\sin\left(T_{k_j}\omega_\mu\right) \tag{6.54}$$

for $\mu = 1\ldots(p-1)$, where p is the number of masses. The next constraint comes from ensuring that the final boundary condition on position is satisfied. This can be found by using the final value theorem. The final value theorem can be represented as

$$x_s(T_f) = \lim_{s\to 0} G(s)G_p(s) \tag{6.55}$$

where $G_p(s)$ is the transfer function of the sth mass position output which can be derived from Equation (6.49). For rest-to-rest maneuvers, the final position of each mass must be the same. $L'H\hat{o}pital's$ rule is used to solve Equation (6.55) since the numerator and denominator of $G(s)G_p(s)$ have two zeros and two poles at the origin, respectively. The next m constraints are the velocity constraints at the time when the friction sign changes, which are shown in Equation (6.56).

$$v_j(t) = \mathscr{L}^{-1}\left[G(s)G_p(s)\right]\bigg|_{T_{k_j}} = 0, \quad j = 1\ldots m. \tag{6.56}$$

There are a total of $2p+m$ constraints with $n+m+1$ unknowns (i.e., the switch and final times). Nonlinear optimizers are used to solve for the unknown switch and final times while minimizing the final time subject to Equations (6.52) through (6.56).

6.1.2.3   Optimality Condition

The new time-optimal control problem with the linearizing net input formulation is written as:

$$\min \int_{0}^{T_f} dt$$

$$\text{subject to}$$

$$\dot{x} = Ax + B(u+(-1)^j f_c), \quad T_{j-1} < t < T_j \tag{6.57}$$
$$\dot{x}_s(T_{k_j}) = 0$$
$$x(0) = x_0 \text{ and } x(T_f) = x_f$$
$$-U \le u \le U$$

for $j = 1\ldots m$ and T_0 which corresponds to the initial time is equal to zero. It is assumed in Equation (6.57) that the frictional mass starts to maneuver from rest in the positive direction with positive velocity. Equation (6.57) has additional interior-point constraints when the velocity of the frictional body is zero compared to the initial problem statement in Equation (6.43). The Lagrangian of this problem can be written as [8]

$$L = \underline{v}^T \underline{N} + \int_0^{T_f} (\mathbf{H} - \underline{\lambda}^T \underline{\dot{x}}) dt \tag{6.58}$$

where,

Hamiltonian $\quad \mathbf{H} = 1 + \underline{\lambda}^T \left(A\underline{x} + B(u + (-1)^j f_c) \right), \quad T_{j-1} < t < T_j$

Interior-point constraint $\underline{N} = \left[\dot{x}_s(T_{k_1}) \; \dot{x}_s(T_{k_2}) \; \dots \; \dot{x}_s(T_{k_m}) \right]^T = \underline{0}$

$$\tag{6.59}$$

for $j = 1 \dots m$ and the Lagrangian multipliers are defined as

$$v_j \begin{cases} \geq 0 \; t = T_{k_j} \\ = 0 \; \text{elsewhere} \end{cases} \quad j = 1 \dots m \tag{6.60}$$

Since the Hamiltonian is not explicitly a function of time, it is equal to zero for all time t. As in the system shown in Equation (6.43), the switching function for the new problem (with linearizing net input) is also given as $B^T \underline{\lambda}(t)$. We also know that for optimality, the switching function must cross zero at the switching times such that

$$B^T \underline{\lambda}(T_{\ell_i}) = 0, \quad i = 1 \dots n \tag{6.61}$$

The costate equation from the optimality condition becomes

$$\underline{\dot{\lambda}} = -\frac{\partial \mathbf{H}}{\partial \underline{x}} = -A^T \lambda, \quad \text{for all time except } t = T_{k_j} \quad j = 1 \dots m. \tag{6.62}$$

When $t = T_{k_j}$, the costates should satisfy the following equations.

$$\begin{cases} \underline{\lambda}(T_{k_j}^-) = \underline{\lambda}(T_{k_j}^+) + v_j \dfrac{\partial N_j}{\partial \underline{x}(T_{k_j})} \\ \mathbf{H}(T_{k_j}^-) = \mathbf{H}(T_{k_j}^+) - v_j \dfrac{\partial N_j}{\partial T_{k_j}} \end{cases}, \quad j = 1 \dots m \tag{6.63}$$

Because $\underline{N}$ is not explicitly a function of time, the Hamiltonian should be continuous such that $\mathbf{H}(T_{k_j}^-) = \mathbf{H}(T_{k_j}^+)$. Therefore, jump discontinuity in the costate has to be chosen to satisfy the continuous Hamiltonian requirements, which can be written in the form

$$v_j \frac{\partial N_j}{\partial \underline{x}(T_{k_j})} = \gamma_j B^T \underline{\lambda}(T_{k_j}^-) \tag{6.64}$$

Once v_j's are determined, we can find an expression for $\underline{\lambda}(T_{\ell_i})$, in terms of the initial costates, $\underline{\lambda}(0)$, from Equations (6.62) and (6.63). Then we can build Equation (6.61) in terms of $\underline{\lambda}(0)$. As an example, assume that the first time that the velocity goes to zero is T_{k_1}, then the costates at $T_{k_1}^-$ are given by Equation (6.65).

$$\underline{\lambda}(T_{k_1}^-) = e^{-A^T T_{k_1}^-} \underline{\lambda}(0). \tag{6.65}$$

The costates at $t = T_{k_1}^+$ are given as

$$\underline{\lambda}(T_{k_1}^+) = (I + J_1)\underline{\lambda}(T_{k_1}^-) = (I + J_1)e^{-A^T T_{k_1}^-}\underline{\lambda}(0) \qquad (6.66)$$

where J_1 is a matrix where $(p+s)^{th}$ row is $-\gamma_j B^T$. Then the costates can be integrated until the next switch with this new initial conditions. This procedure is repeated until the $n \times 2p$ matrix $\mathcal{M}$ is formed for n switch times such that

$$\begin{bmatrix} B^T \underline{\lambda}(T_{\ell_1}) \\ B^T \underline{\lambda}(T_{\ell_2}) \\ \dots \\ B^T \underline{\lambda}(T_{\ell_n}) \end{bmatrix} = \mathcal{M}\underline{\lambda}(0) = \underline{0} \qquad (6.67)$$

$\underline{\lambda}(0)$ is found by the null space of $\mathcal{M}$ which satisfies the Hamiltonian requirements such that $H(t = 0) = 0$. After calculating $\underline{\lambda}(0)$ from Equation (6.67), the costates can then be integrated forward and the resulting switching function must cross zero at the switching time to ensure optimality.

6.1.2.4   Numerical Example

The example problem considered is illustrated in Figure 6.7. The control input and Coulomb friction force is acting on the first mass. The time-optimal control problem can be stated as

$$\min \quad \int_0^{T_f} dt$$

subject to

$$\dot{\underline{x}}(t) = \begin{bmatrix} 0 & 0 & 1 & 0 \\ 0 & 0 & 0 & 1 \\ -1 & 1 & 0 & 0 \\ 0.5 & -0.5 & 0 & 0 \end{bmatrix} \underline{x}(t) + \begin{bmatrix} 0 \\ 0 \\ 1 \\ 0 \end{bmatrix} \left(u - f_c \mathrm{sgn}(\dot{x}_1) \right) \qquad (6.68)$$

$$\underline{x}(0) = \begin{bmatrix} 0 & 0 & 0 & 0 \end{bmatrix}^T, \quad \underline{x}(T_f) = \begin{bmatrix} 1 & 1 & 0 & 0 \end{bmatrix}^T, \quad \text{and} \quad -1 \leq u \leq 1$$

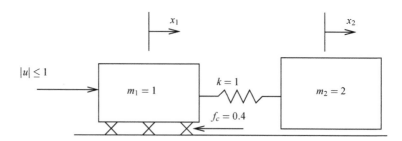

Figure 6.7: Two-Mass-Spring System Subject to Friction

Assuming $n = 3$, $m = 2$, $k = [2\ 4]$, $\ell = [1\ 3\ 5]$, and the flexible poles are given by $s = \pm\omega i$ where $\omega = \sqrt{1.5}$. The net control input is parameterized as shown in Figure 6.6. Then the optimization problem for numerical computation becomes:

$$\min\quad T_f \tag{6.69}$$

subject to the switch time constraints,

$$0 \le T_1 \le T_2 \le T_3 \le T_4 \le T_5 \le T_f \tag{6.70}$$

rigid-body pole cancelation constraint (Equation 6.52),

$$2T_1 - 2f_c T_2 - 2T_3 + 2f_c T_4 + 2T_5 - (1 + f_c)T_f = 0 \tag{6.71}$$

flexible-body pole cancelation constraints (Equations 6.53 and 6.54),

$$\begin{aligned}(1 - f_c) - 2U\cos(\omega T_1) + 2f_c\cos(\omega T_2) + 2\cos(\omega T_3)\\ -2f_c\cos(\omega T_4) - 2\cos(\omega T_5) + (1 + f_c)\cos(\omega T_f) = 0\end{aligned} \tag{6.72}$$

$$\begin{aligned}-2\sin(\omega T_1) + 2f_c\sin(\omega T_2) + 2\sin(\omega T_3)\\ -2f_c\sin(\omega T_4) - 2\sin(\omega T_5) + (1 + f_c)\sin(\omega T_f) = 0\end{aligned} \tag{6.73}$$

final displacement constraint (Equation 6.55),

$$\frac{1}{4}\left(-2T_1^2 + 2f_c T_2^2 + 2T_3^2 - 2f_c T_4^2 - 2T_5^2 + (1 + f_c)T_f^2\right) = 1 \tag{6.74}$$

and first-mass velocity constraints (Equation 6.56),

$$\frac{1}{m_1 + m_2}\left[(1 - f_c)\left(\frac{m_2}{m_1\omega}\sin\left(\omega T_2\right) + T_2\right) - 2\left(\frac{m_2}{m_1\omega}\sin\left(\omega(T_{21})\right) + (T_{21})\right)\right] = 0 \tag{6.75}$$

$$\begin{aligned}\frac{1}{m_1 + m_2}&\left[(1 - f_c)\left(\frac{m_2}{m_1\omega}\sin\left(\omega T_4\right) + T_4\right) - 2\left(\frac{m_2}{m_1\omega}\sin\left(\omega(T_{41})\right) + (T_{41})\right)\right.\\ &\left.+2f_c\left(\frac{m_2}{m_1\omega}\sin\left(\omega(T_{42})\right) + (T_{42})\right) + 2\left(\frac{m_2}{m_1\omega}\sin\left(\omega(T_{43})\right) + (T_{43})\right)\right] = 0\end{aligned} \tag{6.76}$$

where T_{ab} stands for $T_a - T_b$ in Equation (6.75) and Equation (6.76). The constrained nonlinear optimizer of MATLAB is used to solve this problem. The time-delay filter is found to be

$$G(s) = 0.6 - 2e^{-1.8164s} + 0.8e^{-2.1177s} \tag{6.77}$$

$$+ 2e^{-3.1498s} - 0.8e^{-3.5918s} - 2e^{-4.4047s} + 1.4e^{-5.2298s} \tag{6.78}$$

With the resulting control input, the evolution of the states are plotted in Figure 6.8. The costates and switching curve can be computed with the resulting control profile

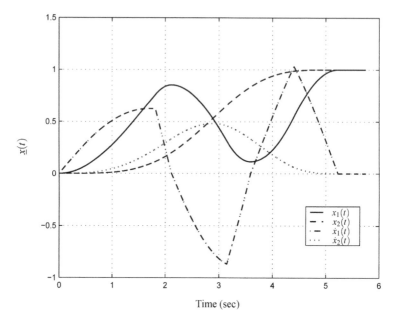

Figure 6.8: Time-Optimal Responses the Two-Mass-Spring System with Coulomb Friction

to verify the optimality. The Hamiltonian for the two-mass-spring system is written as

$$
\mathbf{H} = \begin{cases}
1 + \lambda^T \left(A\underline{x} + B(u - f_c) \right) & 0 < t < T_2 \\
1 + \lambda^T \left(A\underline{x} + B(u + f_c) \right) & T_2 < t < T_4 \\
1 + \lambda^T \left(A\underline{x} + B(u - f_c) \right) & T_4 < t < T_f
\end{cases}
\tag{6.79}
$$

The interior-point constraints are

$$
N_1 = \dot{x}_1(T_2) = 0 \quad \text{and} \quad N_2 = \dot{x}_1(T_4) = 0
\tag{6.80}
$$

Since the Hamiltonian should be continuous, we have $\mathbf{H}(T_2^+) = \mathbf{H}(T_2^-)$ and $\mathbf{H}(T_4^+) = \mathbf{H}(T_4^-)$. The switching function is $B^T \underline{\lambda} = \lambda_3$, and there is a discontinuity only in λ_3. The continuous Hamiltonian requirement yields

$$
1 + \left[\lambda_1(T_2)\ \lambda_2(T_2)\ \lambda_3(T_2^-)\ \lambda_4(T_2) \right] \left(A\underline{x}(T_2) + B(-1 - f_c) \right)
$$
$$
= 1 + \left[\lambda_1(T_2)\ \lambda_2(T_2)\ \lambda_3(T_2^+)\ \lambda_4(T_2) \right] \left(A\underline{x}(T_2) + B(-1 + f_c) \right)
\tag{6.81}
$$

$$
1 + \left[\lambda_1(T_4)\ \lambda_2(T_4)\ \lambda_3(T_4^-)\ \lambda_4(T_4) \right] \left(A\underline{x}(T_4) + B(1 + f_c) \right)
$$
$$
= 1 + \left[\lambda_1(T_4)\ \lambda_2(T_4)\ \lambda_3(T_4^+)\ \lambda_4(T_4) \right] \left(A\underline{x}(T_4) + B(1 - f_c) \right)
\tag{6.82}
$$

Assuming $\lambda_3(T_2^+) = \lambda_3(T_2^-) + \gamma_1 \lambda_3(T_2^-)$ and $\lambda_3(T_4^+) = \lambda_3(T_4^-) + \gamma_2 \lambda_3(T_4^-)$, γ_1 and γ_2 are found to be

$$
\gamma_1 = \frac{-2f_c}{-1 + f_c - k\left(x_1(T_2) - x_2(T_2)\right)} \quad \text{and} \quad \gamma_2 = \frac{2f_c}{1 - f_c - k\left(x_1(T_4) - x_2(T_4)\right)}
\tag{6.83}
$$

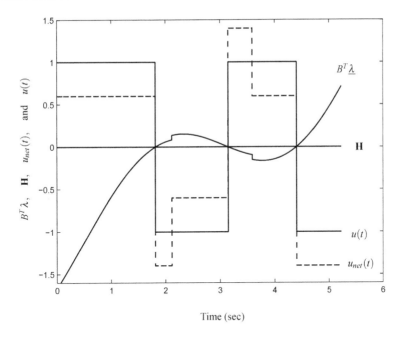

Figure 6.9: Linearizing Control Input, Switching Function, and Hamiltonian

Now the costates at the switch times are found in terms of $\underline{\lambda}(0)$ such that

$$\begin{aligned}
\underline{\lambda}(T_1) &= e^{-A^T T_1}\underline{\lambda}(0)\\
\underline{\lambda}(T_2^-) &= e^{-A^T (T_2-T_1)}\underline{\lambda}(T_1)\\
\underline{\lambda}(T_2^+) &= (I+J_1)\underline{\lambda}(T_2^-)\\
\underline{\lambda}(T_3) &= e^{-A^T (T_3-T_2)}\underline{\lambda}(T_2^+)\\
\underline{\lambda}(T_4^-) &= e^{-A^T (T_4-T_3)}\underline{\lambda}(T_3)\\
\underline{\lambda}(T_4^+) &= (I+J_2)\underline{\lambda}(T_4^-)\\
\underline{\lambda}(T_5) &= e^{-A^T (T_5-T_4)}\underline{\lambda}(T_4^+)
\end{aligned} \tag{6.84}$$

where, J_1 and J_2 is defined as

$$J_1 = \begin{bmatrix} 0 & 0 & 0 & 0\\ 0 & 0 & 0 & 0\\ 0 & 0 & \gamma_1 & 0\\ 0 & 0 & 0 & 0 \end{bmatrix} \quad \text{and} \quad J_2 = \begin{bmatrix} 0 & 0 & 0 & 0\\ 0 & 0 & 0 & 0\\ 0 & 0 & \gamma_2 & 0\\ 0 & 0 & 0 & 0 \end{bmatrix} \tag{6.85}$$

$\mathcal{M}$ is now given to satisfy $\mathcal{M}\underline{\lambda}(0) = \underline{0}$ such that

$$\mathcal{M} = \begin{bmatrix} B^T e^{-A^T T_1}\\ B^T e^{-A^T (T_3-T_2)}(I+J_1)e^{-A^T (T_2-T_1)}e^{-A^T T_1}\\ B^T e^{-A^T (T_5-T_4)}(I+J_2)e^{-A^T (T_4-T_3)}e^{-A^T (T_3-T_2)}(I+J_1)e^{-A^T (T_2-T_1)}e^{-A^T T_1} \end{bmatrix} \tag{6.86}$$

The initial costates that satisfy the constraints $\mathscr{M}\underline{\lambda}(0) = \underline{0}$ and $\mathbf{H}(t = 0) = 0$ can be shown to be

$$\underline{\lambda}(0) = -\frac{\text{Null}(\mathscr{M})}{[\text{Null}(\mathscr{M})]^T B(1 - f_c)} \tag{6.87}$$

where, $\text{Null}(\mathscr{M})$ is the null space of $\mathscr{M}$. The linearizing input, as well as the switching curve is shown in Figure 6.9 for the example problem. The actual input is found by subtracting the friction force from the net input.

6.1.2.5   Variation of Control Structure as a Function of Final Position

In the previous example, the velocity of the first mass is zero at T_2 and T_4. As the final position increases for the previous example, the time gap between T_3 and T_4 becomes smaller until they coincide. The control profile for this transition shown

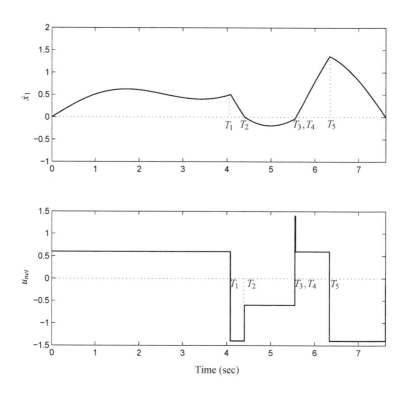

Figure 6.10: Velocity and Control Input Profiles at the Transition

in Figure 6.10 indicates that the velocity crossing coincides with the control switch that will result in a four-switch control profile. The transition displacement for this example is $d \approx 3.4$. After the transition displacement, going back to original profile does not provide a feasible solution. Also, keeping the four-switch transition profile shown in Figure 6.10 after the transition displacement, creates an overconstrained problem. Therefore, the velocity of the first mass should cross zero before the new control switch activates. This implies that now T_3 is the time when the velocity is zero and T_4 is the time when the control switches. At T_3^+, the velocity becomes positive and therefore the friction force becomes $-f_c$. However, the linearizing input with the new friction value drives the first mass to the negative direction if the net input is less than zero. Once the first mass crosses the zero velocity, the net input changes its sign again because of the change in friction value to $+f_c$. This *chattering*

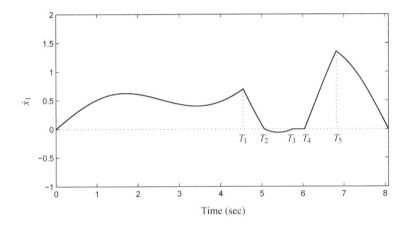

Figure 6.11: Velocity and Control Input Profiles after Transition Displacement

effect will continue until the sum of all the forces to the frictional body overcomes the friction force. Another interpretation of this region is by introducing stiction instead of chattering. Since the sum of all the applied forces to the first mass is not large enough to overcome static friction, the velocity of the first mass stays zero until the new control switch is activated to overcome static friction. One way to formulate an optimal control problem with stiction is adding an additional constraint to the problem such that the velocity of the first mass is zero for this time interval. For the example problem, the predicted velocity profile of the first mass is plotted in Figure 6.11 illustrating that the velocity of the first mass is zero until the new switch activates. It is also possible that the spring force itself can overcome the friction force before the new control switch occurs, which will require an additional constraint satisfying the constraint that the spring force is equal to the Coulomb friction value at the exit of the zero velocity domain. With the velocity profile in Figure 6.11, a new optimal control problem can be formulated as

$$\min \quad \int_0^{T_f} dt$$

subject to
$$
\begin{aligned}
\dot{x} &= A\underline{x} + B(u - f_c), && 0 < t < T_2 \\
\dot{x} &= A\underline{x} + B(u + f_c), && T_2 < t < T_3 \\
\dot{x} &= A\underline{x} + Bu, && T_3 < t < T_4 \\
\dot{x} &= A\underline{x} + B(u - f_c), && T_4 < t < T_f \\
\dot{x}_1 &= 0, && t = T_2 \text{ and } T_3 < t < T_4 \\
\underline{x}(0) &= \underline{x}_0 \text{ and } \underline{x}(T_f) = \underline{x}_f \\
-1 &\le u \le 1
\end{aligned}
\tag{6.88}
$$

Equation (6.88) has a state equality constraint on the velocity of the first mass. Because of this constraint, the friction force during stiction is considered zero. Bryson and Ho [8] have developed a method to handle the state equality constraints in the optimal control problem. From [8], Equation (6.88) can be restated as

$$\min \quad \int_0^{T_f} dt$$

subject to
$$
\begin{aligned}
\dot{x} &= A\underline{x} + B(u - f_c), && 0 < t < T_2 \\
\dot{x} &= A\underline{x} + B(u + f_c), && T_2 < t < T_3 \\
\dot{x} &= A\underline{x} + Bu, && T_3 < t < T_4 \\
\dot{x} &= A\underline{x} + B(u - f_c), && T_4 < t < T_f \\
\ddot{x}_1 &= \frac{k}{m_1}(x_2 - x_1) + u = 0, && T_3 < t < T_4 \\
\dot{x}_1(T_2) &= \dot{x}_1(T_3) = 0 \\
\underline{x}(0) &= \underline{x}_0 \text{ and } \underline{x}(T_f) = \underline{x}_f \\
-1 &\le u \le 1
\end{aligned}
\tag{6.89}
$$

Because the friction force is zero for the zero velocity, the control for $T_3 < t < T_4$ should satisfy the equality constraint in Equation (6.89) such that

$$
u = \frac{k}{m_1}(x_1 - x_2), \qquad T_3 < t < T_4 \tag{6.90}
$$

It is interesting to see that although the control input needed for the first mass to be immobile for $T_3 < t < T_4$ is defined by Equation (6.90), because of the assumption of stiction during the interval, it is possible to use a constant control input value of $u = -1$. Now, the control profile can be parameterized from the predicted velocity profile including the state constraint effect of stiction. The time-delay filter cannot be used for this profile because of the feedback control effect in the chattering interval which alters the system equation. However, the control profile can be used to integrate the system forward in time using the given initial guesses of the switch times to obtain the states at the switch and final times:

$$x(T_1) = e^{AT_1}x_0 + \int_0^{T_1} e^{A(t-\tau)}B(1-f_c)d\tau$$

$$x(T_2) = e^{A(T_2-T_1)}x(T_1) + \int_{T_1}^{T_2} e^{A(t-\tau)}B(-1-f_c)d\tau$$

$$x(T_3) = e^{A(T_3-T_2)}x(T_2) + \int_{T_2}^{T_3} e^{A(t-\tau)}B(-1+f_c)d\tau$$

$$x(T_4) = e^{A^c(T_4-T_3)}x(T_3)$$ \hfill (6.91)

$$x(T_5) = e^{A(T_5-T_4)}x(T_4) + \int_{T_4}^{T_5} e^{A(t-\tau)}B(1-f_c)(\tau)d\tau$$

$$x(T_f) = e^{A(T_6-T_5)}x(T_5) + \int_{T_5}^{T_6} e^{A(t-\tau)}B(-1-f_c)(\tau)d\tau$$

where, A^c is the closed-loop system equation in the interval at $t \in [T_3, T_4]$ using Equation (6.90).

$$A^c = \begin{bmatrix} 0 & 0 & 1 & 0 \\ 0 & 0 & 0 & 1 \\ 0 & 0 & 0 & 0 \\ \dfrac{k}{m_2} & -\dfrac{k}{m_2} & 0 & 0 \end{bmatrix}$$ \hfill (6.92)

For the computational accuracy and convenience, it is very useful to use the following property [9] to compute the states at the switch times:

$$\exp\left(\begin{bmatrix} A & B \\ 0 & 0 \end{bmatrix} t\right) = \begin{bmatrix} e^{At} & \int_0^t e^{A(t-\tau)}Bd\tau \\ 0 & 1 \end{bmatrix}$$ \hfill (6.93)

Now the problem can be solved for switch times numerically by minimizing the final time T_6 with the constraints

$$\begin{aligned} \dot{x}_1(T_2) &= 0 \\ \dot{x}_1(T_3) &= 0 \\ x(T_f) &= [1 \quad 1 \quad 0 \quad 0]^T \end{aligned}$$ \hfill (6.94)

The resulting control profile is used to verify the optimality of the control input by inspecting the optimality conditions. The Lagrangian for this problem can be formulated as [8]

$$L = \psi N_1 + \pi N_2 + \int_{t_0}^{T_f} (\mathbf{H} - \lambda^T \dot{x})dt$$ \hfill (6.95)

where,

Hamiltonian

$$\begin{aligned} \mathbf{H} &= 1 + \lambda^T (Ax + B(u - f_c)), & 0 < t < T_2 \\ \mathbf{H} &= 1 + \lambda^T (Ax + B(u + f_c)), & T_2 < t < T_3 \\ \mathbf{H} &= 1 + \lambda^T (Ax + Bu) + \mu C, & T_3 < t < T_4 \\ \mathbf{H} &= 1 + \lambda^T (Ax + B(u - f_c)), & T_4 < t < T_f \end{aligned}$$ \hfill (6.96)

Interior-point constraint $N_1 = \dot{x}_1(T_2) = 0$
$\qquad\qquad\qquad\qquad\quad N_2 = \dot{x}_1(T_3) = 0$

Equality constraint $\quad C = \ddot{x}_1(t) = k(x_2 - x_1) + u = 0$

and the Lagrangian multipliers are defined as:

$$\mu \begin{cases} \geq 0 \; t \in [T_3, T_4] \\ = 0 \text{ elsewhere} \end{cases}, \quad \psi \begin{cases} \geq 0 \; t = T_2 \\ = 0 \text{ elsewhere} \end{cases} \quad \text{and} \quad \pi \begin{cases} \geq 0 \; t = T_3 \\ = 0 \text{ elsewhere} \end{cases} \quad (6.97)$$

The input u is considered to be within the saturation limit for $T_3 < t < T_4$ because of the stiction assumption with $u = -1$. Then, μ can be found from the following relationship:

$$\frac{\partial H}{\partial u} = 0 = B^T \underline{\lambda} + \mu \quad \text{or} \quad \mu = -B^T \underline{\lambda}, \qquad T_3 < t < T_4 \qquad (6.98)$$

Therefore, the costate equation is found from Equation (6.98) and the necessary optimality condition.

$$\dot{\underline{\lambda}} = -\frac{\partial H}{\partial \underline{x}} = \begin{cases} \left(-A^T + \dfrac{\partial C}{\partial \underline{x}} B^T \right) \underline{\lambda} & T_3 < t < T_4 \\ -A^T \underline{\lambda} & \text{elsewhere except } t = T_2 \text{ and } t = T_3 \end{cases} \qquad (6.99)$$

At the interior points, $t = T_2$ and $t = T_3$, the costates becomes discontinuous by the following equations.

$$\begin{cases} \underline{\lambda}(T_2^-) = \underline{\lambda}(T_2^+) + \psi \dfrac{\partial N_1}{\partial \underline{x}(T_2)} \\ H(T_2^-) = H(T_2^+) - \psi \dfrac{\partial N_1}{\partial T_2} \end{cases} \text{ and } \begin{cases} \underline{\lambda}(T_3^-) = \underline{\lambda}(T_3^+) + \pi \dfrac{\partial N_2}{\partial \underline{x}(T_3)} \\ H(T_3^-) = H(T_3^+) - \pi \dfrac{\partial N_2}{\partial T_3} \end{cases} \qquad (6.100)$$

Because the interior point constraints are not explicit functions of time, the Hamiltonian is continuous such that $H(T_2^-) = H(T_2^+)$ and $H(T_3^-) = H(T_3^+)$. Therefore, ψ and π in Equation (6.100) are chosen to satisfy the continuous Hamiltonian requirement, which yields the following equations:

$$\gamma_1 = \frac{-2f_c}{-1 + f_c - k(x_1(T_2) - x_2(T_2))} \quad \text{and} \quad \lambda_3(T_3^-) = 0 \qquad (6.101)$$

where γ_1 satisfies $\lambda_3(T_2^+) = \lambda_3(T_2^-) + \gamma_1 \lambda_3(T_2^-)$. Costates can be integrated forward in time with the initial $\underline{\lambda}(0)$ using Equations (6.99) and (6.101).

$$\begin{aligned} \underline{\lambda}(T_1) &= e^{-A^T T_1} \underline{\lambda}(0) \\ \underline{\lambda}(T_2^-) &= e^{-A^T (T_2 - T_1)} \underline{\lambda}(T_1) \\ \underline{\lambda}(T_2^+) &= (I + J_1) \underline{\lambda}(T_2^-) \\ \underline{\lambda}(T_3^-) &= e^{-A^T (T_3 - T_2)} \underline{\lambda}(T_2^+) \\ \underline{\lambda}(T_3^+) &= (I + J_2) \underline{\lambda}(T_3^-) \\ \underline{\lambda}(T_4) &= e^{(-A^T + \frac{\partial C}{\partial \underline{x}})(T_4 - T_3)} \underline{\lambda}(T_3^+) \\ \underline{\lambda}(T_5) &= e^{-A^T (T_5 - T_4)} \underline{\lambda}(T_4) \end{aligned} \qquad (6.102)$$

Where, I is an 4×4 identity matrix and

$$J_1 = \begin{bmatrix} 0 & 0 & 0 & 0 \\ 0 & 0 & 0 & 0 \\ 0 & 0 & \gamma_1 & 0 \\ 0 & 0 & 0 & 0 \end{bmatrix} \quad \text{and} \quad J_2 = \begin{bmatrix} 0 & 0 & 0 & 0 \\ 0 & 0 & 0 & 0 \\ 0 & 0 & \infty & 0 \\ 0 & 0 & 0 & 0 \end{bmatrix} \tag{6.103}$$

Because $\underline{\lambda}(T_3^-)$ should be zero, infinite value has to be multiplied to $\underline{\lambda}(T_3^-)$ in J_2 to

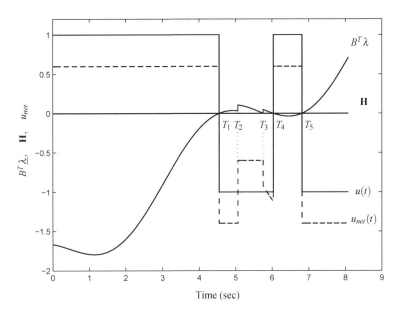

Figure 6.12: Linearizing Control Input, Switching Function, and Hamiltonian

have finite jump discontinuities in the costates. For numerical computations, a large number is used in J_2 instead of ∞. Since the switching curve should cross zero at the actual switch times, T_1, T_4, and T_5, the following should be satisfied:

$$\mathcal{M} \underline{\lambda}(0) = \underline{0} \tag{6.104}$$

where,

$$\mathcal{M} = \begin{bmatrix} B^T e^{-A^T T_1} \\ B^T e^{(-A^T + \frac{\partial C}{\partial \underline{x}})(T_4 - T_3)}(I + J_2)e^{-A^T(T_3 - T_2)}(I + J_1)e^{-A^T(T_2 - T_1)}e^{-A^T T_1} \\ B^T e^{-A^T(T_5 - T_4)}e^{(-A^T + \frac{\partial C}{\partial \underline{x}})(T_4 - T_3)}(I + J_2)e^{-A^T(T_3 - T_2)}(I + J_1)e^{-A^T(T_2 - T_1)}e^{-A^T T_1} \end{bmatrix} \tag{6.105}$$

Therefore, $\underline{\lambda}(0)$ is selected to satisfy Equation (6.104) and Hamiltonian requirement, $H(t = 0) = 0$, such that

$$\underline{\lambda}(0) = -\frac{\text{Null}(\mathcal{M})}{[\text{Null}(\mathcal{M})]^T B(1 - f_c)} \tag{6.106}$$

In Figure 6.12, the linearizing control input along with the switching curve is plotted. The Hamiltonian is also plotted which is zero for all time. The actual control input is computed by subtracting the frictional force from the linearizing control input. Because the friction force is zero when the velocity of the first mass is zero, the linearizing input is equal to the actual input in the stiction region. However, the actual control input becomes $u = -1$ for the stiction interval, since the compensation of the spring force is not necessary to stay stuck. Therefore, the actual control input becomes a three switch bang-bang. The resulting net control input satisfies the necessary optimality conditions as shown in Figure 6.12.

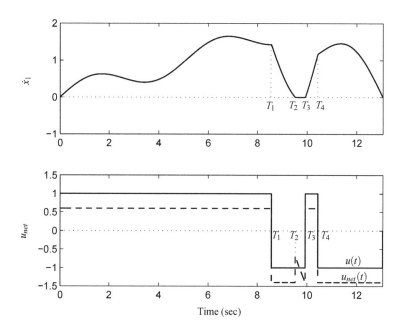

Figure 6.13: Velocity and Control Input Profiles

As the final displacement is increased further, the profile of the control shows another transition where the velocity of the first mass can be illustrated as shown in the upper plot in Figure 6.13. The control profile can be solved in a manner similar to the previous development, when the velocity of the first mass is forced to zero for $T_2 < t < T_3$. The corresponding control input is also plotted in Figure 6.13. The net control input now has four control switches without the negative velocity maneuvers of the first mass. The actual control input stays as a three-switch bang-bang. If the final displacement is increased further, the velocity of the first mass will stay positive during the maneuver. Then, the net control input can be parameterized with the three-switch bang-bang profile as shown in Figure 6.14, which is a Coulomb friction biased control. The time-delay filter approach can be used for this profile since the

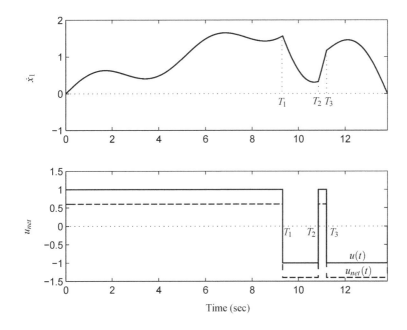

Figure 6.14: Velocity and Control Input Profiles

linearizing input does not depend on the states. The resulting velocity response of the first mass for this net input is shown in Figure 6.14. The summary of the control input transition profile for the example problem is shown in Figure 6.15.

The switch times are plotted as a function of final displacement showing that control switches are smooth curves for final displacement changes. The actual control switches are plotted with solid lines and velocity switches are plotted with dotted lines in Figure 6.15. The linearizing net input and actual control input profiles are also shown in Figure 6.15 for different displacements.

This section focused on the design of time-optimal control profiles for systems subject to Coulomb friction which result in an open-loop control solution. There are legacy systems that are plagued by the effects of friction and which use feedback control for tracking reference inputs. These feedback controlled systems can maneuver the systems states to the proximity of the desired position, but the effect of friction is manifested in the form of steady state error. The following section deals with the

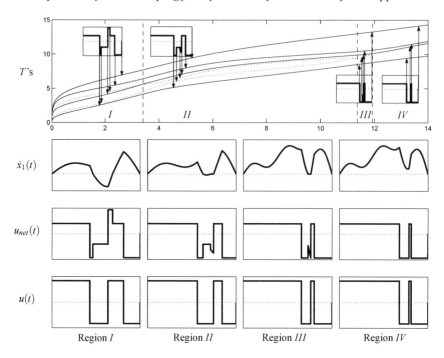

Figure 6.15: Switch Time vs. Displacement with Velocity and Control Input Profiles

design of simple open-loop control profiles to compensate for the small steady-state errors.

6.2 Pulse-Width Pulse-Amplitude Control

For small motions of systems subject to friction, Yang and Tomizuka [10] proposed a technique that uses the pulse-width control (PWC). Here, it is assumed that a standard feedback control is used to move the system to the vicinity of the desired final position. The presence of inaccurate knowledge of friction precludes the accurate tracking of the desired position. In such a setting, an adaptive PWC technique can be used to accurately move the system to the desired position. This technique will be presented as a precursor to the development of the pulse-width pulse-amplitude control (PWPAC) technique.

6.2.1 Rigid Body

This section will review the pulse-width control technique proposed by Yang and Tomizuka [10] for a rigid-body system subject to Coulomb friction. The benefit of estimating the coefficient of friction and using it in conjunction with the pulse-width control will be illustrated. This will be followed by the development of the pulse-amplitude pulse-width control technique that will be shown to reduce the terminal position error compared to pulse-width control.

6.2.1.1 Pulse-Width Control

Consider a rigid-body subject to Coulomb friction as shown in Figure 6.1. The equation of motion of this system is

$$m\ddot{x} + f_c sgn(\dot{x}) = u \tag{6.107}$$

where m is the system mass, f_c is the Coulomb friction coefficient, x is the position of the mass and u is the control input. Consider the response of this system to a pulse of width t_p and height u_p. The response of such a system initially at rest is

$$x(t) = \begin{cases} \frac{1}{2m}t^2(u_p - f_c) & \text{for } 0 < t < t_p \\ \left(\frac{1}{2m}t_p^2 + \frac{t_p}{m}(t - t_p)\right)(u_p - f_c) - \frac{1}{2m}(t - t_p)^2 f_c & \text{for } t_p < t < t_p + \frac{t_p(u_p - f_c)}{f_c} \end{cases} \tag{6.108a}$$

$$\dot{x}(t) = \begin{cases} \frac{1}{m}t(u_p - f_c) & \text{for } 0 < t < t_p \\ \frac{t_p}{m}(u_p - f_c) - \frac{1}{m}(t - t_p)f_c & \text{for } t_p < t < t_p + \frac{t_p(u_p - f_c)}{f_c} \end{cases} \tag{6.108b}$$

Equation (6.108b) states that the velocity of the rigid body at the end of the pulse $(t = t_p)$ is $\frac{1}{m}t_p(u_p - f_c)$. After the duration of the pulse, the frictional force forces the coasting system to rest at $t = t_p + \frac{t_p(u_p - f_c)}{f_c}$. It can be seen from Equation (6.108b) that the system comes to rest at

$$t = t_p + (u_p - f_c)\frac{t_p}{f_c} = \frac{u_p t_p}{f_c} \tag{6.109}$$

with a final displacement (x_f) of

$$x_f = \frac{u_p(u_p - f_c)}{2m f_c}t_p^2 \tag{6.110}$$

where it is assumed that $u_p > f_c$. The friction model used for the determination of the displacement is a simple one which does not capture the stiction phenomenon, Stribeck effect, and so forth. Therefore, the final displacement of the real system will differ from the desired displacement. Subsequent pulses are required to move the system to within a prespecified tolerance of the final desired position.

For an estimated coefficient of friction $\hat{f}_c$, the pulse width, for a desired displacement of x_f is given by the equation:

$$t_p = \sqrt{x_f \frac{2m\hat{f}_c}{u_p(u_p - \hat{f}_c)}}, \tag{6.111}$$

while the actual displacement of the system is:

$$d_1 = \frac{u_p(u_p - f_c)}{2mf_c} x_f \frac{2m\hat{f}_c}{u_p(u_p - \hat{f}_c)} = \frac{(u_p - f_c)\hat{f}_c}{f_c(u_p - \hat{f}_c)} x_f = \Xi x_f \tag{6.112}$$

where $\Xi = \frac{(u_p - f_c)\hat{f}_c}{f_c(u_p - \hat{f}_c)}$. The error in the displacement $x_f - d_1$ is used to determine the pulse width for the subsequent pulse which results in the displacement:

$$d_2 = \Xi(x_f - \Xi x_f) = \Xi x_f - \Xi^2 x_f. \tag{6.113}$$

It can be shown that the displacement of the nth pulse is:

$$d_n = \Xi x_f - \Xi^2 x_f + \Xi^3 x_f - \ldots (-1)^{n+1}\Xi^n x_f \tag{6.114}$$

The total displacement of the system is:

$$x = \sum_{i=1}^{n} d_i = \sum_{i=1}^{n}(-1)^{i+1}\,({}^nC_i)\Xi^i x_f = (1 - (1 - \Xi)^n)x_f \tag{6.115}$$

where ${}^nC_i$ represents the number of combinations of n variables taken i at a time and is given by the equation:

$${}^nC_i = \frac{n!}{(n - i)!\,i!} \tag{6.116}$$

where $n!$ refers to n factorial. For x to converge to x_f, Ξ is required to lie in the range:

$$-1 \leq (1 - \Xi) = (1 - \frac{(u_p - f_c)\hat{f}_c}{f_c(u_p - \hat{f}_c)}) \leq 1 \tag{6.117}$$

which results in the constraints:

$$\hat{f}_c \leq \frac{2u_p f_c}{u_p + f_c} \tag{6.118}$$

$$\hat{f}_c < u_p \text{ and } f_c < u_p \tag{6.119}$$

Figure 6.16 illustrates the region within which $\hat{f}_c$ should lie for the repeated application of the pulse-width control of friction, to converge to the desired displacement. Figure 6.16 is generated for a true friction coefficient of 0.3 and for a peak pulse amplitude of 1.

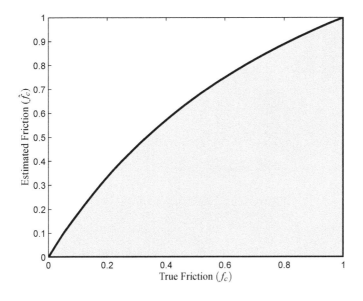

Figure 6.16: Stability Region

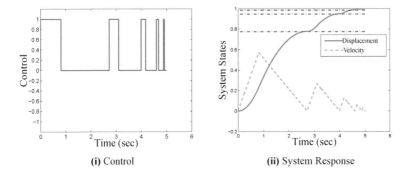

(i) Control   (ii) System Response

Figure 6.17: Underestimated Friction Results

Assuming a true friction coefficient of 0.3 and an estimated friction of 0.25, simulation results of the PWC with a pulse amplitude of 1 are illustrated in Figure 6.17.

It is clear from Figure 6.17(ii) that the underestimation of the coefficient of friction results in a net displacement that is smaller than anticipated for each pulse. This results in a series of positive pulses with monotonically decreasing pulse widths as shown in Figure 6.17(i).

Assuming a true friction coefficient of 0.3 and an estimated friction of 0.35, simulation results of the PWC with a pulse amplitude of 1 are illustrated in Figure 6.18.

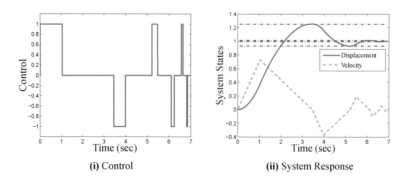

Figure 6.18: Overestimated Friction Results

It is clear from Figure 6.18(ii) that the overestimation of the coefficient of friction results in a net displacement which is larger than anticipated for each pulse. This results in a series of pulses whose signs change and whose widths monotonically decrease as shown in Figure 6.18(i).

Figure 6.19 presents the phase plots of the pulse-width control for the case where the friction is under- and overestimated. When friction is overestimated, significant overshoot can result which might be undesirable when displacement constraints are imposed on the motion of the system.

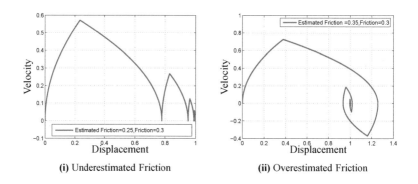

Figure 6.19: Phase Plane Responses

Equation (6.118) results in an upper bound on the estimated friction $\hat{f}_c$ of 0.4615 for a true friction of 0.3 and a peak pulse amplitude of 1. Assuming a friction estimate of 0.5 results in an unstable response as shown in Figure 6.20, confirming the bounds given in Equation (6.118).

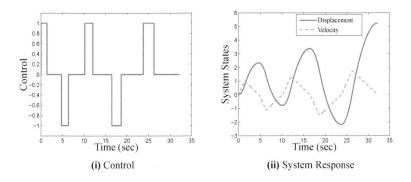

Figure 6.20: Unstable Friction Results

6.2.1.2 Adaptive Pulse-Width Control

The friction model used for the determination of the pulse width is a simple one, which does not capture the stiction, Stribeck effect, and so forth. Therefore, the final displacement will differ from the desired displacement. Subsequent pulses are required to move the system to within a prespecified tolerance of the final desired position. This prompts the use of an algorithm for the real-time estimation of the friction coefficient ($\hat{f}_c$) which can subsequently be used to solve for the pulse width to drive the system to the final position. Assuming that only the pulse width can be modified for each implementation of the pulse, Yang and Tomizuka [10] suggested representing the system displacement dynamics as:

$$x(k+1) = x(k) + d(k+1) \tag{6.120}$$

and the displacement error as:

$$e(k) = x_f - x(k). \tag{6.121}$$

The error dynamics can now be represented as:

$$e(k+1) = e(k) - d(k+1) \tag{6.122}$$

where the incremental displacement is defined as:

$$d(k+1) = \underbrace{\frac{u_p(u_p - f_c)}{2mf_c}}_{b} t_p^2 sgn(e(k)) = bt_p^2 sgn(e(k)) = b\omega(k) \tag{6.123}$$

where

$$\omega(k) = t_p^2 sgn(e(k)). \tag{6.124}$$

If b which is a function of the coefficient of friction is treated as a known quantity, the feedback control law

$$\omega(k) = \frac{1}{b}K_c e(k) \qquad (6.125)$$

results in the error dynamics:

$$e(k+1) = (1 - K_c)e(k) \qquad (6.126)$$

which from discrete time stability constraint requires $0 < K_c < 2$ which will place the discrete-time pole inside the unit circle.

Since b is unknown, the following control law is proposed:

$$\omega(k) = \frac{1}{\hat{b}}K_c e(k) \qquad (6.127)$$

which implies

$$t_p(k+1) = \sqrt{\omega(k)} = \sqrt{\frac{K_c}{\hat{b}(k)}|e(k)|} \qquad (6.128)$$

where $\hat{b}$ is the estimate of b.

The recursive least-squares algorithm is used to estimate the unknown parameters. The adaptation algorithm is given by the equations in Reference [11]:

$$\pi(k) = P(k-1)u(k) \qquad (6.129)$$

$$P(k) = \lambda^{-1}P(k-1) - \lambda^{-1}K(k)u^H(k)P(k-1) \qquad (6.130)$$

$$K(k) = \frac{\pi(k)}{\lambda + u^H(k)\pi(k)} \qquad (6.131)$$

$$\xi = d(k) - \hat{b}^H(k-1)u(k) \qquad (6.132)$$

$$\hat{b}(k) = \hat{b}(k-1) + K(k)\xi^H(k) \qquad (6.133)$$

$$u(k) = [t_p^2] \qquad (6.134)$$

The superscript H stands for the Hermitian transpose. In this case, all values are real and thus H can be replaced by T. In [11], K is described as the time-varying gain vector and P as the inverse correlation matrix. The λ in the algorithm represents the forgetting factor. Generally, the forgetting factor is selected to be smaller than but close to 1 and is set to 1 if all data is to be equally weighted. If the forgetting factor is not required, λ should be set to 1.

For the initiation of the adaptation algorithm, initial conditions for $P(0)$ and $\hat{b}(0)$ are required. $\hat{b}(0)$ should be chosen as accurately as possible and the inverse covariance matrix $P(0)$ is set to $P(0) = \delta^{-1}I$. δ should be large if the sensors are noisy and small if otherwise.

Assuming a true friction coefficient of 0.3 and an initial estimate of friction of 0.25, simulation results with a pulse amplitude of 1 are illustrated in Figure 6.21.

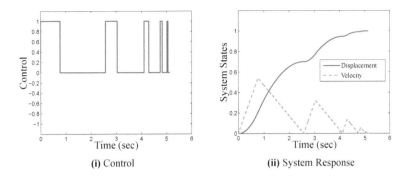

(i) Control   **(ii)** System Response

Figure 6.21: Underestimated Friction Results

It is clear from Figure 6.21(ii) that the underestimation of the coefficient of friction results in a net displacement which is smaller than anticipated for each pulse. This results in a series of positive pulses with monotonically decreasing pulse widths as shown in Figure 6.21(i). The maneuver time is 5.143 seconds, compared to 5.0200 for the nonadaptive pulse-width controller, which is about the same.

Assuming a true friction coefficient of 0.3 and an estimated friction of 0.35, simulation results with a pulse amplitude of 1 are illustrated in Figure 6.22.

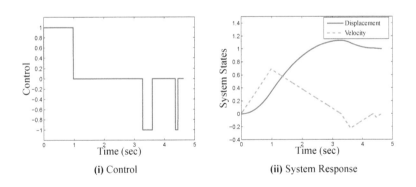

(i) Control   **(ii)** System Response

Figure 6.22: Overestimated Friction Results

It is clear from Figure 6.22(ii) that the overestimation of the coefficient of friction results in a net displacement which is larger than anticipated for each pulse. This results in a series of pulses whose signs change and whose widths monotonically decrease as shown in Figure 6.22(i). The maneuver time is 4.648 seconds, compared to 6.9440 seconds for the nonadaptive pulse-width controller, which is a significant

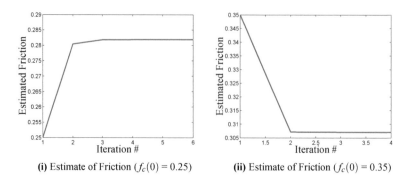

(i) Estimate of Friction ($f_c(0) = 0.25$)   **(ii)** Estimate of Friction ($f_c(0) = 0.35$)

Figure 6.23: Adaptive PWC

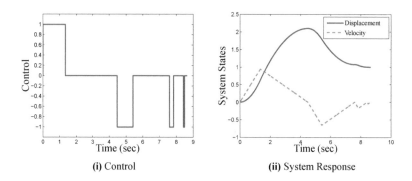

(i) Control   **(ii)** System Response

Figure 6.24: Unstable Friction Estimate

improvement.

Figure 6.23 illustrates the evolution of the estimate of the friction coefficient and it is clear that they tend toward the true friction coefficient of 0.3.

The nonadaptive system resulted in an unstable response when a friction estimate of 0.5 was used for a system with a true friction of 0.3. Figure 6.24 illustrates the stable response generated by the adaptive friction estimator used in conjunction with the pulse-width controller. The controller is initiated with a friction estimate of 0.5.

6.2.1.3   Pulse-Amplitude Pulse-Width Control (PAPWC)

The PAPWC permits adaptation of the amplitude of the pulse in addition to pulse width. This additional degree of freedom is exploited in digital implementation of the PAPWC since the pulse width of the PWC might not coincide with an integer multiple of the sampling interval. Since the desired displacement remains the same for the PWC and the PAPWC, altering the pulse width results in an adjustment of the pulse amplitude as well.

The pulse width needs to be rounded to the higher integer multiple of the sampling time T, i.e., $T_x = nT > t_p$, so that the corresponding pulse amplitude is smaller than or equal to the maximum permitted pulse amplitude u_p. The desired displacement d must be the same for both t_p and T_x, which results in:

$$d(t_p) = d(T_x) \tag{6.135}$$

$$bt_p^2 \, sgn(e) = b^* T_x^2 \, sgn(e) \tag{6.136}$$

$$\frac{u_p(u_p - f_c)}{2mf_c} t_p^2 \, sgn(e) = \frac{u_p^*(u_p^* - f_c)}{2mf_c} T_x^2 \, sgn(e) \tag{6.137}$$

The constant b must change to b^* to satisfy Equation (6.135). This can only be done by varying the pulse amplitude u_p^*, since this is the only free parameter in b^*. Solving Equation (6.137) results in [12]:

$$u_p^* = 0.5 f_c + 0.5 \sqrt{f_c^2 + 4u_p(u_p - f_c)\frac{t_p^2}{(T_x)^2}} \tag{6.138}$$

Instead of using t_p and u_p, the control pulse is specified by T_x and u_p^*. Since the coefficient of friction f_c is an unknown quantity, an adaptation algorithm is necessary for the implementation of the PAPWC scheme.

Rewriting the variable b as:

$$\hat{b} = \begin{bmatrix} \frac{1}{2mf_c} & \frac{1}{2m} \end{bmatrix} \begin{Bmatrix} u_p^2 \\ -u_p \end{Bmatrix} = \begin{bmatrix} A_1 & A_2 \end{bmatrix} \begin{Bmatrix} u_p^2 \\ -u_p \end{Bmatrix} \tag{6.139}$$

the variables that need to be estimated for are A_1 and A_2. From the estimates $\hat{A}_1$ and $\hat{A}_2$, an estimate of f_c can be derived by dividing $\hat{A}_2$ by $\hat{A}_1$. Using $\hat{f}_c$ in Equation (6.138), the pulse amplitude is calculated.

The PAPWC algorithm requires two parameters to be estimated and the algorithm is given by the equations:

$$\pi(k) = P(k-1)u(k) \tag{6.140}$$

$$P(k) = \lambda^{-1}P(k-1) - \lambda^{-1}K(k)u^H(k)P(k-1) \tag{6.141}$$

$$K(k) = \frac{\pi(k)}{\lambda + u^H(k)\pi(k)} \tag{6.142}$$

$$\xi = d(k) - \hat{a}^H(k-1)u(k) \tag{6.143}$$

$$\hat{a}(k) = \hat{a}(k-1) + K(k)\xi^H(k) \tag{6.144}$$

$$u(k) = \begin{bmatrix} f_p^2 t_p^2 & -f_p t_p^2 \end{bmatrix}^T \tag{6.145}$$

Assuming as in the section on adaptive pulse-width control, a true friction coefficient of 0.3 and an initial estimate of the coefficient of friction of 0.25 and 0.35, numerical simulations are used to illustrate the benefit of the adaptive pulse-amplitude pulse-width control. Figures 6.25 and 6.26 illustrate the control profiles and the corresponding evolution of the system states. It is clear from Figures 6.25 and 6.26 that the maneuver time is 4.641 and 4.93 seconds, for the maneuvers which correspond to an underestimation and overestimation of the coefficient of friction, respectively. This compares to 5.020 and 4.648 seconds, for the adaptive pulse-width control.

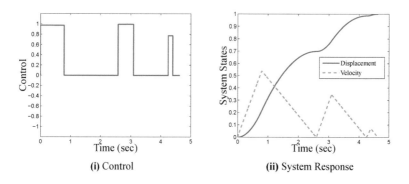

(i) Control　　　　　　　**(ii)** System Response

Figure 6.25: Adaptive PAPWC

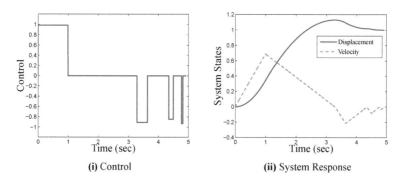

(i) Control　　　　　　　**(ii)** System Response

Figure 6.26: Adaptive PAPWC

Figure 6.27 illustrates the evolution of the estimate of the friction coefficient and it is clear that they tend toward the true friction coefficient of 0.3.

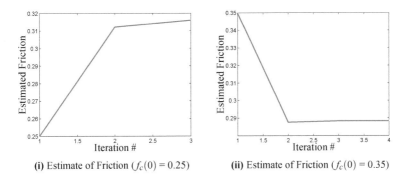

(i) Estimate of Friction ($f_c(0) = 0.25$) **(ii)** Estimate of Friction ($f_c(0) = 0.35$)

Figure 6.27: Adaptive PAPWC

A true comparison of the adaptive PWC and PAPWC requires that identical sampling times be selected. In the section on adaptive PWC, it was assumed that the pulse width could be implemented without any round-off of the maneuver time to an integer multiple of the sampling time. Figures 6.28 and 6.29 illustrate the time evolution of the system states and the control profiles when the controller is initiated with an under- and an overestimate of the coefficient of friction respectively. The solid lines correspond to the adaptive PWC and the dashed line to the adaptive PAPWC. It is clear from Figure 6.28, that the adaptive PWC results in a *limit cycle*. In both scenarios, the adaptive PAPWC outperforms the adaptive PWC. A sampling time of $T_s = 0.05$ was selected for the simulation.

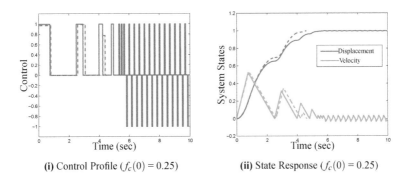

(i) Control Profile ($f_c(0) = 0.25$) **(ii)** State Response ($f_c(0) = 0.25$)

Figure 6.28: Adaptive PWC, PAPWC

The smallest displacement of the mass when subject to a pulse-width control profile is:

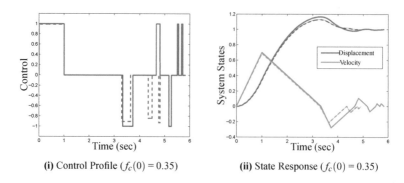

(i) Control Profile ($f_c(0) = 0.35$)   (ii) State Response ($f_c(0) = 0.35$)

Figure 6.29: Adaptive PWC, PAPWC

$$d_{min}(T_s) = \frac{u_p(u_p - f_c)}{2mf_c}T_s^2 = 0.0029, \tag{6.146}$$

for $m = 1$, $T_s = 0.05$, $u_p = 1$, and $f_c = 0.3$. The smallest displacement of the PAPWC is given by the equation:

$$d_{min}(T_s) = \frac{f_s(f_s - f_c)}{2mf_c}T_s^2 = 0.000167, \tag{6.147}$$

where f_s, the static friction is assumed to be 0.4. When the desired resolution is less than the smallest possible displacement, there exists the potential for *limit cycles*.

6.2.2   Benchmark Problem

This section will extend the pulse-amplitude pulse-width control technique for point-to-point control of the benchmark floating oscillator system subject to friction. A three pulse control is parameterized which assumes that the velocity of the mass subject to friction will not change sign over the duration of the maneuver. The pole-zero cancelation approach developed in Chapter 2 will be exploited in the design of the control profile.

6.2.2.1   Problem Formulation

The floating oscillator under the influence of friction is illustrated in Figure 6.30, where m_1, m_2 are the first and second mass, k is the spring constant, u the control input, f the friction force and x_1, x_2 are the positions of the first and second mass. The equation of motion of the system can be written as

$$M\ddot{\underline{x}} + K\underline{x} = D(u - f) \tag{6.148}$$

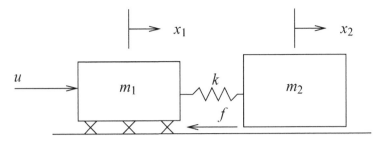

Figure 6.30: Floating Oscillator under Friction

where M, K, D, and $\underline{x}$ are

$$M = \begin{bmatrix} m_1 & 0 \\ 0 & m_2 \end{bmatrix}, \quad K = \begin{bmatrix} k & -k \\ -k & k \end{bmatrix}, \quad D = \begin{bmatrix} 1 \\ 0 \end{bmatrix}, \quad \underline{x} = \begin{bmatrix} x_1 \\ x_2 \end{bmatrix} \qquad (6.149)$$

The friction force is modeled as a nonlinear function of the velocity which accounts for static and Coulomb friction. The friction model can be represented as:

$$f(\underline{x}, u) = \begin{cases} f_c sgn(\dot{x}_1) & \text{if } \dot{x}_1 \neq 0 \\ f_s sgn(u_s) & \text{if } \dot{x}_1 = 0 \text{ and } u_s > f_s \\ u_s & \text{if } \dot{x}_1 = 0 \text{ and } u_s \leq f_s \end{cases} \qquad (6.150)$$

where f_s is the static friction, f_c is the Coulomb friction and u_s is the sum of the forces applied to the first mass, which is

$$u_s = u + k(x_2 - x_1). \qquad (6.151)$$

If the velocity of the first mass remains larger than zero during the entire maneuver, the friction force for a rest-to-rest maneuver becomes

$$f = f_c \left[1 - H(t - T_f) \right] \qquad (6.152)$$

where $H(\cdot)$ is the Heaviside step function and T_f is the final time. With this friction model, Equation (6.148) becomes

$$M\ddot{\underline{x}} + K\underline{x} = D\left\{ u - f_c \left[1 - H(t - T_f) \right] \right\}. \qquad (6.153)$$

It is more convenient to study the floating oscillator system if the decoupled equations of motion are used. Define the decoupled states as $\underline{z} = [\theta, q]^T$, where θ and q denote the rigid and flexible body states of the system. The transformation matrix V can be formed from the eigenvectors of the system, which decouples the system with the relationship $\underline{x} = V\underline{z}$. The decoupled system model is

$$\begin{bmatrix} \ddot{\theta} \\ \ddot{q} \end{bmatrix} + \begin{bmatrix} 0 & 0 \\ 0 & \omega^2 \end{bmatrix} \begin{bmatrix} \theta \\ q \end{bmatrix} = \begin{bmatrix} \dfrac{1}{m_1 + m_2} \\ -\dfrac{1}{m_1 + m_2} \end{bmatrix} \left\{ u - f_c \left[1 - H(t - T_f) \right] \right\} \qquad (6.154)$$

where V and ω are

$$V = \begin{bmatrix} 1 & -\dfrac{m_2}{m_1} \\ 1 & 1 \end{bmatrix}, \quad \omega = \sqrt{\dfrac{k(m_1 + m_2)}{m_1 m_2}} \tag{6.155}$$

For a rest-to-rest maneuver problem, the boundary conditions are

$$\begin{aligned} x_1(0) = x_2(0) = 0 && x_1(T_f) = x_2(T_f) = d \\ \dot{x}_1(0) = \dot{x}_2(0) = 0 && \dot{x}_1(T_f) = \dot{x}_2(T_f) = 0 \end{aligned} \tag{6.156}$$

where, d is the desired position at T_f and the corresponding boundary conditions of the decoupled states are

$$\begin{aligned} \theta(0) = \dot{\theta}(0) = 0 && \theta(T_f) = d, \ \dot{\theta}(T_f) = 0 \\ q(0) = \dot{q}(0) = 0 && q(T_f) = \dot{q}(T_f) = 0. \end{aligned} \tag{6.157}$$

6.2.2.2 Pole-Zero Cancelation

In our development, a three pulse profile is initially assumed as shown in Figure 6.31(a). The control profile results in an active control period of time T_3 at which point the system is traveling with nonzero velocity [13]. Following the implementation of the third pulse, the system is allowed to coast to rest. The pulse widths are

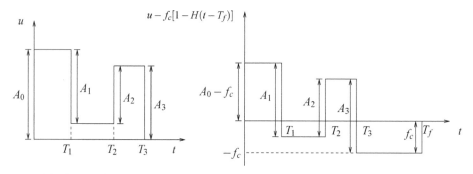

(a) Three Pulse Input   (b) Coulomb Friction Biased Input

Figure 6.31: Input Profile

selected by the user, and can correspond to the sampling time or an integral multiple of the sampling time for the digital implementation of the controller. The pulse amplitudes are subsequently determined to satisfy the boundary conditions. Since the velocity state of the first mass is assumed to remain positive over the duration of the maneuver, the new input to the linear system is biased by the magnitude of Coulomb friction as shown in Figure 6.31(b). The Coulomb friction biased input can be parameterized as

$$u(t) - f_c[1 - H(t - T_f)] = (A_0 - f_c) - A_1 H(t - T_1) \tag{6.158}$$
$$+ A_2 H(t - T_2) - A_3 H(t - T_3) + f_c H(t - T_f).$$

This biased input $u(t) - f_c[1 - H(t - T_f)]$ can be generated by a time-delay filter with a transfer function

$$G(s) = (A_0 - f_c) - A_1 e^{-sT_1} + A_2 e^{-sT_2} - A_3 e^{-sT_3} + f_c e^{-sT_f} \tag{6.159}$$

when it is subject to a unit step input. Since the control input in Figure 6.31(a) should be zero for $t \geq T_3$, we arrive at the constraint

$$A_0 - A_1 + A_2 - A_3 = 0. \tag{6.160}$$

In order to eliminate the vibration at the end of the maneuver, a pair of zeros of the transfer function (Equation 6.159), should cancel the flexible mode poles of the system [14]. To cancel the flexible mode poles, the real and imaginary parts of the transfer function $G(s)$ evaluated at $s = j\omega$, should be equated to zero resulting in the equations:

$$A_0 - A_1 \cos \omega T_1 + A_2 \cos \omega T_2 - A_3 \cos \omega T_3 = f_c(1 - \cos \omega T_f) \tag{6.161}$$

$$-A_1 \sin \omega T_1 + A_2 \sin \omega T_2 - A_3 \sin \omega T_3 = -f_c \sin \omega T_f \tag{6.162}$$

The displacement of the rigid-body at the final time is sum of the rigid-body displacement at $t = T_3$ and the coasting displacement:

$$\theta(T_f) = \theta(T_3) + \frac{m_1 + m_2}{2f_c} \left[\dot{\theta}(T_3)\right]^2 = d. \tag{6.163}$$

$\theta(T_3)$ and $\dot{\theta}(T_3)$ are found by solving the rigid-body differential equation:

$$\theta(T_3) = \frac{1}{2(m_1 + m_2)} \left[A_0(T_3)^2 - A_1(T_3 - T_1)^2 + A_2(T_3 - T_2)^2 - f_c(T_3)^2\right] \tag{6.164}$$

$$\dot{\theta}(T_3) = \frac{1}{m_1 + m_2} \left[A_0 T_3 - A_1(T_3 - T_1) + A_2(T_3 - T_2) - f_c T_3\right]. \tag{6.165}$$

The final time can be found by adding the coasting time to T_3. Since, satisfying Equations (6.161) and (6.162) is equivalent to the flexible states being forced to zero at the final time, the coasting time is found by solving the rigid body equation for $t > T_3$, by equating the velocity of the rigid body to be zero at $t = T_f$. The resulting final time is

$$T_f = T_3 + \frac{\dot{\theta}(T_3)(m_1 + m_2)}{f_c} = \frac{1}{f_c}(A_1 T_1 - A_2 T_2 + A_3 T_3) \tag{6.166}$$

The solution of the multi-pulse controller is determined by solving Equations (6.160), through (6.163) which results in four nonlinear equations in four unknowns $A_{0,1,2,3}$. An iterative approach for solving for the final time and the pulse amplitudes will be presented in the next section.

6.2.2.3   Iterative Solution Approach

If the flexible motion states $q(T_f)$ and $\dot{q}(T_f)$ are forced to zero at the final time, residual vibration will be eliminated. Since the final time in Equation (6.166) is a function of pulse amplitudes, Equations (6.161) and (6.162) are difficult to solve. To address this problem, the states of the flexible mode at $t = T_3$ that will force the flexible motion to be zero at the final time are derived. Solutions of the flexible mode equation at $t = T_3$ are

$$q(T_3) = -\frac{1}{\omega^2(m_1 + m_2)}[-f_c - A_0 \cos \omega T_3 + A_1 \cos \omega(T_3 - T_1)$$

$$(6.167)$$

$$-A_2 \cos \omega(T_3 - T_2) + A_3 + f_c \cos \omega T_3]$$

$$\dot{q}(T_3) = -\frac{1}{\omega(m_1 + m_2)}[A_0 \sin \omega T_3 - A_1 \sin \omega(T_3 - T_1)$$

$$(6.168)$$

$$+A_2 \sin \omega(T_3 - T_2) - f_c \sin \omega T_3].$$

The equation of motion of the flexible mode for the coasting period with initial conditions $q(T_3)$ and $\dot{q}(T_3)$ is

$$\ddot{q}_c + \omega^2 q_c = \frac{f_c[1 - H(t - T_c)]}{m_1 + m_2} \qquad (6.169)$$

where q_c is the flexible mode state for the coasting period and the coasting time $T_c = T_f - T_3$. The solution to Equation (6.169) is

$$q_c(t) = \frac{f_c}{m_1 + m_2}\left[\left(\frac{1}{\omega^2} - \frac{\cos \omega t}{\omega^2}\right) - \left(\frac{1}{\omega^2} - \frac{\cos \omega(t - T_c)}{\omega^2}\right)H(t - T_c)\right]$$

$$(6.170)$$

$$+ q(T_3) \cos \omega t + \frac{\dot{q}(T_3) \sin \omega t}{\omega}$$

$$\dot{q}_c(t) = \frac{f_c}{m_1 + m_2}\left(\frac{\sin \omega t}{\omega}\right) - q(T_3)\omega \sin \omega t + \dot{q}(T_3)\cos \omega t. \qquad (6.171)$$

At $t = T_c$, the flexible motion should be eliminated. By substituting T_c into Equations (6.170) and (6.171) and equating them to zero, the flexible states at $t = T_3$ which will force the flexible states to zero at the final time are

$$\begin{bmatrix} q(T_3) \\ \dot{q}(T_3) \end{bmatrix} = \begin{bmatrix} -\dfrac{f_c(\cos \omega T_c - 1)}{\omega^2(m_1 + m_2)} \\ -\dfrac{f_c \sin \omega T_c}{\omega(m_1 + m_2)} \end{bmatrix}. \qquad (6.172)$$

The flexible states at $t = T_3$ shown in Equation (6.172) will force the flexible motion to be eliminated at the end of the maneuver if T_c is known. However, the total maneuver time is a function of pulse amplitudes and therefore, an iterative approach is used to find the total maneuver time and T_c. Rewriting the constraint Equations (6.160), (6.167), and (6.168) in terms of A_1, A_2, and A_3, the constraint equations in matrix form become

$$
\begin{bmatrix}
-1 & 1 & -1 \\
-\cos\omega(T_3 - T_1) & \cos\omega(T_3 - T_2) & -1 \\
-\sin\omega(T_3 - T_1) & \sin\omega(T_3 - T_2) & 0
\end{bmatrix}
\begin{bmatrix}
A_1 \\
A_2 \\
A_3
\end{bmatrix}
$$

$$
= \begin{bmatrix}
-1 \\
-\cos\omega T_3 \\
-\sin\omega T_3
\end{bmatrix} A_0 + \begin{bmatrix}
0 \\
f_c\cos\omega T_3 - f_c + \omega^2(m_1 + m_2)q(T_3) \\
f_c\sin\omega T_3 - \omega(m_1 + m_2)\dot{q}(T_3)
\end{bmatrix}.
$$

$$\tag{6.173}$$

To find initial values for the input pulse amplitudes and final time, solve Equation (6.173) for A_1, A_2, and A_3 in terms of A_0 by letting $q(T_3) = \dot{q}(T_3) = 0$. By substituting A_1, A_2, and A_3 into the rigid body constraint in Equation (6.163), A_0 can be determined. Once the pulse amplitudes are found, the final time is found by substituting the pulse amplitudes into Equation (6.166). With this initial T_f, flexible states at $t = T_3$ are computed using Equation (6.172). The new pulse amplitudes and total maneuver time is calculated with these new flexible states at $t = T_3$. This procedure is repeated until the flexible states and final time converge.

Instead of the iterative approach proposed in this section, an optimization problem can be posed to minimize T_c subject to the constraints given by Equations (6.163), (6.166), (6.172), and (6.173), where the variables to be solved for are A_0, A_1, A_2, A_3, and T_c. We have verified the solution generated by the iterative algorithm and it is coincident with that of the optimization algorithm.

6.2.2.4 Bounds on Control Pulse

The multiple pulse input development so far allows the user to select pulse widths for the prescribed displacement. However, bounds on pulse amplitudes should be examined because of the positive velocity constraint. The first pulse amplitude A_0 should be greater than the static friction, f_s. Also, the pulse amplitude should be less than the permitted peak input amplitude, u_p. Therefore, the bounds on A_0 become

$$f_s < A_0 \leq u_p \tag{6.174}$$

The velocity of the first mass can be written as

$$
\dot{x}_1(t) = \frac{1}{m_1 + m_2}[(A_0 - f_c)g(t) - A_1 g(t - T_1)H(t - T_1) + A_2 g(t - T_2)H(t - T_2)
$$
$$
- A_3 g(t - T_3)H(t - T_3)] \tag{6.175}
$$

where $g(t)$ is a function defined by the following equation:

$$g(t) = t + \frac{m_2\sin\omega t}{m_1\omega} \tag{6.176}$$

For $0 < t < T_1$, Equation (6.175) reduces to

$$\dot{x}_1(t) = \frac{A_0 - f_c}{m_1 + m_2} g(t) \qquad (0 < t < T_1) \tag{6.177}$$

Rathbun [15] showed that selection of a specific mass ratio guarantees positive velocity of the first mass for $0 < t < T_1$. Since A_0 is greater than f_c, Equation (6.177) is always positive if $g(t)$ is positive. Rewriting the equation for $g(t)$ yields

$$g(t) = t\left(1 + \frac{m_2}{m_1}\frac{\sin \omega t}{\omega t}\right) = t\left[1 + \frac{m_2}{m_1}\text{sinc}(\omega t)\right] > 0 \tag{6.178}$$

Knowing that $-0.2172 < \text{sinc}(\omega t) < 1$, Equation (6.178) results in bounds on the ratio of the first and the second mass which guarantees positive velocity, which is

$$\frac{m_1}{m_2} > 0.2172. \tag{6.179}$$

For systems which violate this mass ratio, multi-pulse control profiles can also be designed. However, the model of the system will be represented by a series of piecewise linear systems which makes solving the problem significantly more difficult. The local maximum and minimum of $g(t)$ coincides with the time when the velocity of the first mass reaches its local maximum and minimum. The time derivative of $g(t)$ is

$$\dot{g}(t) = 1 + \frac{m_2}{m_1}\cos \omega t \tag{6.180}$$

It is patent that $\dot{g}(t)$ is always positive and hence $g(t)$ is a nondecreasing function if

$$\frac{m_1}{m_2} \geq 1 \tag{6.181}$$

If $0.2172 < \dfrac{m_1}{m_2} < 1$, $g(t)$ is always positive but not a monotonically increasing function. Similarly, the velocity of the first mass for $T_1 < t < T_2$ becomes

$$\dot{x}_1(t) = \frac{A_0 - f_c}{m_1 + m_2} g(t) - \frac{A_1}{m_1 + m_2} g(t - T_1) \qquad (T_1 < t < T_2). \tag{6.182}$$

The first mass velocity will reach its minimum or maximum values for $T_1 < t < T_2$ when

$$\ddot{x}_1(t) = \frac{1}{m_1 + m_2}\left[A_0 - A_1 - f_c + \frac{(A_0 - f_c)m_2}{m_1}\cos \omega t - \frac{A_1 m_2}{m_1}\cos \omega(t - T_1)\right] = 0 \tag{6.183}$$

Solving Equation (6.183) for t, the velocity of the first mass will have extreme values when

$$t_{min,max} = \frac{1}{\omega}\left[\sin^{-1}\left(-\frac{C}{D}\right) - \phi + 2n\pi\right] \tag{6.184}$$

where, C, D, and ϕ are defined as

$$C = A_0 - A_1 - f_c$$

$$D = \sqrt{\left[\frac{m_2(A_0 - f_c - A_1 \cos \omega T_1)}{m_1}\right]^2 + \left[\frac{-m_2 A_1 \sin \omega T_1}{m_1}\right]^2} \qquad (6.185)$$

$$\phi = \tan^{-1}\left(-\frac{A_0 - f_c - A_1 \cos \omega T_1}{A_1 \sin \omega T_1}\right)$$

for Equation (6.184). The $t_{min,max}$ might have none or multiple solutions, for different integer value of n because $t_{min,max}$ should lie in the interval $[T_1 \quad T_2]$. Therefore, the minimum velocity of the first mass in this interval becomes

$$\min(\dot{x}_1) = \begin{cases} \min[\dot{x}_1(T_1), \dot{x}_1(T_2), \dot{x}_1(t_{min,max})] & \text{if } 0.2172 < m_1/m_2 < 1 \\ \min[\dot{x}_1(T_1), \dot{x}_1(T_2)] & \text{if } m_1/m_2 \geq 1 \end{cases}$$

$$(6.186)$$

Therefore, the constraint on the pulse amplitudes for positive velocity for $T_1 < t < T_2$ can be derived from Equation (6.182), which is

$$A_1 < (A_0 - f_c)\frac{g(t_{min})}{g(t_{min} - T_1)} \qquad (6.187)$$

where, t_{min} is the time when the velocity of the first mass is at its minimum. If this condition is violated, stiction of the first mass will occur. The control profile should be modified because the constant friction assumption is no longer valid. For $T_2 < t < T_3$, positive A_2 will guarantee positive velocity of the first mass provided the velocity is positive for $t < T_2$. The velocity of the first mass for $T_3 < t < T_f$ is

$$\dot{x}_1(t) = \dot{\theta}(T_3) - \frac{f_c}{m_1 + m_2}\left[(t - T_3) + \frac{m_2 \sin \omega(t - T_f)}{m_1 \omega}\right] \qquad (6.188)$$

According to Equation (6.188), the velocity of the first mass never goes to zero if the mass ratio satisfies the condition shown in Equation (6.179). Equation (6.188) is a monotonically decreasing function if $m_1/m_2 < 1$.

6.2.2.5 Numerical Simulation

Numerical simulations are used to illustrate the performance of the proposed controllers. The parameter values used in the simulation are shown in Table 6.1. The first simulation is performed with the initial and final states of

$$\begin{aligned} x_1(0) = x_2(0) = 0 \quad & x_1(T_f) = x_2(T_f) = 0.1 \\ \dot{x}_1(0) = \dot{x}_2(0) = 0 \quad & \dot{x}_1(T_f) = \dot{x}_2(T_f) = 0. \end{aligned} \qquad (6.189)$$

The pulse widths are chosen such that $T_1 = 0.1$ seconds, $T_2 = 0.2$ seconds, and $T_3 = 0.3$ seconds. If the condition number of the matrix on the left-hand side of Equation (6.173) is large, or if the resulting solution is one which violates Equation (6.174), one has to select a different pulse width. The controller for this problem

Table 6.1: Parameters Used in the Simulation

Symbol	Description	Value
m_1	Mass 1	$80\ kg$
m_2	Mass 2	$100\ kg$
ω	Natural Frequency	$50\ rad/sec$
u_p	Peak Input	$1000\ N$
f_s	Static Friction	$137\ N$
f_c	Coulomb Friction	$111\ N$

is shown in Figure 6.32 and the corresponding response of the system is plotted in Figure 6.33. The solid line and the dashed line represent the states of the first and second masses, respectively. It is evident that the first mass velocity is always pos-

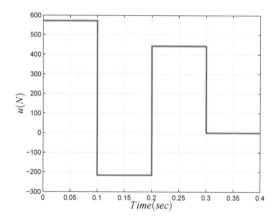

Figure 6.32: Three-Pulse Control Input ($d = 0.1$)

itive and therefore unidirectional friction force is applied to the system during the maneuver. The system begins to coast at $t = 0.3$ seconds and the undesirable vibration is eliminated at the final time. The response plot of the decoupled states are also shown in Figure 6.34. The flexible states at $t = 0.3$ are forced such that at the final time the flexible states become zero.

Figure 6.35 illustrates the variation of the pulse input amplitudes as a function of commanded displacement. The pulse widths are selected to be, $T_1 = 0.04$ seconds, $T_2 = 0.08$ seconds, and $T_3 = 0.12$ seconds. For this specific pulse width, the velocity of the mass goes to zero during the maneuver for $d = 0.0039$ m, which corresponds to the lower displacement bound with the chosen pulse widths. For a displacement of

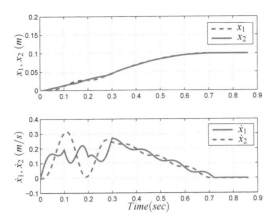

Figure 6.33: Response of the System ($d = 0.1m$)

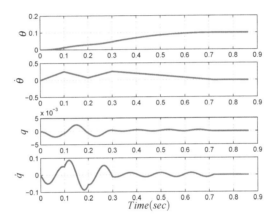

Figure 6.34: Response of the System, Decoupled States ($d = 0.1m$)

$d = 0.278$ m, the control input saturates which corresponds to the upper displacement bound for the prescribed pulse width. The upper and lower bounds can be changed by selecting pulses of different width.

The relationship of the pulse input amplitudes for different natural frequency values is illustrated in Figure 6.36. It is clear from Figure 6.36 that there are no feasible solutions in the frequency range of 76 to 79 rad/sec. This is because the matrix in Equation (6.173) becomes singular when the switching time is chosen such that:

$$\omega(T_3 - T_2) = 2n\pi \quad \text{or} \quad \omega(T_3 - T_1) = 2n\pi \tag{6.190}$$

where, n is a positive integer. Therefore different pulse widths should be selected for designing a controller if the pulse widths chosen make the condition number of the

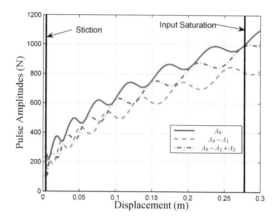

Figure 6.35: Pulse Amplitudes vs. Displacement

matrix in Equation (6.173) very large.

Incorrect estimates of the friction and frequency of the system result in a steady state error and residual vibration at the final time. Figure 6.37 represents the steady state error of the first mass with respect to variations in the spring constant and Coulomb friction. It can be seen that the final displacement is more sensitive to the friction variation compared to variations in the frequency. Robustness to the frequency variation can be obtained with additional pulses which permits placing additional zeros of the time-delay filter at the estimated pole locations which correspond to the vibratory modes. The friction variation can be handled by successive application of the proposed control profile until the desired position is reached. Sta-

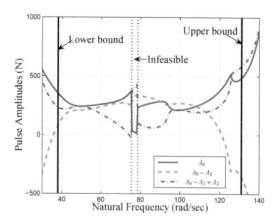

Figure 6.36: Pulse Amplitudes vs. Natural Frequency

bility of the iterative operation is guaranteed for bounded friction and spring constant variations if the steady state error remains less than 100%. The performance of the controller can be enhanced if the uncertain/varying parameters are adapted during the iterative application of the input. Figure 6.38 shows the variation of the resid-

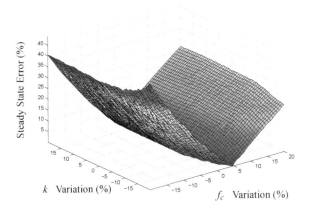

Figure 6.37: Steady-State Error with Friction and Spring Constant Variations

ual energy with respect to the variations in friction and spring constant. The total residual energy includes the energy from the residual vibration of the second mass in addition to the norm of the steady-state error of the first mass. It can be seen that if the friction coefficient is overestimated, the residual energy increases at a rate greater than when the friction coefficient is underestimated.

6.3 Summary

This chapter focused on the development of open-loop controllers for systems with friction. As in the previous chapters, the two systems considered to illustrate the controllers are a rigid body system and the floating oscillator benchmark problem which includes one rigid body mode and one flexible mode. The first problem considered is the design of time-optimal control profiles. A simple technique which eliminates the frictional nonlinearity by integrating the friction force into a linearizing control profile is proposed. The resulting control profile is not bang-bang, but it drives a linear system permitting the formulation of a parameter optimization problem for the rest-to-rest maneuver of a linear system.

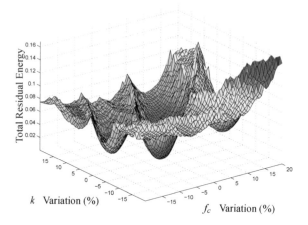

Figure 6.38: Total Residual Energy with Friction and Spring Constant Variations

A pulse-width controller which is useful for small motions is also presented in this chapter. A closed-form expression describing the relationship between the pulse width and the displacement as a function of the coefficient of friction is determined. The effect of uncertainty on the stability of the repeated application of the pulse-width controller is studied and bounds on the overestimation of the coefficient of friction is solved for which is useful for the analysis of the controlled system. To improve the performance of the pulse-width controller, an estimator is included into the control algorithm which results in reduction in the maneuver time compared to the nonadaptive pulse-width controller. To account for constraints on the pulse width which results from the sampling time, a pulse-width pulse-amplitude controller is developed which in conjunction with the estimator results in improved performance. The technique to design pulse-width pulse-amplitude controller for rigid-body systems is extended to control flexible structures subject to friction and is illustrated on the benchmark floating oscillator problem.

Exercises

6.1 Design a time-delay filter for a second order system subject to Coulomb friction:

$$\ddot{x} + 0.1\dot{x} + x + 0.1\text{sgn}(\dot{x}) = u$$

to move the system states where the initial and final states are:

$$x(0) = 0, \dot{x}(0) = 0, \quad x(t_f) = 1, \dot{x}(t_f) = 0$$

Hint: Since the velocity does not change sign for the posicast control, the friction force acts as a bias.

6.2 Design a time-optimal control for a rigid body system subject to Coulomb friction:

$$\ddot{x} + 0.1\dot{x} + 0.1\text{sgn}(\dot{x}) = u$$

to move the system states where the initial and final states are:

$$x(0) = 0, \dot{x}(0) = -1, \quad x(t_f) = 1, \dot{x}(t_f) = 0$$

The control input is constrained to:

$$-1 \le u(t) \le 1.$$

Show that the time-optimal control is a single-switch bang-bang profile with a switch time of T = 2.2699 and a final time of t_f = 3.2853.

6.3 A high-precision gantry robot is used for accurately riveting panels on wings of an aircraft. The end effector is located on a cantilever resulting in vibratory motion. The model for a single-axis gantry robot is:

$$\ddot{\theta} + 0.1\text{sgn}(\dot{\theta}) + 0.5\ddot{\phi} = u$$

$$0.5\ddot{\theta} + 0.5\ddot{\phi} + 500\phi = 0$$

Design a three-pulse pulse-amplitude control profile to move the system states from:

$$\theta(0) = 0, \dot{\theta}(0) = 0, \phi(0) = 0, \dot{\phi}(0) = 0$$

to

$$\theta(t_f) = 0.01, \dot{\theta}(t_f) = 0, \phi(t_f) = 0, \dot{\phi}(t_f) = 0$$

Illustrate that the three-pulse profile that maneuvers the system from the initial position of rest to the final position of rest is given by the profile illustrated in Figure 6.39.

6.4 For the undamped second-order frictional system given by the equation:

$$\ddot{x} + \omega^2 x + f_c \text{sgn}(\dot{x}) = u,$$

show that the application of a pulse of amplitude A applied for a time interval of T_1 will bring the velocity state to rest at a time

$$T_2 = -\frac{1}{\omega}\left(-\omega T_1 + \arctan\frac{\sin(\omega T_1)(A - f_c)}{A\cos(\omega T_1) - A - f_c\cos(\omega T_1)}\right)$$

and the corresponding displacement at time T_2 is:

$$\frac{A - f_c}{\omega^2}(\cos(\omega(T_1 - T_2))(1 - \cos(\omega T_1)) - \sin(\omega(T_1 - T_2))\sin(\omega T_1)) - \frac{f_c}{\omega^2}(1 - \cos(\omega(T_1 - T_2)))$$

Determine the pulse profile to move a system with a natural frequency $\omega=1$, from the initial to the final states given by:

$$x(0) = 0, \dot{x}(0) = 0, \quad x(T_2) = 1, \dot{x}(T_2) = 0.$$

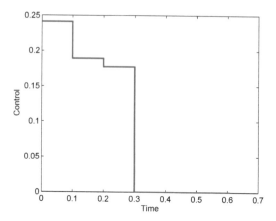

Figure 6.39: Three-Pulse Control Profile

References

[1] K. Peng, G. Cheng, B.M. Chen, and T.H. Lee. Improvement on a hard disk drive servo system using friction and disturbance compensation. In *Proceedings. 2003 IEEE/ASME International Conference on Advanced Intelligent Mechatronics, 2003. AIM 2003.*, volume 2, pages 1160–1165, July 2003.

[2] K. Takaishi, T. Imamura, Y. Mizoshita, S. Hasegawa, T. Ueno, and T. Yamada. Microactuator control for disk drive. *IEEE Transactions on Magnetics.*, 32(3):1863–1866, May 1996.

[3] J. R. Ryoo, K. B. Jin, T.-Y. Doh, and M. J. Chung. New fine seek control for optical disk drives. In *American Control Conference, 1999. Proceedings of the 1999*, volume 5, pages 3635–3639, 1999.

[4] V. M. Hung and U. J. Na. Adaptive neural fuzzy control for robot manipulator friction and disturbance compensator. In *Control, Automation and Systems, 2008. ICCAS 2008. International Conference on*, pages 2569–2574, October 2008.

[5] F. Bruni, F. Caccavale, C. Natale, and L. Villani. Experiments of impedance control on an industrial robot manipulator with joint friction. In *Control Applications, 1996., Proceedings of the 1996 IEEE International Conference on*, pages 205–210, September 1996.

[6] A. Piazzi and A. Visioli. Optimal dynamic-inversion-based control of an overhead crane. 149(5):405–411, September 2002.

[7] J.-J. Kim, R. Kased, and T. Singh. Time-optimal control of flexible systems subject to friction. *Optimal Control: Applications and Methods*, 29(4):257–277, 2008.

[8] Arthur E. Bryson and Yu-Chi Ho. *Applied Optimal Control: Optimization, Estimation and Control*. Hemisphere Publishing, 1975.

[9] C. F. van Loan. Computing integrals involving the matrix exponential. *IEEE Trans. on Auto. Cont.*, (3):395–404, 1978.

[10] S. Yang and M. Tomizuka. Adaptive pulse width control for precise positioning under the influence of stiction and Coulomb friction. *Journal of Dynamic Systems, Measurements, and Control*, 110:221–227, September 1988.

[11] S. Haykin. *Adaptive Filter Theory*. Prentice Hall, 2002.

[12] J. J. M. van de Wijdeven and T. Singh. Adaptive pulse amplitude pulse width control of systems subject to Coulomb and viscous friction. In *ACC*, Denver, CO, 2003.

[13] J-J. Kim and T. Singh. Controller design for flexible systems with friction: Pulse amplitude control. *ASME Journal of Dynamic Systems, Measurement and Control*, 127(3):336–344, September 2005.

[14] T. Singh and S. R. Vadali. Robust time delay control. *ASME J. of Dyn. Systems, Measurements, and Cont.*, 115(2):303–306, 1993.

[15] D. B. Rathbun. *Pulse Modulation Control for Flexible Systems Under the Influence of Nonlinear Friction*. PhD thesis, University at Washington, Seattle, WA, 2001.

7

Numerical Approach

The advancement and perfection of mathematics are intimately connected with the prosperity of the State.

Napoleon (1769–1821); military and political leader of France

THIS chapter describes different techniques to solve for the parameters of input shaper/time-delay filters (IS/TDF) and saturating controllers. Except for a very small class of systems, closed form solutions for IS/TDF and saturating controllers cannot be derived. One has to therefore resort to numerical techniques to determine the optimal control profiles. This chapter describes three techniques to determine the IS/TDF and saturating controllers. The first entails parameterizing the control profile and solving for the optimal parameters via a gradient based constrained optimization algorithm. The second technique exploits the strengths of linear programming in conjunction with a *bisection* algorithm to determine the optimal IS/TDF and saturating control profiles. The third exploits semidefinite optimization techniques to optimize a linear cost subject to linear matrix inequalities (LMI). All three techniques are illustrated on simple examples. MATLAB program listing which encode the two algorithms are also provided. This chapter deals with input shaper/time-delay filters, saturating controllers, minimax controllers that desensitize the control profile to modeling errors, and also presents results for benchmark examples.

7.1 Parameter Optimization

Input shapers/time-delay filters can be parameterized in terms of delay times and gains. For instance, the IS, or the nonrobust time-delay filter can be represented as:

$$A_1\delta(t - T_1) + A_2\delta(t - T_2) \quad \text{(Impulse Sequence)} \tag{7.1}$$

$$A_1 e^{-sT_1} + A_2 e^{-sT_2} \quad \text{(Time-Delay Filter)} \tag{7.2}$$

where A_1, A_2, T_1, and T_2 are the parameters to be determined. The constraints on the parameters are that the impulse sequence force the residual energy of an underdamped harmonic oscillator

$$\ddot{x} + 2\zeta\omega\dot{x} + \omega^2 x = u, \tag{7.3}$$

subject to the impulse sequence to be zero at time T_2. The corresponding constraint on the time-delay filter is that a pair of zeros of the time-delay filter cancel the poles of the underdamped harmonic oscillator which are:

$$s = -\zeta\omega \pm j\omega\sqrt{1 - \zeta^2}. \tag{7.4}$$

The residual energy of the system resulting from a sequence of impulses is

$$V(\omega, \zeta) = e^{-\zeta\omega T_n}\sqrt{C(\omega, \zeta)^2 + S(\omega, \zeta)^2} \tag{7.5}$$

where

$$C(\omega, \zeta) = \sum_{i=1}^{n} A_i e^{\zeta\omega T_i}\cos(\omega\sqrt{1 - \zeta^2}T_i) \tag{7.6}$$

$$S(\omega, \zeta) = \sum_{i=1}^{n} A_i e^{\zeta\omega T_i}\sin(\omega\sqrt{1 - \zeta^2}T_i) \tag{7.7}$$

and n is the number of impulses. Assuming $T_1 = 0$, for a two-impulse input shaper, the zero residual energy and pole-cancelation constraints can both be reduced to:

$$A_1 + A_2 e^{\zeta\omega T_2}\cos(\omega\sqrt{1 - \zeta^2}T_2) = 0 \tag{7.8a}$$

$$A_2 e^{\zeta\omega T_2}\sin(\omega\sqrt{1 - \zeta^2}T_2) = 0 \tag{7.8b}$$

Finally, the requirement that the steady-state output of the time-delay filter be equal to the reference input results in the constraint

$$A_1 + A_2 = 1. \tag{7.9}$$

This results in the solution

$$A_1 = \frac{\exp(\frac{\zeta\pi}{\sqrt{1-\zeta^2}})}{1 + \exp(\frac{\zeta\pi}{\sqrt{1-\zeta^2}})}, \quad A_2 = \frac{1}{1 + \exp(\frac{\zeta\pi}{\sqrt{1-\zeta^2}})}, \quad T_2 = \frac{\pi}{\omega\sqrt{1 - \zeta^2}} \tag{7.10}$$

For systems with multiple modes, or with multi-inputs, closed-form solutions for the IS/TDF cannot be derived. Numerical optimization is necessary in such situations. The IS/TDF with N delays can be represented as

$$\sum_{i=1}^{N} A_i \delta(t - T_i) \quad \text{(IS)} \tag{7.11a}$$

$$\sum_{i=1}^{N} A_i \exp(-sT_i) \quad \text{(TDF)} \tag{7.11b}$$

with the constraints

$$\sum_{i=1}^{N} A_i = 1 \tag{7.12}$$

and

$$\sum_{i=1}^{N} A_i exp(-\zeta_j \omega_j T_i) \cos(T_i \omega_j \sqrt{1 - \zeta_j^2}) = 0, \forall j \quad (IS) \tag{7.13}$$

$$\sum_{i=1}^{N} A_i exp(-\zeta_j \omega_j T_i) \sin(T_i \omega_j \sqrt{1 - \zeta_j^2}) = 0, \forall j \quad (IS) \tag{7.14}$$

$$\Re\left(\sum_{i=1}^{N} A_i exp(-s_j T_i)\right) = 0, \forall s_j \quad (TDF) \tag{7.15}$$

$$\Im\left(\sum_{i=1}^{N} A_i exp(-s_j T_i)\right) = 0, \forall s_j \quad (TDF) \tag{7.16}$$

where $j = 1, 2, ..., m$, where m is the number of poles to be canceled. The optimization problem to be solved is:

$$\min J = T_N \tag{7.17a}$$

subject to $\tag{7.17b}$

$$\Re\left(\sum_{i=1}^{N} A_i exp(-s_j T_i)\right) = 0, \quad \forall s_j, j = 1, 2, .., m \tag{7.17c}$$

$$\Im\left(\sum_{i=1}^{N} A_i exp(-s_j T_i)\right) = 0, \quad \forall s_j, j = 1, 2, .., m \tag{7.17d}$$

$$\sum_{i=1}^{N} A_i = 1 \tag{7.17e}$$

$$A_i \geq 0 \tag{7.17f}$$

$$T_i - T_{i+1} < 0 \quad \forall i = 1, 2, ..., N - 1. \tag{7.17g}$$

MATLAB Script 7.1 returns the cost and the gradient of the cost with respect to the optimization variables. The inclusion of analytical gradients are optional. In the absence of analytical gradients, MATLAB uses finite difference to determine the gradients.

MATLAB Program 7.1:

```
function [f,df] = cost(x,poles);

nDelays = floor(length(x)/2);
```

```
T = x(1:nDelays);                      % Time Delays
A = x(nDelays+1:end);                  % Delay Gains

f = T(end); df = zeros(length(x)); df(nDelays) = 1;
```

MATLAB Script 7.2 returns the equality and inequality constraints.

MATLAB Program 7.2:

```
function [g,geq] = constraints(x,poles);

nDelays = floor(length(x)/2);
T = [0 x(1:nDelays)];                  % Time Delays
A = x(nDelays+1:end);                  % Delay Gains

% Check if repeated poles have to be canceled
uniq_poles = unique(poles);
for indx = 1:length(uniq_poles),
    rpoles(indx) = length(find(poles==uniq_poles(indx)));
end;

counter = 1;
for ind = 1:length(uniq_poles),
    for rindx = 1:rpoles(ind),
        geq_temp = 0;
        for Del_ind = 1:length(T),
            geq_temp = geq_temp + A(Del_ind)*exp(-poles(ind)*...
            T(Del_ind))*(T(Del_ind)^(rindx-1));
        end;
    geq(counter)=real(geq_temp);
    geq(counter+1)=imag(geq_temp);
    counter = counter+2;
    end;
end;
geq(counter) = sum(A)-1;

counter = 1;
for ind = 1:length(T)-1,
    g(counter) = T(ind)-T(ind+1);
    counter= counter+1;
end;

for ind = 1:length(A),
    g(counter) = -A(ind);
    counter=counter+1;
end;
```

An IS/TDF for a single mode is designed using the program described above. Consider a system with a pair of undamped poles at $s = \pm j$. The following MATLAB script is executed:

MATLAB Program 7.3:

```
x0 = [3 1 0.7]; poles = j;

x = fmincon('cost',x0,[],[],[],[],[],[],'constraints',[],poles)
```

where $x0$ is the initial guess of the parameters to be optimized for. The vector $x0$ contains three terms corresponding to the parameters T, A_1, and A_2 of a single time-delay filter. The variable *poles* is the listing of the poles to be canceled. Only one of the pair of complex conjugate poles need to be listed. The MATLAB function for constrained minimization is "fmincon." The output of the optimizer is

```
x =

    3.1416    0.5000    0.5000
```

which coincides with the solution given by Equations (2.14) and (2.17) which corresponds to the solution of the *Posicast* controller. If it is desired to desensitize the IS/TDF to errors in estimated natural frequency, it has been shown that one needs to place multiple zeros of the time-delay filter at the estimated location of the poles to be canceled. This will requires additional delays in the filter transfer function. The MATLAB script for this problem is:

MATLAB Program 7.4:

```
x0 = [3 6 0.3 0.7 0.4]; poles = [j;j];

x = fmincon('cost',x0,[],[],[],[],[],[],'constraints',[],poles)
```

The output of the optimizer is

```
x =

    3.1416    6.2832    0.2500    0.5000    0.2500
```

which is the same as the solution given by Equation (2.41) determined in Chapter 2. This also corresponds to the ZVD input shaper by Singer and Seering [1].

Saturating controllers refer to control profiles where the control constraint is active for all or part of the maneuver. Such control profiles correspond to time-optimal, fuel-time optimal, fuel-constrained time-optimal control profiles besides others. Saturating control profiles can be parameterized by switch times of the control profiles. For instance, the time-optimal solution for rest-to-rest maneuver of a double integrator is a single switch bang-bang control profile. The control profile is thus, parameterized in terms of the switch time and the maneuver time.

A saturating control profile can be parameterized as

$$u(s) = \frac{1}{s} \sum_{i=1}^{N} A_i exp(-sT_i) \tag{7.18}$$

where each A_i belongs to the set

$$A_i = \begin{bmatrix} -2 & -1 & 1 & 2 \end{bmatrix}. \tag{7.19}$$

The saturating control profile can be represented as the output of a time-delay filter subject to a step input. The transfer function of the corresponding time-delay filter is:

$$G_c(s) = \sum_{i=1}^{N} A_i exp(-sT_i). \tag{7.20}$$

Equation (7.18) in conjunction with Equation (7.19) can generate bang-bang and bang-off-bang control profiles, if $u(t)$ is constrained to lie between -1 and 1. The constraints for parameter optimization problem for rest-to-rest maneuver can be derived by requiring that a set of zeros of the time-delay filter cancel the poles of the system. This results in the equations:

$$\Re \left(\sum_{i=1}^{N} A_i exp(-s_j T_i) \right) = 0 \ \forall s_j \tag{7.21a}$$

$$\Im \left(\sum_{i=1}^{N} A_i exp(-s_j T_i) \right) = 0 \ \forall s_j \tag{7.21b}$$

If R repeated poles need to be canceled, the corresponding constraint equations are:

$$\Re \left(\sum_{i=1}^{N} (T_i)^{k-1} A_i exp(-s_j T_i) \right) = 0 \ \forall s_j \tag{7.22a}$$

$$\Im \left(\sum_{i=1}^{N} (T_i)^{k-1} A_i exp(-s_j T_i) \right) = 0 \ \forall s_j \tag{7.22b}$$

where k varies from 1 to R. The final constraint is derived from the system transfer function

$$\frac{y(s)}{u(s)} = G(s) \tag{7.23}$$

so as to satisfy the rigid-body boundary condition. The constraint can be written as:

$$y_f = \lim_{s \to 0} G(s) G_c(s) = \lim_{s \to 0} \left(G(s) \sum_{i=1}^{N} A_i exp(-sT_i) \right). \tag{7.24}$$

Since the transfer function of the time-delay filter $G_c(s)$ has zeros at location of the poles of the system, the limit value is indeterminate. Repeatedly apply *L'Hôpital's* rule until the limit is determinate. That is the final constraint for the parameter optimization problem.

Example 7.1: Design a time-optimal controller for the rest-to-rest maneuver of the benchmark problem:

$$\ddot{x}_1 + x_1 - x_2 = u \tag{7.25a}$$

$$\ddot{x}_2 - x_1 + x_2 = 0 \tag{7.25b}$$

where the boundary conditions are

$$x_1(0) = x_2(0) = \dot{x}_1(0) = \dot{x}_2(0) = 0 \tag{7.26a}$$

$$x_1(t_f) = x_2(t_f) = 1, \dot{x}_1(t_f) = \dot{x}_2(t_f) = 0 \tag{7.26b}$$

Parameterizing the optimal control profile as

$$u(s) = \frac{1}{s}\left(1 - 2e^{-sT_1} + 2e^{-sT_2} - 2e^{-sT_3} + e^{-sT_4}\right), \tag{7.27}$$

the transfer function of the system where the output is the displacement of the second mass is

$$G(s) = \frac{x_2}{u} = \frac{1}{s^2(s^2 + 2)}. \tag{7.28}$$

Using Equation (7.24), we have

$$x_2 = \lim_{s \to 0} \frac{1}{s^2(s^2 + 2)}\left(1 - 2e^{-sT_1} + 2e^{-sT_2} - 2e^{-sT_3} + e^{-sT_4}\right) = \frac{0}{0}. \tag{7.29}$$

Using *L'Hôpital's* rule twice we have,

$$x_2 = \lim_{s \to 0} \frac{1}{12s^2 + 4}\left(-2T_1^2 e^{-sT_1} + 2T_2^2 e^{-sT_2} - 2T_3^2 e^{-sT_3} + T_4^2 e^{-sT_4}\right) \tag{7.30}$$

which results in the constraint

$$-2T_1^2 + 2T_2^2 - 2T_3^2 + T_4^2 = 4; \tag{7.31}$$

the parameter optimization problem can be stated as

$$\min\ J = T_4 \tag{7.32a}$$

subject to $\tag{7.32b}$

$$\left(1 - 2e^{-sT_1} + 2e^{-sT_2} - 2e^{-sT_3} + e^{-sT_4}\right) = 0 \text{ for } s = 0 \tag{7.32c}$$

$$\left(-2T_1 e^{-sT_1} + 2T_2 e^{-sT_2} - 2T_3 e^{-sT_3} + T_4 e^{-sT_4}\right) = 0 \text{ for } s = 0 \tag{7.32d}$$

$$\Re\left(1 - 2e^{-sT_1} + 2e^{-sT_2} - 2e^{-sT_3} + e^{-sT_4}\right) = 0 \text{ for } s = j\sqrt{2} \tag{7.32e}$$

$$\Im\left(1 - 2e^{-sT_1} + 2e^{-sT_2} - 2e^{-sT_3} + e^{-sT_4}\right) = 0 \text{ for } s = j\sqrt{2} \qquad (7.32\text{f})$$

$$-2T_1^2 + 2T_2^2 - 2T_3^2 + T_4^2 = 4 \qquad (7.32\text{g})$$

$$A_i \geq 0 \qquad (7.32\text{h})$$

$$T_i - T_{i+1} < 0 \quad \forall i. \qquad (7.32\text{i})$$

The MATLAB code for Example 7.1 is:

MATLAB Program 7.5:

```
function [f] = cst(x);

T1 = x(1); T2 = x(2); T3 = x(3); T4 = x(4);

f = x(4);
```


MATLAB Program 7.6:

```
function [g,geq] = constr(x);

T1 = x(1); T2 = x(2); T3 = x(3); T4 = x(4);

s = j*sqrt(2);

geq(1) = -2*T1+2*T2-2*T3+T4;

geq(2) =
real(1-2*exp(-s*T1)+2*exp(-s*T2)-2*exp(-s*T3)+exp(-s*T4));

geq(3) =
imag(1-2*exp(-s*T1)+2*exp(-s*T2)-2*exp(-s*T3)+exp(-s*T4));

geq(4) = -2*T1^2+2*T2^2-2*T3^2+T4^2-4;

g(1) = -T1; g(2) = -(T2-T1); g(3) = -(T3-T2); g(4) = -(T4-T3);
```

MATLAB Scripts 7.5 and 7.6 return the cost and the inequality and equality constraints respectively. The MATLAB function "fmincon" is used to solve this constrained optimization problem.

7.1.1 Minimax Control

MATLAB provides a function "fminimax" which solves a minimax optimization problem. To illustrate this, consider the design of a minimax time-delay filter for a single spring-mass-dashpot system

$$\ddot{y} + 0.2\dot{y} + ky = ku \tag{7.33}$$

with uncertainty in the coefficient of stiffness k. Consider the design of a two-time-delay filter whose transfer function is:

$$\frac{u(s)}{r(s)} = A_0 + A_1 \exp(-sT_1) + A_2 \exp(-sT_2). \tag{7.34}$$

To ensure that the steady state value of $u(s)$ is the same as the reference input $r(s)$, we require

$$A_0 + A_1 + A_2 = 1. \tag{7.35}$$

The minimax optimization problem can be stated as the determination of A_0, A_1, A_2, T_1, and T_2, parameters of the time-delay filter so as to

$$\min_{T_i,A_i} \max_k \frac{1}{2}\dot{y}^2 + \frac{1}{2}k(y-1)^2 \tag{7.36}$$

evaluated at T_2. The MATLAB script 7.7 lists the functions which returns a variable "f" which is 21 dimensional vector of the residual energies evaluated for the uncertain stiffness ranging from 0.7 to 1.3.

MATLAB Program 7.7:

```
function f = fmin(x);

T1 = x(1);
T2 = x(2);
A0 = x(3);
A1 = x(4);
A2 = x(5);

Avec = [A0 A1 A2];
Tvec = [0 T1 T2];

kvec = linspace(0.7,1.3,21);

for indk = 1:length(kvec);
    k = kvec(indk);
    A = [0 1;-k -0.2];
    B = [0;k];
    P = [A B;zeros(1,3)];
    xftemp = zeros(2,1);
    for ind = 1:length(Avec),
        Phi = expm(P*(Tvec(end)-Tvec(ind)));
        xftemp = xftemp + Avec(ind)*Phi(1:2,3);
    end;
```

```
      f(indk) = (0.5*(xftemp(1)-1)^2*k + 0.5*xftemp(2)^2);
end;
```

MATLAB Script 7.8 is the main script which calls the script 7.7 and saves the optimal solution in the variable "xopt."

MATLAB Program 7.8:

```
x0 = [pi 2*pi 0.25 0.5 0.25];
Aeq = [0 0 1 1 1];
Beq = [1];

options = optimset('Display','iter');

xopt = fminimax(@fun,x0,[],[],Aeq,Beq,[],[],[], options)
```

The output of the optimizer is:

```
xopt =

    3.1688      6.3406      0.3450      0.4730      0.1820
```

which is illustrated in Figure 7.1 via the dashed line, where the maximum magnitude of the residual energy corresponding to the minimax solution over the uncertain domain is significantly smaller than that of the robust time-delay filter which locates a pair of zeros at the nominal location of the poles of the system.

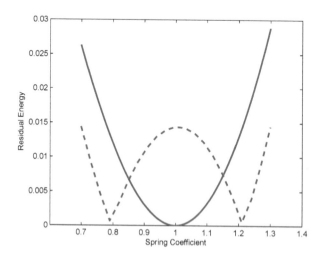

Figure 7.1: Variation of Residual Energy

7.1.2 Analytic Gradients

Gradient-based optimizers require exact values or estimates of the gradient of the cost with respect to the parameters being optimized for to calculate the updates of the parameters. Often optimization software determine the gradients using finite difference. However, if analytical gradients can be provided to the software, it will expedite the convergence of the algorithm. In this section, the determination of the analytic gradients for the control profiles which have been parameterized to generate staircase profiles or bang-bang and bang-off-bang profiles is described.

Consider the cost function which corresponds to the residual energy of a system at the end of the maneuver

$$J(T_n) = \frac{1}{2} \left(\dot{x}^T M \dot{x} + x^T K x \right) \tag{7.37}$$

where T_N corresponds to the final time. Here we consider a linear mechanical system whose equations of motion are given as

$$M\ddot{x} + C\dot{x} + Kx = Du \tag{7.38}$$

where M, C, and K correspond to the mass, damping, and stiffness matrices, respectively. The D matrix is the control-influence matrix. The second-order model can be represented in state-space form as

$$\dot{z} = Az + Bu, \ z = \begin{Bmatrix} x \\ \dot{x} \end{Bmatrix}, \tag{7.39}$$

where

$$A = \begin{bmatrix} 0 & I \\ -M^{-1}K & -M^{-1}C \end{bmatrix}, \text{ and } B = \begin{bmatrix} 0 \\ M^{-1}D \end{bmatrix} \tag{7.40}$$

For a staircase control profile parameterized as

$$u = \sum_{i=0}^{N} A_i \mathcal{H}(t - T_i), \text{ with } T_0 = 0, \tag{7.41}$$

the parameters to be optimized for include A_i and T_i for $i \in [1 \ N]$. The gradients of the cost with respect to A_i and T_i are

$$\frac{dJ}{dA_i} = \left(\dot{x}^T M \frac{d\dot{x}}{dA_i} + x^T K \frac{dx}{dA_i} \right) \bigg|_{T_N} \tag{7.42}$$

$$\frac{dJ}{dT_i} = \left(\dot{x}^T M \frac{d\dot{x}}{dT_i} + x^T K \frac{dx}{dT_i} \right) \bigg|_{T_N} \tag{7.43}$$

which require the state sensitivities at the final time. The sensitivity state equations are

$$\frac{d\dot{z}}{dA_i} = A\frac{dz}{dA_i} + B\mathcal{H}(t - T_i) \tag{7.44}$$

$$\frac{d\dot{z}}{dT_i} = A\frac{dz}{dT_i} - B\delta(t - T_i) \tag{7.45}$$

where $\delta(.)$ is the Dirac delta function. Equation (7.44) which describes the dynamics of the sensitivity of the states with respect to the parameter A_i is essentially the response of the system to a unit step input applied at time T_i. The van Loan identity can be used to determine the terminal time sensitivity state without numerical integration.

The sensitivity of the states with respect to the parameters T_i can be derived easily since Equation (7.45) which describes their dynamics is driven by an impulse input. The solution of Equation (7.45) can be represented as

$$\frac{dz}{dT_i} = -\int_0^{T_N} e^{A(T_N - \tau)} A_i B\delta(\tau - T_i)\, d\tau \tag{7.46}$$

which can be rewritten as

$$\frac{dz}{dT_i} = -e^{A(T_N - T_i)} A_i B \tag{7.47}$$

The derivatives $\frac{dx}{dT_i}$ and $\frac{d\dot{x}}{dT_i}$ can be extracted from Equation (7.47).

The sensitivity of the states to the final time T_N is a little more complicated since it appears in the integrand and in the limit of Equation (7.49). The system response to the staircase input (Equation (7.41))

$$\dot{z} = Az + B\sum_{i=0}^{N} A_i \mathcal{H}(t - T_i) \tag{7.48}$$

for zero initial conditions evaluated at the final time (T_N) is

$$z(T_N) = \int_0^{T_N} e^{A(T_N - \tau)} B\sum_{i=0}^{N} A_i \mathcal{H}(\tau - T_i)\, d\tau \tag{7.49}$$

and the sensitivity with respect to the final time is

$$\frac{dz(T_N)}{dT_N} = \int_0^{T_N} e^{A(T_N - \tau)} AB\sum_{i=0}^{N} A_i \mathcal{H}(\tau - T_i)\, d\tau$$
$$\underbrace{-\int_0^{T_N} e^{A(T_N - \tau)} ABA_N \delta(\tau - T_N)\, d\tau}_{BA_N} + \sum_{i=0}^{N} BA_i \mathcal{H}(T_N - T_i) \tag{7.50}$$

It can be seen that the addend corresponding to $i = N$ of the last term of Equation (7.50) cancels the second term on the right-hand side of Equation (7.50). The first integral of Equation (7.50) can be calculated since the input is piecewise constant.

$$
\int_0^{T_N} e^{A(T_N-\tau)} AB \sum_{i=0}^{N} A_i \mathcal{H}(\tau - T_i)\, d\tau = \sum_{i=0}^{N} A_i \left(e^{A(T_N-T_i)} - I \right) B\mathcal{H}(T_N - T_i)
$$

$$
= \sum_{i=0}^{N-1} A_i \left(e^{A(T_N-T_i)} - I \right) B\mathcal{H}(T_N - T_i) \tag{7.51}
$$

since the contribution of addend corresponding to $i = N$ is zero. Equation (7.50) can now be represented as

$$
\frac{dz(T_N)}{dT_N} = \sum_{i=0}^{N-1} A_i \left(e^{A(T_N-T_i)} - I \right) B\mathcal{H}(T_N - T_i) + \sum_{i=0}^{N-1} BA_i \mathcal{H}(T_N - T_i)
$$

$$
= \sum_{i=0}^{N-1} A_i e^{A(T_N-T_i)} B\mathcal{H}(T_N - T_i) \tag{7.52}
$$

Since Equation (7.52) is evaluated at T_N, the Heaviside function is equal to 1 for all i. Comparing Equation (7.52) to Equation (7.47), we have the relationship

$$
\frac{dz(T_N)}{dT_N} = -\sum_{i=0}^{N-1} \frac{dz(T_N)}{dT_i} \tag{7.53}
$$

which states that once the state sensitivities with respect to T_i have been determined for i from 1 to $N - 1$, the sensitivity to the final time is given by the negative of the sum of the sensitivities to the rest of the parameters T_i.

MATLAB optimization functions "fmincon," "fminimax" are gradient-based optimization algorithms and as with any gradient-based algorithms, these often converge to local optima. The initial guess strongly influences whether the gradient-based optimization algorithm converges to the global optima. Computationally expensive fixes to the problem are available, e.g., Monte Carlo initial guesses, gridding initial guesses, simulated annealing, and so forth. Posing a problem as a convex optimization problem eliminate the problem of converging to a local optima. Linear programming and semidefinite programming are two techniques which ensure convergence to a global optima. The next section presents a technique for the determination of optimal control profile using linear programming. This technique requires converting the continuous-time problem to a discrete-time problem. As the sampling interval decreases, the solution of the discrete-time problem converges to the continuous-time problem.

7.2   Linear Programming

A *linear program* is a problem which can be represented in the form:

$$\text{minimize} \quad c^T z \tag{7.54a}$$
$$\text{subject to} \quad A_{eq}z = b_{eq} \tag{7.54b}$$
$$A z \geq b \tag{7.54c}$$

where z is the vector of parameters to be solved for. The *simplex method* for linear programming that has been called one of the top 10 algorithms of the millennium by the *IEEE Computer Society* [2], is a powerful algorithm that has been used extensively to solve a variety of problems. Leonid Vitalyevich Kantorovich (who earned his doctorate at the age of 18) and George Dantzig are considered the fathers of linear programming (LP). In 1939, Kantorovich posed an LP problem to solve a problem dealing with manufacturing schedules. Dantzig proposed the simplex algorithm for solving a linear programming problem when working on a research project for the U.S. Air Force, which related to coordinating supplies for troops in World War II. Kantorovich was jointly awarded the Nobel Prize in Economics in 1975 for his work on optimal allocation of scarce resources.

Linear programming can be exploited in the design of controllers for linear systems. Consider the discrete-time state-space representation of a dynamical system with a sampling time T_s:

$$\underline{x}(k+1) = G\underline{x}(k) + Hu(k) \text{ where } k = 1, 2, ..., N \tag{7.55}$$

where $\underline{x}$ is the state vector and u is the control input. The state response for the control input $u(k)$ is

$$\underline{x}(k+1) = G^k\underline{x}(1) + \sum_{i=1}^{k} G^{k-i}Hu(i) \tag{7.56}$$

where $\underline{x}(1)$ represents the initial state of the system. To solve control problems with specified initial and final states, in addition to the final time (T_f), the final state constraint can be represented as

$$\underline{x}(N+1) = G^N\underline{x}(1) + \sum_{i=1}^{N} G^{N-i}Hu(i) \tag{7.57}$$

where the maneuver time $T_f = (N+1)T_s$, i.e., the maneuver time is discretized into N intervals. Equation 7.57 can be rewritten in the standard equality constraint form:

$$[G^{N-1}H \quad G^{N-2}H \quad \cdots \quad GH \quad H] \begin{bmatrix} u_1 \\ u_2 \\ \cdot \\ \cdot \\ \cdot \\ u_N \end{bmatrix} = \underline{x}_{N+1} - G^N \underline{x}_1 \qquad (7.58)$$

The limits on the control which are specified as

$$u_{lb} \leq u(k) \leq u_{ub} \qquad (7.59)$$

where u_{lb} and u_{ub} are the lower and upper bounds on the control, respectively, and can be included in the problem formulation. Constraints on the states or combination of states which have to be satisfied over the duration of the maneuver can be included in the problem formulation. The state constraints can be represented as:

$$CG^N \underline{x}(1) + \sum_{i=1}^{N-1} CG^{N-i} H u(i) \leq 0 \quad \forall i \in [0 \ N] \qquad (7.60)$$

where C is the output matrix. Note that while states and inputs can be specified, constrained or treated as unconstrained variables within the linear program formulation (Equation 7.54), this is not the case with the maneuver terminal time. T_N must be specified. This limits how minimum time control is to be managed, as described in the next section.

7.2.1 Minimum Time Control

Since the linear programming approach requires the maneuver time to be specified, the determination of the time-optimal controller using this approach requires that an initial estimate of the maneuver time be used to determine whether it supports a feasible solution. If it does not, the candidate optimal maneuver time is increased, else it is iteratively decreased till the maneuver time cannot be reduced without violating the constraints. A MATLAB-based Bisection algorithm (Program 7.9) is used to find the lower bound for the feasible region of the maneuver time. This program requires the name of a file which returns the *exitflag* which indicates whether the *linprog* command resulted in a feasible solution. This program requires a lower (X_L) and upper (X_U) bound of the maneuver time. One can select 0 for the lower bound and a large time for the upper bound if one has no idea of the maneuver time. The bracketing interval is halved every iteration and after N iterations, the optimal is guaranteed to lie in the interval $\frac{X_U - X_L}{2^N}$. In other words for a desired error tolerance of ε, the algorithm requires

$$N = \log_2\left(\frac{X_U - X_L}{\varepsilon}\right) \qquad (7.61)$$

iterations irrespective of the function considered. The complexity of each iteration however, is determined by the function.

The minimum time formulation can be used to solve for the time-optimal control profiles for systems with rigid-body modes which results in a bang-bang profile for *Normal* systems. It can also be used for the determination of *input shaped* reference profiles for systems without rigid-body modes.

First, we will consider the problem of design of pre-filtered reference profile to drive the system states to their final values in finite time. Consider the undamped second-order system

$$\begin{Bmatrix} \dot{x} \\ \ddot{x} \end{Bmatrix} = \begin{bmatrix} 0 & 1 \\ -\omega^2 & 0 \end{bmatrix} \begin{Bmatrix} x \\ \dot{x} \end{Bmatrix} + \begin{bmatrix} 0 \\ \omega^2 \end{bmatrix} u \qquad (7.62)$$

where the initial and final states of the system are:

$$\begin{Bmatrix} x \\ \dot{x} \end{Bmatrix}(0) = \begin{bmatrix} 0 \\ 0 \end{bmatrix} \quad \begin{Bmatrix} x \\ \dot{x} \end{Bmatrix}(t_f) = \begin{bmatrix} 1 \\ 0 \end{bmatrix}. \qquad (7.63)$$

To pose a linear programming problem, the continuous-time model is discretized which results in the equation:

$$\begin{Bmatrix} \dot{x} \\ \ddot{x} \end{Bmatrix}(k+1) = \underbrace{\begin{bmatrix} \cos(\omega T) & \frac{\sin(\omega T)}{\omega} \\ -\omega \sin(\omega T) & \cos(\omega T) \end{bmatrix}}_{G} \begin{Bmatrix} x \\ \dot{x} \end{Bmatrix}(k) + \underbrace{\begin{bmatrix} -\cos(\omega T) + 1 \\ \omega \sin(\omega T) \end{bmatrix}}_{H} u(k). \qquad (7.64)$$

From Equation (7.63), the final value of the control input $u(t_f)$ can be determined which results in the constraint:

$$u(t_f) = 1. \qquad (7.65)$$

MATLAB Program 7.9: Bisection Algorithm

```
function   [XL,XR] = bisection(fun,XL,XR)

% This program determines the lower bound of the feasible region

% display line
    fprintf('\n\n%s\t%s\t%s\n','Iter.','X_L','X_U');

% evaluate feasibility at lower, upper & mid-point of variable
    Flag_L = feval(fun,XL);
    Flag_U = feval(fun,XR);
    Flag_M = feval(fun,(XL+XR)/2);

% refine until convergence criteria is satisfied:
    epsilon_t = 1e-3;
    counter = 0;
    max_fun_eval = 1e4;
    while abs(XR-XL) > epsilon_t
```

```
counter = counter + 1;
if counter > max_fun_eval
    disp ( 'Maximum Number of Function Evaluations exceeded!')
    break
end
fprintf ([ '%0.0f\t%0.2f\t%0.2f\n'] , counter ,XL,XR);
if Flag_M < 0
    XL = (XL+XR)/2;
    Flag_L = Flag_M;
    Flag_M = feval (fun ,(XL+XR)/2);
elseif Flag_M > 0
    XR = (XL+XR)/2;
    Flag_R = Flag_M;
    Flag_M = feval (fun ,(XL+XR)/2);
else
    disp ('Linprog Reached Maximum Number of Iterations!')
end;
end
```

Example 7.2:

Design a minimum time control profile for the system

$$\ddot{x} + x = u \tag{7.66}$$

to satisfy the boundary conditions

$$x(0) = \dot{x} = 0, \text{ and } x(t_f) = 1, \dot{x}(t_f) = 0, \tag{7.67}$$

where u has to satisfy the constraint

$$0 \leq u \leq 1. \tag{7.68}$$

The discrete time state-space representation of the simple harmonic oscillator (Equation 7.66) is

$$\begin{Bmatrix} x(k+1) \\ \dot{x}(k+1) \end{Bmatrix} = \begin{bmatrix} \cos(T) & \sin(T) \\ -\sin(T) & \cos(T) \end{bmatrix} \begin{Bmatrix} x(k) \\ \dot{x}(k) \end{Bmatrix} + \begin{bmatrix} -\cos(T)+1 \\ \sin(T) \end{bmatrix} u(k) \tag{7.69}$$

where T is the sampling interval. The boundary conditions are

$$\underline{x}_1 = \begin{Bmatrix} 0 \\ 0 \end{Bmatrix} \text{ and } \underline{x}_{N+1} = \begin{Bmatrix} 1 \\ 0 \end{Bmatrix}. \tag{7.70}$$

Since, the desired maneuver is a rest-to-rest maneuver, we require

$$\underline{x}_{N+1+i} = \underline{x}_{N+1} \quad \forall i = 1,2,3,...,\infty \tag{7.71}$$

Substituting Equation (7.71), with $i = 1$, in Equation (7.69), we have

$$\begin{Bmatrix} 1 \\ 0 \end{Bmatrix} = \begin{bmatrix} \cos(T) & \sin(T) \\ -\sin(T) & \cos(T) \end{bmatrix} \begin{Bmatrix} 1 \\ 0 \end{Bmatrix} + \begin{bmatrix} -\cos(T)+1 \\ \sin(T) \end{bmatrix} u \qquad (7.72)$$

or

$$\begin{bmatrix} 1-\cos(T) & -\sin(T) \\ \sin(T) & 1-\cos(T) \end{bmatrix} \begin{Bmatrix} 1 \\ 0 \end{Bmatrix} = \begin{bmatrix} -\cos(T)+1 \\ \sin(T) \end{bmatrix} = \begin{bmatrix} -\cos(T)+1 \\ \sin(T) \end{bmatrix} u \qquad (7.73)$$

which implies that the steady-state value of u is 1. This terminal equality constraint has to be included in the linear programming problem.

MATLAB Program 7.10:

```
function [exitflag] = example1(Tf);
%
% Program to Solve for the Input Shaped Control Profile for a
% Harmonic Oscillator using Linear Programming
%
X0 = [0;0];                    % Inital States
Xf = [1;0];                    % Final States

    No_samp = 101;
    T = Tf/(No_samp-1);            % Sampling Time

    G = [cos(T) sin(T); -sin(T) cos(T)];     % Transition Matrix
    H = [-cos(T)+1;sin(T)];             % Control Influence Matrix

    f = zeros(No_samp,1);             % Cost function coefficient
    % Equality Constraint calculation
    Aeq1 = [];
    for ind = 1:(No_samp)
        Aeq1 = [G^(ind-1)*H Aeq1];
    end;
    beq1 = Xf - G^No_samp*X0;

    % Equality constraint which requires final input values be 1
    Aeq2 = zeros(1,No_samp);
    Aeq2(1,end) = 1;
    beq2 = 1;

    Aeq = [Aeq1;Aeq2];
    beq = [beq1;beq2];

    vlb = zeros(No_samp,1);
    vub = 1*ones(No_samp,1);
    u0 = 0.5*ones(size(f));
    % Minimum Time Input Shaper
```

```
[u,fval,exitflag] = linprog(f,[],[],Aeq,beq,vlb,vub,u0);
if exitflag > 0
    t = linspace(0,Tf,No_samp);
    C = eye(2); D = [0;0];
    sys = ss(G,H,C,D,T);
    figure(1)
    [ys,ts,xs] = lsim(sys,u);
    subplot(211), stairs(ts,ys(:,1));
    xlabel('Time'); ylabel('Displacement');
    subplot(212), stairs(ts,ys(:,2));
    xlabel('Time'); ylabel('Velocity');
    figure(2)
    stairs(t,u);
    xlabel('Time'); ylabel('Velocity');
end;
```

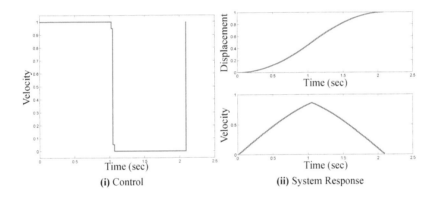

(i) Control **(ii)** System Response

Figure 7.2: Linear Programming Solution

Figure 7.2 illustrates the input and the corresponding system response for a minimum time cost function.

Solving the problem with the constraint that the rate of change of the control input be greater than zero results in the classic *input-shaper* solution, as shown in Figure 7.3.

To account for uncertainties in the plant model, a linear programming problem is formulated where additional constraints which are a function of the sensitivity of the system states to these variables are included. Consider the spring-mass system

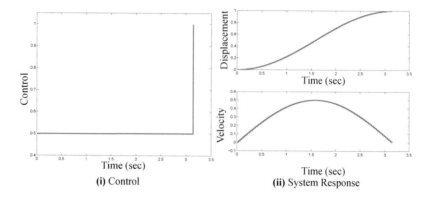

(i) Control   **(ii) System Response**

Figure 7.3: Linear Programming Solution

$$\left\{\begin{matrix} \dot{x} \\ \ddot{x} \end{matrix}\right\} = \begin{bmatrix} 0 & 1 \\ -\omega^2 & 0 \end{bmatrix} \left\{\begin{matrix} x \\ \dot{x} \end{matrix}\right\} + \begin{bmatrix} 0 \\ \omega^2 \end{bmatrix} u \tag{7.74}$$

where the initial and final states of the system are:

$$\left\{\begin{matrix} x \\ \dot{x} \end{matrix}\right\}(0) = \begin{bmatrix} 0 \\ 0 \end{bmatrix} \quad \left\{\begin{matrix} x \\ \dot{x} \end{matrix}\right\}(t_f) = \begin{bmatrix} 1 \\ 0 \end{bmatrix}. \tag{7.75}$$

The sensitivity of the system states to the natural frequency ω are

$$\left\{\begin{matrix} \frac{dx}{d\omega} \\ \frac{d\dot{x}}{d\omega} \end{matrix}\right\} = \begin{bmatrix} 0 & 1 \\ -\omega^2 & 0 \end{bmatrix} \left\{\begin{matrix} \frac{dx}{d\omega} \\ \frac{d\dot{x}}{d\omega} \end{matrix}\right\} - \begin{bmatrix} 0 \\ 2\omega \end{bmatrix} x. \tag{7.76}$$

The boundary conditions can be derived from Equation (7.76) by forcing the rate of change of the sensitivity states to zero, resulting in the boundary conditions:

$$\left\{\begin{matrix} \frac{dx}{d\omega} \\ \frac{d\dot{x}}{d\omega} \end{matrix}\right\}(0) = \begin{bmatrix} 0 \\ 0 \end{bmatrix} \quad \left\{\begin{matrix} \frac{dx}{d\omega} \\ \frac{d\dot{x}}{d\omega} \end{matrix}\right\}(t_f) = \begin{bmatrix} -\frac{2}{\omega}x(t_f) \\ 0 \end{bmatrix}. \tag{7.77}$$

The augmented system model is

$$\left\{\begin{matrix} \dot{x} \\ \ddot{x} \\ \frac{dx}{d\omega} \\ \frac{d\dot{x}}{d\omega} \end{matrix}\right\} = \begin{bmatrix} 0 & 1 & 0 & 0 \\ -\omega^2 & 0 & 0 & 0 \\ 0 & 0 & 0 & 1 \\ -2\omega & 0 & -\omega^2 & 0 \end{bmatrix} \left\{\begin{matrix} x \\ \dot{x} \\ \frac{dx}{d\omega} \\ \frac{d\dot{x}}{d\omega} \end{matrix}\right\} + \begin{bmatrix} 0 \\ \omega^2 \\ 0 \\ 0 \end{bmatrix} u \tag{7.78}$$

with the associated boundary conditions

$$\left\{\begin{matrix} x \\ \dot{x} \\ \frac{dx}{d\omega} \\ \frac{d\dot{x}}{d\omega} \end{matrix}\right\}(0) = \begin{bmatrix} 0 \\ 0 \\ 0 \\ 0 \end{bmatrix} \quad \left\{\begin{matrix} x \\ \dot{x} \\ \frac{dx}{d\omega} \\ \frac{d\dot{x}}{d\omega} \end{matrix}\right\}(t_f) = \begin{bmatrix} 1 \\ 0 \\ -\frac{2}{\omega} \\ 0 \end{bmatrix}. \tag{7.79}$$

The continuous time model is converted to discrete time model and the optimal input shaping control profile is determined as above.

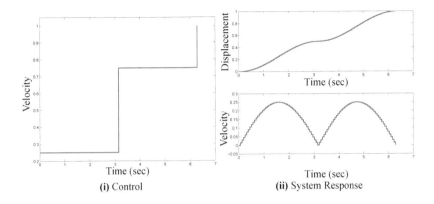

Figure 7.4: Linear Programming Solution

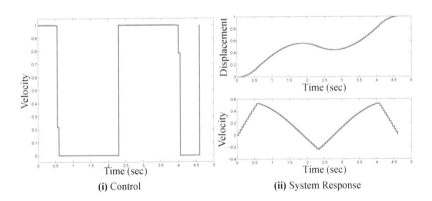

Figure 7.5: Linear Programming Solution

Figure 7.4 illustrates the desensitized input profile which can be seen to be identical to the ZVD input-shaped control or the output of the robust time-delay filter. Eliminating the constraint that the shaped control profile cannot decrease, results in the robust time-optimal control input as shown in Figure 7.5.

7.2.2 Minimum Fuel Control

The minimum fuel cost function requires minimizing the cost function

$$J = \int_0^{t_f} |u| \, dt \propto \sum_{i=1}^{N} |u_i| \tag{7.80}$$

where the maneuver time t_f is specified. The cost function is not in the form of the standard linear programming problem (Equations 7.54a through 7.54c). The minimum fuel optimal control problem can be reformulated by defining auxiliary variables u_+ and u_- where both these variable are greater than or equal to zero. The terminal state constraint equation can be rewritten as

$$\begin{bmatrix} G^{N-1}H & -G^{N-1}H & G^{N-2}H & -G^{N-2}H & \cdots & H & -H \end{bmatrix} \begin{bmatrix} u_{1+} \\ u_{1-} \\ u_{2+} \\ u_{2-} \\ \cdot \\ \cdot \\ \cdot \\ u_{N+} \\ u_{N-} \end{bmatrix} = \underline{x}_{N+1} - G^N \underline{x}_1 \tag{7.81}$$

The minimum fuel optimal control problem can now be stated as

$$\min \; J = \sum_{i=1}^{N} u_{i+} + u_{i-} = \begin{bmatrix} 1 & 1 & \cdots & 1 & 1 \end{bmatrix} \begin{bmatrix} u_{1+} \\ u_{1-} \\ u_{2+} \\ u_{2-} \\ \cdot \\ \cdot \\ \cdot \\ u_{N+} \\ u_{N-} \end{bmatrix} \tag{7.82a}$$

subject to $\hspace{8cm}$ (7.82b)

$$\begin{bmatrix} G^{N-1}H & -G^{N-1}H & \cdots & H & -H \end{bmatrix} \begin{bmatrix} u_{1+} \\ u_{1-} \\ u_{2+} \\ u_{2-} \\ \cdot \\ \cdot \\ \cdot \\ u_{N+} \\ u_{N-} \end{bmatrix} = \underline{x}_{N+1} - G^N \underline{x}_1 \tag{7.82c}$$

$$0 \leq u_{i+} \leq 1 \quad \forall i \tag{7.82d}$$

$$0 \leq u_{i-} \leq 1 \quad \forall i \tag{7.82e}$$

MATLAB Program 7.11: Minimum Fuel Optimal Control

```
function [exitflag] = example3(Tf);
% Minimum Fuel Control
% Program to Solve for the rest–to–rest Control Profile for a
% 2–Mass–system, using Linear Programming
%
%
X0 = [0;0;0;0;];              % Initial States
Xf = [1;1;0;0];              % Final States
M = [1 0;0 1]; K = [1 −1;−1 1]; Ac = [zeros(2) eye(2); −inv(M)*K
zeros(2)]; Bc = [zeros(2,1);inv(M)*[1;0]]; Cc = eye(4); Dc =
zeros(4,1); sysc = ss(Ac,Bc,Cc,Dc);

    No_samp = 101;           % Sampling Time
    T = Tf/(No_samp−1);

    sysd = c2d(sysc,T);
    [G,H,C,D] = ssdata(sysd);
    H2 = [H −H];

    f = ones(2*No_samp,1);            % Cost function coefficient

    % Equality Constraint calculation
    Aeq = [];
    for ind = 1:(No_samp)
        Aeq = [G^(ind−1)*H2 Aeq];
    end;
    beq = Xf − G^No_samp*X0;

    vlb = 0*ones(size(f));
    vub = 1*ones(size(f));
    u0 = 0.5*ones(size(f));
    % Minimum Fuel Controller
    [u,fval,exitflag] = linprog(f,[],[],Aeq,beq,vlb,vub,u0);

    if exitflag > 0
        u1 = u(1:2:end);
        u2 = u(2:2:end);
        usim = [u1(:)−u2(:)];
        t = linspace(0,Tf,No_samp);
        figure(1)
        [yy,tt,xx] = lsim(sysd,usim);
        subplot(211), plot(tt,xx(:,1:2));
        xlabel('Time'); ylabel('Displacement');
```

```
    subplot(212),  stairs(t,xx(:,3:4));
    xlabel('Time');  ylabel('Velocity');
    figure(2)
    subplot(211),  stairs(t,u1);
    xlabel('Time');  ylabel('Control u_+');
    subplot(212),  stairs(t,u2);
    xlabel('Time');  ylabel('Control u_-');
    figure(3)
    plot(t,usim)
    xlabel('Time');  ylabel('Control');
end;
```

The minimum fuel optimal control problem is solved for the undamped floating oscillator problem where the maneuver time is assumed to be 6 seconds. The LP problem results in a six-switch optimal control profile which is antisymmetric about the mid-maneuver time.

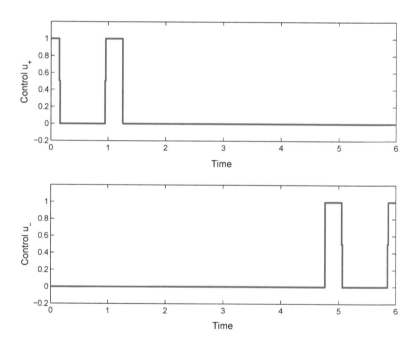

Figure 7.6: Resolved Control Input

Figure 7.6 illustrates the variation of the resolved control input u_+ and u_-. It is clear that the resolved control inputs lie between 0 and 1 and their structures corre-

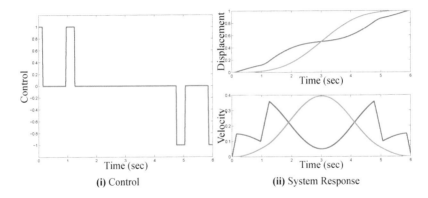

Figure 7.7: Linear Programming Solution

spond to the bang-off-bang structures.

Figure 7.7 illustrates the time-response of the system states and the minimum fuel optimal-control input which is antisymmetric about the mid-maneuver time.

7.2.3 Fuel/Time Optimal Control

The fuel/time optimal control corresponds to a control profile that minimizes the cost function

$$J = \int_0^{t_f} (1 + \alpha|u|) \, dt \tag{7.83}$$

where α is a weighting parameter which specifies the relative importance of the fuel consumed to the maneuver time. This cost function is not conducive to the formulation of an LP problem unlike the minimum fuel optimal controller. A technique similar to the approach used to solve the minimum-time control problem is proposed. Figure 7.8 illustrates the variation of the cost (Equation 7.83) as a function of the maneuver time for $\alpha = 8$ for the undamped floating oscillator benchmark problem.

It is clear from Figure 7.8 that the variation of the weighted fuel/time cost is a multi-modal function of the maneuver time. A single-dimension search strategy such as the golden section algorithm can be used to solve for the maneuver time which corresponds to the minimum of the weighted fuel/time cost function. It should be noted that the global minimum should be bracketed to preclude the possibility of the algorithm converging to a local minima.

MATLAB Program 7.12: Golden Section Algorithm

```
function [Xmin, fmin] = golden1(fun, XL, XR, tol)
% This program determines the minimum of a function by
% bracketing the minimum
```

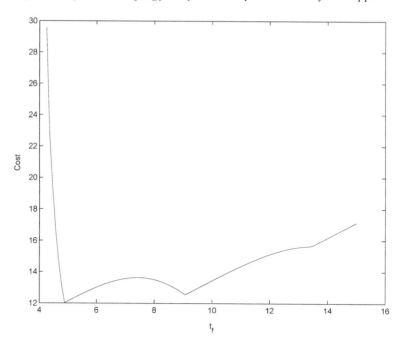

Figure 7.8: Cost Function Variation vs. Maneuver Time

```
% display line
    fprintf('\n\n%s\t%s\t%s\n','Iter.','X_L','X_R');

C = (3-sqrt(5))/2; R = 1-C;

X1 = XL + C*(XR-XL); X2 = XL + R*(XR-XL); f1 = feval(fun,X1); f2 =
feval(fun,X2);

counter = 1; while abs(XR-XL) > tol*(abs(X1)+abs(X2)),
  fprintf(['%0.0f\t%0.2f\t%0.2f\n'],counter,XL,XR);
  if f2 < f1,
    XL = X1;
    X1 = X2;
    X2 = XL + C*(XR-X1);
    f1 = f2;
    f2 = feval(fun,X2);
  else
    XU = X2;
    X2 = X1;
    X1 = X2 + C*(XL-X2);
    f2 = f1;
    f1 = feval(fun,X1);
  end
```

```
    counter = counter+1;
end

if f1 < f2,
    Xmin = X1;
    fmin = f1;
else
    Xmin = X2;
    fmin = f2;
end
```

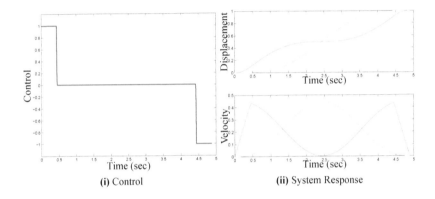

(i) Control   **(ii)** System Response

Figure 7.9: Linear Programming Solution

Figure 7.9 illustrates the two switch optimal control profile and the corresponding system response for a weighting parameter value of $\alpha = 8$. The solution is identical to that presented by Singh [3] where a parameter optimization problem is solved to arrive at the fuel/time optimal control profiles for an undamped two-mass spring system.

7.2.4 Minimax Control

Design of input shaped controllers which are insensitive to modeling uncertainties has been posed in the framework of a minimax problem by Singh [4]. The residual energy of the system over the domain of uncertainty is calculated and the maximum magnitude is minimized. Since the residual energy is a quadratic function, it is not compatible with the constraints of an LP problem.

In this section, we describe how a minimax problem can be solved using the LP framework. Consider the problem

$$\text{minimize} \quad \underset{i=1,2,\ldots,M}{\text{maximum}} \ c_i^T z + d_i \tag{7.84a}$$

$$\text{subject to} \quad A_{eq} z = b_{eq} \tag{7.84b}$$

$$A z \leq b \tag{7.84c}$$

where z is a vector to be solved for and M is the number of linear constraints. Defining a variable f which is equal to the maximum of $c_i^T z + d_i$ for all i, the minimax problem can be stated as

$$\text{minimize} \quad f \tag{7.85a}$$

$$\text{subject to} \quad A_{eq} z = b_{eq} \tag{7.85b}$$

$$A z \leq b \tag{7.85c}$$

$$c_i^T z + d_i \leq f \ \forall i = 1,2,\ldots,M \tag{7.85d}$$

This corresponds to an LP problem where the variables f and z are solved for.

A linear model of a structure with n degrees of freedom, can be represented as

$$M\ddot{x} + C\dot{x} + Kx = Du \tag{7.86}$$

where M, C, and K are the mass, damping and stiffness matrices and D is the control influence matrix which is a function of the location of the actuators. The residual energy of this system is the sum of the kinetic and potential energy at the end of the maneuver (t_f):

$$E(t_f) = \frac{1}{2}\dot{x}^T M\dot{x} + \frac{1}{2}x^T Ky = \frac{1}{2}(\sqrt{M}\dot{x})^T(\sqrt{M}\dot{x}) + \frac{1}{2}(\sqrt{K}x)^T\sqrt{K}y \tag{7.87}$$

which is a common metric to use in the design of robust controllers [4]. For systems with rigid-body modes, a pseudo-potential energy terms has to be added to the kinetic and potential energy terms to guarantee that $E(t_f)$ is positive definite. The resulting cost function is

$$E(t_f) = \frac{1}{2}\dot{x}^T M\dot{x} + \frac{1}{2}x^T Ky + \frac{1}{2}k'(y_r - y_f)^2 \tag{7.88}$$

where y_r refers to the rigid-body mode and y_f the corresponding desired final displacement.

The residual energy corresponds to the ℓ_2 norm and does not satisfy the constraints for an LP problem formulation. In the following development, the ℓ_1 and ℓ_∞ metrics will be first exploited to demonstrate the formulation of an LP problem for the design of desensitized input profiles.

7.2.4.1 ℓ_∞ Optimal Control

For a system with uncertain damping (C) and stiffness (K) matrices, define a metric

$$f = \max_i \left\| \begin{bmatrix} \sqrt{M}\dot{x}^i(t_f) \\ \sqrt{K^i}x^i(t_f) \end{bmatrix} \right\|_\infty \tag{7.89}$$

where C^i and K^i represent the ith embodiment of the uncertain damping and stiffness matrices respectively. Weighting the displacement and the velocities with the matrix square root of the stiffness and mass matrices respectively, result in equal contribution to the residual energy cost, by the individual displacement and velocity states. f which corresponds to the ℓ_∞ norm can be rewritten as

$$f = \max_i \left\{ |\dot{y}_1^i|, |\dot{y}_2^i|, \ldots, |\dot{y}_n^i|, |y_1^i|, |y_2^i|, \ldots, |y_n^i| \right\}. \tag{7.90}$$

where

$$\dot{y}^i = \begin{bmatrix} \dot{y}_1^i \\ \dot{y}_2^i \\ \vdots \\ \dot{y}_n^i \end{bmatrix} = \sqrt{M}\dot{x}^i \quad \text{and} \quad y^i = \begin{bmatrix} y_1^i \\ y_2^i \\ \vdots \\ y_n^i \end{bmatrix} = \sqrt{K_i}x^i. \tag{7.91}$$

Figure 7.10 illustrates the region whose boundaries correspond to the ℓ_∞ norm equal to f, for the variables X_i in two and three dimensions. In two-dimensional space, this space corresponds to a square and in three-dimensional space it corresponds to a cube.

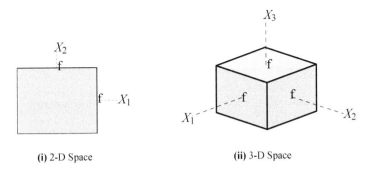

(i) 2-D Space   (ii) 3-D Space

Figure 7.10: Volume Representing $\ell_\infty = f$

The state space representation of the system in continuous time is

$$\left\{ \begin{matrix} \dot{x} \\ \ddot{x} \end{matrix} \right\} = \begin{bmatrix} 0 & I \\ -M^{-1}K & -M^{-1}C \end{bmatrix} \left\{ \begin{matrix} x \\ \dot{x} \end{matrix} \right\} + \begin{bmatrix} 0 \\ M^{-1}Du \end{bmatrix} \tag{7.92}$$

or

$$\dot{X} = AX + Bu \tag{7.93}$$

which can be rewritten in discrete time form as

$$X(k+1) = GX(k) + Hu(k) \tag{7.94}$$

which is the form which will be used in the LP formulation of the minimax optimization problem. The terminal states are

$$X(N) = \left\{ \begin{matrix} x(N) \\ \dot{x}(N) \end{matrix} \right\} \Rightarrow \left\{ \begin{matrix} y(N) = \sqrt{K}x(N) \\ \dot{y}(N) = \sqrt{M}\dot{x}(N) \end{matrix} \right\} \tag{7.95}$$

where N is the index that corresponds to the final time in a discrete-time representation of the system states.

The minimax problem can be stated as

$$\text{minimize} \quad f \tag{7.96a}$$
$$\text{subject to} \tag{7.96b}$$

$$-f \le y_1^i(N) \le f$$
$$-f \le y_2^i(N) \le f$$
$$\vdots$$
$$-f \le y_n^i(N) \le f$$
$$-f \le \dot{y}_1^i(N) \le f \quad \forall i = 1,2,\ldots,Z \tag{7.96c}$$
$$-f \le \dot{y}_2^i(N) \le f$$
$$\vdots$$
$$-f \le \dot{y}_n^i(N) \le f$$
$$u_{min} \le u(m) \le u_{max} \quad \forall m = 1,2,\ldots,N-1 \tag{7.96d}$$

where Z is the number of uncertain models over which the ℓ_∞ norm is minimized. u_{min} and u_{max} correspond to the minimum and maximum permitted control.

Example 7.3: To illustrate the proposed approach, consider the system

$$\ddot{x} + 0.2\dot{x} + kx = ku \tag{7.97}$$

where the spring-stiffness k is uncertain and lies in the range

$$0.7 \le k \le 1.3 \tag{7.98}$$

The optimization problem is to design a control profile that moves the system from rest to a final position of rest while minimizing the maximum magnitude of the residual states over the range of uncertain ks. The boundary conditions are

$$\begin{bmatrix} x(0) \\ \dot{x}(0) \end{bmatrix} = \begin{bmatrix} 0 \\ 0 \end{bmatrix} \text{ and } \begin{bmatrix} x(t_f) \\ \dot{x}(t_f) \end{bmatrix} = \begin{bmatrix} 1 \\ 0 \end{bmatrix}. \tag{7.99}$$

The cost function is

$$\min \left(f = \max_i \left\| \begin{bmatrix} (x^i(t_f) - 1) \\ \dot{x}^i(t_f) \end{bmatrix} \right\|_\infty \right) \tag{7.100}$$

Constraint to guarantee a unidirectional change in the magnitude of the control profile is imposed. This constraint results in control profile which have a staircase form. The resulting LP problem is:

$$\text{minimize } f \tag{7.101a}$$

$$\text{subject to}$$

$$+\dot{x}(N) < f \tag{7.101b}$$

$$-\dot{x}(N) < f \tag{7.101c}$$

$$+\sqrt{k^j}x(N) < f \tag{7.101d}$$

$$-\sqrt{k^j}x(N) < f \tag{7.101e}$$

$$\begin{Bmatrix} x \\ \dot{x} \end{Bmatrix}(N) = G^{N-1}\begin{Bmatrix} x \\ \dot{x} \end{Bmatrix}(1) + \sum_{i=1}^{N-1} G^{N-i}Hu(i) \tag{7.101f}$$

$$0 \le u(m) \le 1 \ \forall m = 1,2,\ldots,N = 501 \tag{7.101g}$$

$$u(m) - u(m+1) \le 0 \ \forall m = 1,2,\ldots,500 \tag{7.101h}$$

The minimax solution generated by optimizing for the parameters of a transfer function of a two-time-delay filter using the gradient based approach presented in the Chapter 5 results in the optimal input shaped control profile

$$u_{minimax} = 0.3452 + 0.4730\mathcal{H}(t - 3.1703) + 0.1818\mathcal{H}(t - 6.3405). \tag{7.102}$$

The maneuver time of 6.3405 seconds is used to solve the LP problem. Discretizing the uncertain domain of k into 21 models, the ℓ_∞ norm minimization problem is solved and the resulting control profile and the system response for the nominal model which corresponds to a stiffness of $k = 1$, are shown in Figure 7.11. Figure 7.12(i) illustrates the variation of residual energy of the ℓ_∞ (* symbols) and the minimax (○ symbols) control profiles, respectively. It is clear that the minimax solution has a smaller maximum magnitude of residual energy over the range of uncertain k. It should be pointed out that the determination of the ℓ_∞ control profile is faster than the determination of the minimax control profile.

Figure 7.12(ii) illustrates the absolute magnitude of the residual displacement and velocity error for the 21 models spanning the entire uncertain space. The * represent the residual state errors for the system driven by the ℓ_∞ control profile and the ○

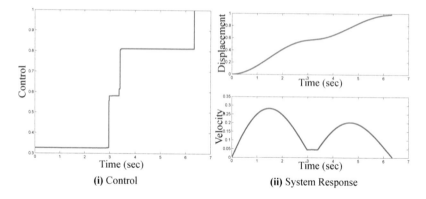

(i) Control   (ii) System Response

Figure 7.11: Linear Programming Solution

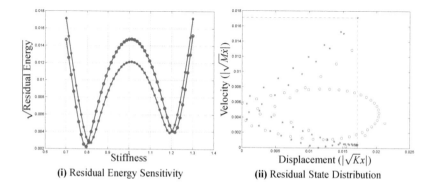

(i) Residual Energy Sensitivity   (ii) Residual State Distribution

Figure 7.12: Energy and Norm Plots

correspond to the residual state errors for the minimax controlled system. The dashed line corresponds to the value of the ℓ_∞ norm minimized by the LP problem. It can be seen that all the $*$ symbols lie within the square bounded by the dashed line.

Since the maneuver time is a user-selected variable, it is of interest to explore the effect of variation of the maneuver time on the residual energy distribution and the ℓ_∞ norm. Three control profiles are derived for maneuver times of $t_f = 6.3405, 7.3405$, and 8.3405 seconds. Figure 7.13 illustrates the reduction in the maximum magnitude of the residual energy over the range of uncertain k as a function of increasing maneuver time. Since the input-shaped control profile is not parameterized in terms of a sequence of impulse sequences (input shaper) or the transfer function of a time-delay filter, the flexibility of the selection of the maneuver time in the LP formulation is conducive to real-time implementation of the control profiles.

Figure 7.14 illustrates the reduction of the ℓ_∞ norm as the maneuver time increases. It can be seen that as the maneuver time varies, the components of the residual energy of the number of uncertain plants that lie on the ℓ_∞ boundary varies between two and

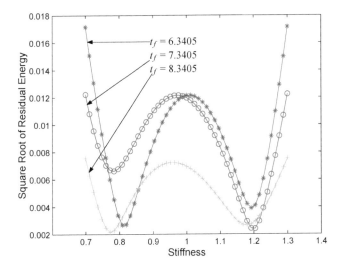

Figure 7.13: Residual Energy

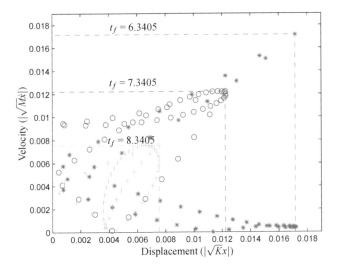

Figure 7.14: Residual State Distribution

three.

Figure 7.15 illustrates the variation of the control profiles as a function of the maneuver time. It can be seen for the selected maneuver times that all three control profiles can be parameterized by time delays.

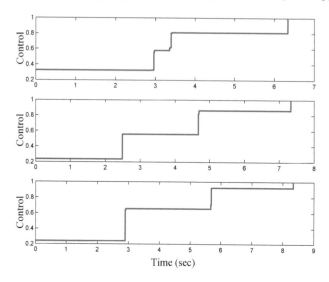

Figure 7.15: Spectrum of Control Profiles

Since it is clear that the insensitivity of the control profile to errors in estimated spring stiffness decreases with maneuver time, the proposed approach provides a technique to trade-off robustness for reduced maneuver time. Figure 7.16 illustrates the variation of the ℓ_∞ norm which bounds the absolute magnitude of the residual tracking error of all the uncertain models, as a function of the maneuver time. It can be seen that there is a rapid reduction of the ℓ_∞ norm until a maneuver time of about 9.5 seconds. A further increase in maneuver time does not significantly reduce the ℓ_∞ norm. This curve can be used by the control designer to trade-off maneuver time for robustness.

Example 7.4: To illustrate the ℓ_∞ robust optimization technique on system with a rigid-body mode, consider the the benchmark floating oscillator system

$$\begin{bmatrix} 1 & 0 \\ 0 & 1 \end{bmatrix} \begin{Bmatrix} \ddot{x}_1 \\ \ddot{x}_2 \end{Bmatrix} + \begin{bmatrix} k & -k \\ -k & k \end{bmatrix} \begin{Bmatrix} x_1 \\ x_2 \end{Bmatrix} = \begin{bmatrix} 1 \\ 0 \end{bmatrix} u \tag{7.103}$$

where the spring stiffness k is uncertain and lies in the range

$$0.7 \le k \le 1.3 \tag{7.104}$$

The state and control matrices of the discrete time model for a sampling time of T

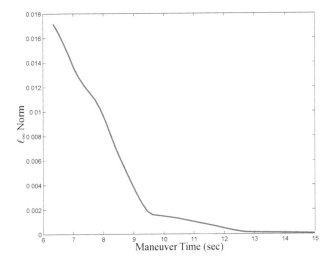

Figure 7.16: Norm Variation vs. Maneuver Time

are

$$
G = \begin{bmatrix}
\frac{(1+\cos(\sqrt{2k}T))}{2} & \frac{(1-\cos(\sqrt{2k}T))}{2} & \frac{T}{2}+\frac{\sqrt{2}}{4\sqrt{k}}\sin(\sqrt{2k}T) & \frac{T}{2}-\frac{\sqrt{2}}{4\sqrt{k}}\sin(\sqrt{2k}T) \\
\frac{(1-\cos(\sqrt{2k}T))}{2} & \frac{(1+\cos(\sqrt{2k}T))}{2} & \frac{T}{2}-\frac{\sqrt{2}}{4\sqrt{k}}\sin(\sqrt{2k}T) & \frac{T}{2}+\frac{\sqrt{2}}{4\sqrt{k}}\sin(\sqrt{2k}T) \\
-\sqrt{\frac{k}{2}}\sin(\sqrt{2k}T) & \sqrt{\frac{k}{2}}\sin(\sqrt{2k}T) & \frac{(1+\cos(\sqrt{2k}T))}{2} & \frac{(1-\cos(\sqrt{2k}T))}{2} \\
\sqrt{\frac{k}{2}}\sin(\sqrt{2k}T) & -\sqrt{\frac{k}{2}}\sin(\sqrt{2k}T) & \frac{(1-\cos(\sqrt{2k}T))}{2} & \frac{(1+\cos(\sqrt{2k}T))}{2}
\end{bmatrix}
\tag{7.105}
$$

$$
H = \begin{bmatrix}
\frac{kt^2-\cos(\sqrt{2}\sqrt{k}t)+1}{4k} \\
\frac{\cos(\sqrt{2}\sqrt{k}t)-1+kt^2}{4k} \\
\frac{2t\sqrt{k}+\sqrt{2}\sin(\sqrt{2}\sqrt{k}t)}{4\sqrt{k}} \\
-\frac{\sqrt{2}\sin(\sqrt{2}\sqrt{k}t)-2t\sqrt{k}}{4\sqrt{k}}
\end{bmatrix}
\tag{7.106}
$$

The optimization problem is to design a control profile that moves the system from rest to a final position of rest while minimizing the maximum magnitude of the residual states over the range of uncertain ks. The boundary conditions are

$$
\begin{bmatrix} x_1(0) \\ x_2(0) \\ \dot{x}_1(0) \\ \dot{x}_2(0) \end{bmatrix} = \begin{bmatrix} 0 \\ 0 \\ 0 \\ 0 \end{bmatrix} \quad \text{and} \quad \begin{bmatrix} x_1(t_f) \\ x_2(t_f) \\ \dot{x}_1(t_f) \\ \dot{x}_2(t_f) \end{bmatrix} = \begin{bmatrix} 1 \\ 1 \\ 0 \\ 0 \end{bmatrix}.
\tag{7.107}
$$

The cost function is

$$\min \left(f = \max_i \left\| \begin{bmatrix} [x_1^i(t_f) - 1)] \\ (x_2^i(t_f) - 1) \\ \dot{x}_1^i(t_f) \\ \dot{x}_2^i(t_f) \end{bmatrix} \right\|_\infty \right) \tag{7.108}$$

Since the stiffness matrix is singular, the system states cannot be weighted by the square root of the stiffness matrices. To account for this, a pseudo-spring stiffness of 1 is attached to the first mass so as to have 0 potential energy at the final position. The resulting LP problem is:

$$\text{minimize} \quad f \tag{7.109a}$$

$$\text{subject to}$$

$$+\dot{x}_1(N) < f \tag{7.109b}$$

$$-\dot{x}_1(N) < f \tag{7.109c}$$

$$+\dot{x}_2(N) < f \tag{7.109d}$$

$$-\dot{x}_2(N) < f \tag{7.109e}$$

$$+\frac{1}{\sqrt{2}} \begin{bmatrix} \sqrt{k^i + k'} & 0 \\ -\frac{k^i}{\sqrt{k^i+k'}} & \sqrt{k^i - \frac{(k^i)^2}{k^i+k'}} \end{bmatrix} \begin{Bmatrix} x_1 \\ x_2 \end{Bmatrix}(N) < \begin{Bmatrix} f \\ f \end{Bmatrix} \tag{7.109f}$$

$$-\frac{1}{\sqrt{2}} \begin{bmatrix} \sqrt{k^i + k'} & 0 \\ -\frac{k^i}{\sqrt{k^i+k'}} & \sqrt{k^i - \frac{(k^i)^2}{k^i+k'}} \end{bmatrix} \begin{Bmatrix} x_1 \\ x_2 \end{Bmatrix}(N) < \begin{Bmatrix} f \\ f \end{Bmatrix} \tag{7.109g}$$

$$-1 \le u(m) \le 1 \quad \forall m = 1, 2, \dots, 501 \tag{7.109h}$$

$$\begin{Bmatrix} x_1 \\ x_2 \\ \dot{x}_1 \\ \dot{x}_2 \end{Bmatrix}(N) = G^{N-1} \begin{Bmatrix} x_1 \\ x_2 \\ \dot{x}_1 \\ \dot{x}_2 \end{Bmatrix}(1) + \sum_{i=1}^{N-1} G^{N-i} Hu(i) \tag{7.109i}$$

where the Cholesky decomposition of the stiffness matrix is used to weigh the displacement states. k' corresponds to the stiffness of the pseudo-spring whose potential energy is zero when the first mass is at the desired final state and is selected to be 1 for the following results.

Figure 7.17 illustrates the control profile and the corresponding evolution of the system states. The solid line represents the states of the mass to which the control is applied and the dashed line represent the states of the noncollocated mass. The control profile for the ℓ_∞ minimax controller is a five-switch bang-bang profile which has a structure identical to that designed to minimize the ℓ_2 norm. The maneuver time is selected to be 5.866, which is the maneuver time of the ℓ_2 minimax optimal controller.

Figure 7.18 illustrates the sensitivity of the residual energy as a function of varying spring stiffness and the residual error states of the two mass for the entire set of uncertain plants which are shown to lie in a box whose ℓ_∞ norm is 0.03285.

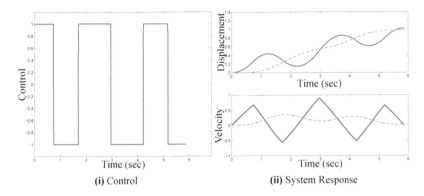

Figure 7.17: Linear Programming Solution

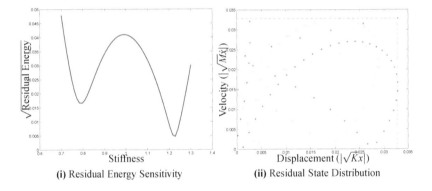

Figure 7.18: Energy and Norm Plots

7.2.4.2 ℓ_1 Optimal Control

Figure 7.19 illustrates the region whose boundaries corresponds to the ℓ_1 norm equal to f. For the two dimension case, the polygon is a cube rotated 45 degrees. For the three-dimension case, the octahedron surfaces correspond to the ℓ_1 norm being a constant.

This section will focus on the design of control profiles which minimize the ℓ_1 norm of the residual states over the span of uncertain models. To illustrate the formulation of the necessary constraints, the two- and three-dimensional examples will be exploited. Consider a system with two states X_1 and X_2 as shown in Figure 7.19(i), the equations which constrain the ℓ_1 norm to be less than f are:

$$X_1 + X_2 < f \tag{7.110}$$
$$X_1 - X_2 < f \tag{7.111}$$

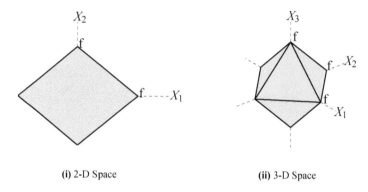

(i) 2-D Space   (ii) 3-D Space

Figure 7.19: Surfaces Representing $\ell_1 = f$

$$-X_1 + X_2 < f \tag{7.112}$$
$$-X_1 - X_2 < f \tag{7.113}$$

For a system with three states X_1, X_2, and X_3, as shown in Figure 7.19(ii), the corresponding constraints are

$$X_1 + X_2 + X_3 < f \tag{7.114}$$
$$X_1 + X_2 - X_3 < f \tag{7.115}$$
$$X_1 - X_2 + X_3 < f \tag{7.116}$$
$$X_1 - X_2 - X_3 < f \tag{7.117}$$
$$-X_1 + X_2 + X_3 < f \tag{7.118}$$
$$-X_1 + X_2 - X_3 < f \tag{7.119}$$
$$-X_1 - X_2 + X_3 < f \tag{7.120}$$
$$-X_1 - X_2 - X_3 < f \tag{7.121}$$

It can be seen from the Equations (7.110) through (7.113) and (7.114) through (7.121), that the coefficients of the linear equations correspond to the vertices of a square and cube, respectively, of edge length equal to $\sqrt{2}$, centered at the origin. For examples, the vertices of a square centered at the origin with edge length $\sqrt{2}$ are $\{1, 1\}$, $\{1, -1\}$, $\{-1, 1\}$, and $\{-1, -1\}$ which are also the coefficients of the linear constraint equations. Thus, the constraint equations for a system of order n can be derived by determining the vertices of a n dimensional hypercube of edge length $\sqrt{2}$, and centered at the origin.

The minimax problem to design control profiles which minimize the ℓ_1 norm of the residual states over the domain of uncertain models can be stated in an LP form as:

$$\text{minimize} \quad f \tag{7.122a}$$

subject to

$$\left[\sum_{k=1}^{n} C_{kj} y_k^i(N) \le f \right]_{j=1,2,\dots,2^n}^{i=1,2,\dots,Z} \tag{7.122b}$$

where $C_{kj} = \pm 1$

$$u_{min} \le u(m) \le u_{max} \quad \forall m = 1,2,\dots,N \tag{7.122c}$$

where n is the order of the dynamical system, N is the number of samples over which the control is parameterized and Z is the number of uncertain models over which the the ℓ_1 norm is minimized. The coefficients $C_{kj} = \pm 1$ correspond to the vertices of an n-dimensional hypercube centered at the origin and side length of $\sqrt{2}$.

Example 7.5:
  A spring-mass-dashpot system with an uncertain spring stiffness modeled as

$$\ddot{x} + 0.2\dot{x} + kx = ku \tag{7.123}$$

with the uncertain stiffness given by Equation (7.104) as in Example 7.3. The ℓ_1 optimal-control problem is given as:

$$\text{minimize} \quad f \tag{7.124a}$$

subject to

$$+\dot{x} + \sqrt{k^i}\, x < f \tag{7.124b}$$

$$+\dot{x} - \sqrt{k^i}\, x < f \tag{7.124c}$$

$$-\dot{x} + \sqrt{k^i}\, x < f \tag{7.124d}$$

$$-\dot{x} - \sqrt{k^i}\, x < f \tag{7.124e}$$

$$0 \le u(m) \le 1 \quad \forall m = 1,2,\dots,501 \tag{7.124f}$$

$$u(m) - u(m+1) \le 0 \quad \forall m = 1,2,\dots,500 \tag{7.124g}$$

where $i = 1,2,\dots,21$ correspond to the 21 discrete values of k over the range of 0.7–1.3. The ℓ_1 optimal control profile and the corresponding evolution of the system states is shown in Figure 7.20 for a maneuver time of $t_f = 6.3405$.

  Figure 7.20(i) illustrates the sensitivity curve for the ℓ_1 (∗) and minimax (∘) optimal control profiles. It is clear that the sensitivity curve corresponding to the minimax solution has a smaller maximum magnitude of the residual energy compared to the ℓ_1 solution. Figure 7.20(i) displays the absolute values of the residual error of the system states for the 21 uncertain models used in the design of the robust controller.

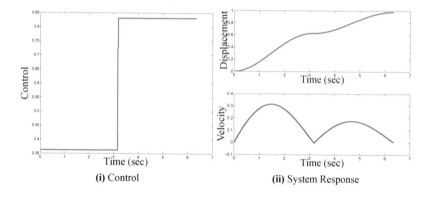

(i) Control **(ii) System Response**

Figure 7.20: Linear Programming Solution

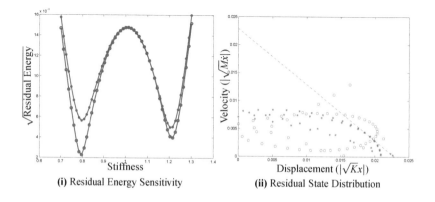

(i) Residual Energy Sensitivity **(ii) Residual State Distribution**

Figure 7.21: Energy and Norm Plots

The $*$ symbols which represent the ℓ_1 solution are bounded by the dashed line which corresponds to the ℓ_1 norm of 0.0229.

Since the LP approach to solving the minimax ℓ_1 problem requires the maneuver time to be specified, the effect of the maneuver time on the robustness can be studied. Figure 7.22 illustrates the sensitivity curves for maneuver times of $t_f = 6.3405$, 7.3405, and 8.3405 seconds. As was evident in Example 7.3, an increase in the maneuver time results in increased robustness of the controllers.

As in Example 7.3, the variation of the ℓ_1 norm as a function of the maneuver time is plotted in Figure 7.25. The ℓ_∞ is plotted as well and it can be seen that the trends are identical and the knee of both curves occur around a maneuver time of 9.5 seconds.

Example 7.6: To compare the minimax solution of the ℓ_∞ and the ℓ_1 cost function for systems with rigid-body modes, the two-mass spring benchmark problem is stud-

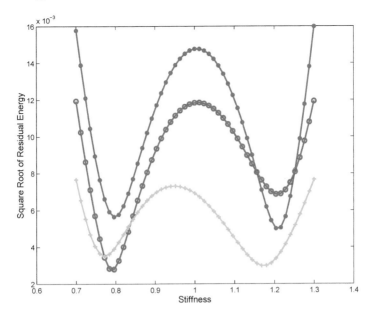

Figure 7.22: Residual Energy

ied. The cost function is to minimize the maximum ℓ_1 norm of the residual states for a set of plants spanning the uncertain space. Defining a state transformation:

$$\begin{Bmatrix} y_1 \\ y_2 \\ \dot{y}_1 \\ \dot{y}_2 \end{Bmatrix}(N) = \frac{1}{\sqrt{2}} \begin{bmatrix} \sqrt{k^i + k^l} & 0 & 0 & 0 \\ -\frac{k^i}{\sqrt{k^i+k^l}} & \sqrt{k^i - \frac{(k^i)^2}{k^i+k^l}} & 0 & 0 \\ 0 & 0 & 1 & 0 \\ 0 & 0 & 0 & 1 \end{bmatrix} \begin{Bmatrix} x_1 \\ x_2 \\ \dot{x}_1 \\ \dot{x}_2 \end{Bmatrix}(N), \qquad (7.125)$$

where k^l is the stiffness of a pseudo-spring to guarantee a positive definite potential energy function about the equilibrium states. An LP problem is formulated:

$$\text{minimize} \quad f \qquad (7.126a)$$

subject to

$$-f < +y_1(N) + y_2(N) + \dot{y}_1(N) + \dot{y}_2(N) < f \qquad (7.126b)$$
$$-f < -y_1(N) + y_2(N) + \dot{y}_1(N) + \dot{y}_2(N) < f \qquad (7.126c)$$
$$-f < -y_1(N) - y_2(N) + \dot{y}_1(N) + \dot{y}_2(N) < f \qquad (7.126d)$$
$$-f < -y_1(N) - y_2(N) - \dot{y}_1(N) + \dot{y}_2(N) < f \qquad (7.126e)$$
$$-f < +y_1(N) - y_2(N) + \dot{y}_1(N) + \dot{y}_2(N) < f \qquad (7.126f)$$
$$-f < +y_1(N) - y_2(N) - \dot{y}_1(N) + \dot{y}_2(N) < f \qquad (7.126g)$$
$$-f < +y_1(N) + y_2(N) - \dot{y}_1(N) + \dot{y}_2(N) < f \qquad (7.126h)$$

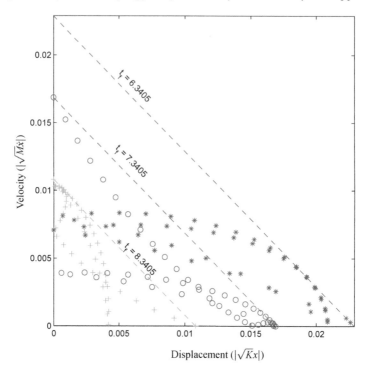

Figure 7.23: Residual State Distribution

$$-f < -y_1(N) + y_2(N) + \dot{y}_1(N) - \dot{y}_2(N) < f \tag{7.126i}$$

$$-1 \le u(m) \le 1 \quad \forall m = 1, 2, \ldots, 1001 \tag{7.126j}$$

$$\begin{Bmatrix} x_1 \\ x_2 \\ \dot{x}_1 \\ \dot{x}_2 \end{Bmatrix}(N) = G^{N-1} \begin{Bmatrix} x_1 \\ x_2 \\ \dot{x}_1 \\ \dot{x}_2 \end{Bmatrix}(1) + \sum_{i=1}^{N-1} G^{N-i} H u(i) \tag{7.126k}$$

Discretizing the maneuver time into 1000 intervals, the LP problem is solved. Figure 7.26 illustrates the ℓ_1 minimax control profile and the evolution of the nominal system states.

The solid line represents the states of the mass to which the control is applied and the dashed line represents the states of the noncollocated mass. The control profile for the ℓ_1 minimax controller is a five-switch bang-bang profile which has a structure that is identical to that designed to minimize the ℓ_2 norm. The maneuver time is selected to be 5.9093 which is the maneuver time of the ℓ_2 minimax optimal controller. Figure 7.27 illustrates the sensitivity plot of the ℓ_1 and the ℓ_2 minimax controllers. The ℓ_2 minimax controller is derived using a nonlinear programming approach and its sensitivity is illustrated by the line with the ∘ marks. It can be seen that the ℓ_2 minimax solution has a performance which is marginally better than that of the ℓ_1 con-

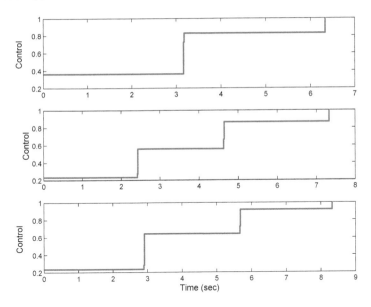

Figure 7.24: Spectrum of Control Profiles

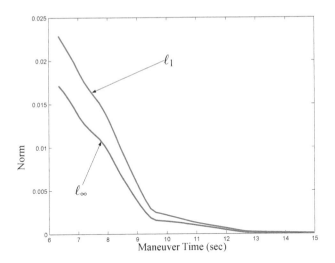

Figure 7.25: Norm Variation vs. Maneuver Time

troller. The absolute magnitude of the residual displacement states are plotted against the residual velocity states for both the ℓ_1 (∗) and ℓ_2 (∘) solutions. The dashed line corresponds to the optimal constant ℓ_1 norm profile determined from the optimization problem to be 0.0859. It appears intriguing that none of the ∗ symbols lie on the dashed line. This is because, we are attempting to represent a four-dimensional (4D)

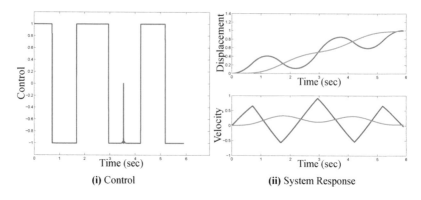

(i) Control   (ii) System Response

Figure 7.26: Linear Programming Solution

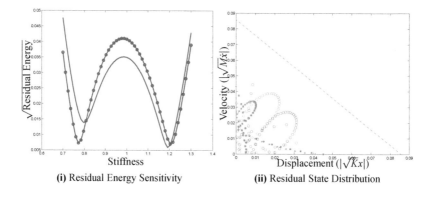

(i) Residual Energy Sensitivity   (ii) Residual State Distribution

Figure 7.27: Energy and Norm Plots

hyper-octahedron with a quadrant. Figure 7.28 illustrates all possible combinations of the absolute magnitude of the system states and it can be seen that some $*$ symbols lie on the constant ℓ_1 norm lines. This set of graphs does not comprehensively cap-ture the entire 4D hyper-octahedron, rather it looks at slices of the hyper-octahedron. It can be seen that all the $\circ$ symbols lie within the region bounded by the dashed lines indicating that the maximum magnitude of residual state errors for the ℓ_1 minimax solution is greater than that of the ℓ_2 minimax solution.

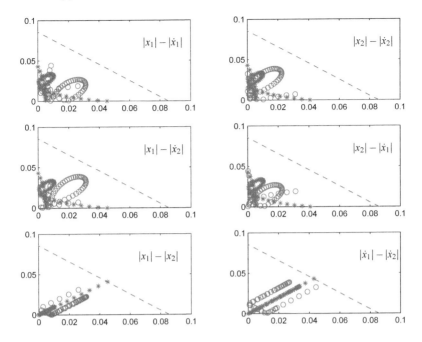

Figure 7.28: Residual State Distribution

7.2.4.3 ℓ_2 Optimal Control

In this section, an approach to use linear programming to solve the minimax problem that minimizes the residual energy of the system which is a ℓ_2 norm is presented. This involves approximating the n-dimensional hypersphere with $(n-1)$-dimensional hyperplanes (simplex). A simplex is the convex hull of $n+1$ points in n dimensional space where the points do not all lie in some $n-1$ dimensional subspace. For the single dimension, the simplex is a line, for two dimension, it is a triangle, for the third dimension it is a tetrahedeon, in the fourth dimension it is a pentatope, and so on. Barycentric coordinates [5] which are also referred to homogeneous coordinates or trilinear coordinates use $n+1$ points to locate a point in n space relative to some n dimensional simplex. Barycentric coordinates sum to unity. By requiring all the coordinates to be positive constraints the points to lie in the simplex.

Figure 7.29 illustrates the approximation of an unit circle with lines and Figure 7.30 illustrates a unit sphere with multiple planes. Considering the three sequential approximation of the circle and sphere with lines and planes as Level 1, 2, and 3, it is clear from Level 3, that the approximation of the ℓ_2 norm by planes is quite good.

An affine combination of a set of points $\mathscr{P}_1$, $\mathscr{P}_2$, $\mathscr{P}_3$, ..., $\mathscr{P}_n$ is the point:

$$w_1\mathscr{P}_1 + w_2\mathscr{P}_2 + w_3\mathscr{P}_3 + \ldots w_n\mathscr{P}_n \qquad (7.127)$$

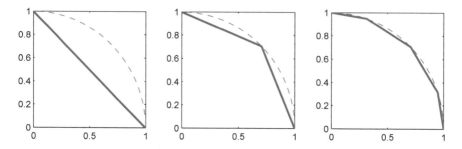

Figure 7.29: Approximation of Circle with Lines

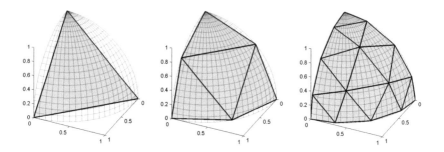

Figure 7.30: Approximation of Sphere with Planes

where

$$w_1 + w_2 + w_3 + \ldots + w_n = 1. \tag{7.128}$$

The coordinates $\begin{bmatrix} w_1 & w_2 & w_3 & \ldots & w_n \end{bmatrix}$, are called the barycentric coordinates of the point given by Equation (7.127). If all the w_i are required to be positive, the barycentric coordinates represent all the points which lie in the Convex Hull of the points $\mathscr{P}_1, \mathscr{P}_2, \mathscr{P}_3, \ldots, \mathscr{P}_n$.

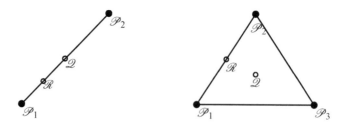

Figure 7.31: Barycentric Examples

Figure 7.31 illustrates the barycentric concept. The points $\mathscr{Q}$ and $\mathscr{R}$ on the line connecting the points $\mathscr{P}_1$ and $\mathscr{P}_2$ are given by the barycentric coordinates $[0.5 \ 0.5]$ and $[0.25 \ 0.75]$, respectively. Similarly, the barycentric coordinates for the points $\mathscr{Q}$ and $\mathscr{R}$ on the triangle formed by the points $\mathscr{P}_1$, $\mathscr{P}_2$, and $\mathscr{P}_3$ are given by the three-tuples $[1/3 \ 1/3 \ 1/3]$ and $[1/2 \ 1/2 \ 0]$, respectively.

The barycentric coordinates means for parameterizing specific points in an n-simplex provides us a simple mechanism to recursively divide the n-simplex into smaller segments. For example, if we start with two points specifying a line

$$\mathscr{P}_1 = \begin{bmatrix} 1 \\ 1 \end{bmatrix} \text{ and } \mathscr{P}_2 = \begin{bmatrix} 3 \\ 5 \end{bmatrix}, \tag{7.129}$$

and we are interested in dividing the line into four segments, we first determine the point with a barycentric coordinate of $[0.5 \ 0.5]$ which corresponds to the point

$$\mathscr{P}_3 = 0.5\mathscr{P}_1 + 0.5\mathscr{P}_2 = 0.5 \begin{bmatrix} 1 \\ 1 \end{bmatrix} + 0.5 \begin{bmatrix} 3 \\ 5 \end{bmatrix} = \begin{bmatrix} 2 \\ 3 \end{bmatrix}. \tag{7.130}$$

The next set of points are determined by recursively calculating the points corresponding to the barycentric coordinate of $[0.5 \ 0.5]$ of the points $\mathscr{P}_1$ and $\mathscr{P}_3$ and $\mathscr{P}_2$ and $\mathscr{P}_3$ which results in the points

$$\mathscr{P}_4 = 0.5\mathscr{P}_1 + 0.5\mathscr{P}_3 = 0.5 \begin{bmatrix} 1 \\ 1 \end{bmatrix} + 0.5 \begin{bmatrix} 2 \\ 3 \end{bmatrix} = \begin{bmatrix} 1.5 \\ 2 \end{bmatrix} \tag{7.131}$$

and

$$\mathscr{P}_5 = 0.5\mathscr{P}_3 + 0.5\mathscr{P}_2 = 0.5 \begin{bmatrix} 2 \\ 3 \end{bmatrix} + 0.5 \begin{bmatrix} 3 \\ 5 \end{bmatrix} = \begin{bmatrix} 2.5 \\ 4 \end{bmatrix}. \tag{7.132}$$

The approach that we will follow to generate hyperplanes to approximate the hypersphere begins by determining the hyperplane which corresponds to the polytope which corresponds to the constant ℓ_1 norm. For example, the constant $\ell_1 = 1$, polytopes for the two-dimensional space are given by the lines:

$$x_1 + x_2 = 1 \tag{7.133}$$
$$-x_1 - x_2 = 1 \tag{7.134}$$
$$-x_1 + x_2 = 1 \tag{7.135}$$
$$x_1 - x_2 = 1 \tag{7.136}$$

which can be parameterized in terms of lines connecting the points

$$\mathscr{P}_1 = \begin{bmatrix} 1 \\ 0 \end{bmatrix}, \mathscr{P}_2 = \begin{bmatrix} 0 \\ 1 \end{bmatrix}, \text{ for the line} \qquad +x_1 + x_2 = 1,$$

$$\mathscr{P}_4 = \begin{bmatrix} 0 \\ -1 \end{bmatrix}, \mathscr{P}_3 = \begin{bmatrix} -1 \\ 0 \end{bmatrix}, \text{ for the line} \qquad -x_1 - x_2 = 1,$$

$$\mathscr{P}_3 = \begin{bmatrix} -1 \\ 0 \end{bmatrix}, \mathscr{P}_2 = \begin{bmatrix} 0 \\ 1 \end{bmatrix}, \text{ for the line} \qquad -x_1 + x_2 = 1,$$

$$\mathscr{P}_1 = \begin{bmatrix} 1 \\ 0 \end{bmatrix}, \mathscr{P}_4 = \begin{bmatrix} 0 \\ -1 \end{bmatrix}, \text{ for the line} \qquad +x_1 - x_2 = 1.$$

Consider the line connecting $\mathscr{P}_1$ and $\mathscr{P}_2$ (Equation 7.133) which approximates the arc of an unit radius circle in the first quadrant. The barycentric coordinates $\begin{bmatrix} 0.5 & 0.5 \end{bmatrix}$ of the points $\mathscr{P}_1$ and $\mathscr{P}_2$ correspond to the point:

$$\mathscr{P}_5 = 0.5\mathscr{P}_1 + 0.5\mathscr{P}_2 = \begin{bmatrix} 0.5 \\ 0.5 \end{bmatrix}. \tag{7.137}$$

Normalizing this point such that its Euclidean norm is unity results in the point:

$$\mathscr{P}_5 = \begin{bmatrix} \frac{1}{\sqrt{2}} \\ \frac{1}{\sqrt{2}} \end{bmatrix}. \tag{7.138}$$

The line which connects $\mathscr{P}_1$ and $\mathscr{P}_5$, and $\mathscr{P}_5$ and $\mathscr{P}_2$ are given by the equations:

$$0.9239x_1 + 0.3827x_2 = 0.9239 \tag{7.139}$$
$$0.3827x_1 + 0.9239x_2 = 0.9239. \tag{7.140}$$

Similarly, the lines defined by Equations (7.134) through (7.136) can be divided to result in a total of eight lines to approximate the circle.

Similarly, for the three-dimensional space, the planes can be parameterized in terms of the point:

$$\mathscr{P}_1 = \begin{bmatrix} 1 \\ 0 \\ 0 \end{bmatrix}, \mathscr{P}_2 = \begin{bmatrix} 0 \\ 1 \\ 0 \end{bmatrix}, \qquad \mathscr{P}_3 = \begin{bmatrix} 0 \\ 0 \\ 1 \end{bmatrix}, \text{ for } x_1 + x_2 + x_3 = 1$$

$$\mathscr{P}_1 = \begin{bmatrix} 1 \\ 0 \\ 0 \end{bmatrix}, \mathscr{P}_2 = \begin{bmatrix} 0 \\ 1 \\ 0 \end{bmatrix}, \qquad \mathscr{P}_6 = \begin{bmatrix} 0 \\ 0 \\ -1 \end{bmatrix}, \text{ for } x_1 + x_2 - x_3 = 1$$

$$\mathscr{P}_1 = \begin{bmatrix} 1 \\ 0 \\ 0 \end{bmatrix}, \mathscr{P}_5 = \begin{bmatrix} 0 \\ -1 \\ 0 \end{bmatrix}, \qquad \mathscr{P}_3 = \begin{bmatrix} 0 \\ 0 \\ 1 \end{bmatrix}, \text{ for } x_1 - x_2 + x_3 = 1$$

$$\mathscr{P}_1 = \begin{bmatrix} 1 \\ 0 \\ 0 \end{bmatrix}, \mathscr{P}_5 = \begin{bmatrix} 0 \\ -1 \\ 0 \end{bmatrix}, \qquad \mathscr{P}_6 = \begin{bmatrix} 0 \\ 0 \\ -1 \end{bmatrix}, \text{ for } x_1 - x_2 - x_3 = 1$$

$$\mathscr{P}_4 = \begin{bmatrix} -1 \\ 0 \\ 0 \end{bmatrix}, \mathscr{P}_2 = \begin{bmatrix} 0 \\ 1 \\ 0 \end{bmatrix}, \qquad \mathscr{P}_3 = \begin{bmatrix} 0 \\ 0 \\ 1 \end{bmatrix}, \text{ for } -x_1 + x_2 + x_3 = 1$$

$$\mathscr{P}_4 = \begin{bmatrix} -1 \\ 0 \\ 0 \end{bmatrix}, \mathscr{P}_2 = \begin{bmatrix} 0 \\ 1 \\ 0 \end{bmatrix}, \qquad \mathscr{P}_6 = \begin{bmatrix} 0 \\ 0 \\ -1 \end{bmatrix}, \text{ for } -x_1 + x_2 - x_3 = 1$$

$$\mathscr{P}_4 = \begin{bmatrix} -1 \\ 0 \\ 0 \end{bmatrix}, \mathscr{P}_5 = \begin{bmatrix} 0 \\ -1 \\ 0 \end{bmatrix}, \qquad \mathscr{P}_3 = \begin{bmatrix} 0 \\ 0 \\ 1 \end{bmatrix}, \text{ for } -x_1 - x_2 + x_3 = 1$$

$$\mathscr{P}_4 = \begin{bmatrix} -1 \\ 0 \\ 0 \end{bmatrix}, \mathscr{P}_5 = \begin{bmatrix} 0 \\ -1 \\ 0 \end{bmatrix}, \qquad \mathscr{P}_6 = \begin{bmatrix} 0 \\ 0 \\ -1 \end{bmatrix}, \text{ for } -x_1 - x_2 - x_3 = 1$$

These points, which define patches are used to divide each patch into smaller patches using the barycentric coordinate approach. These result in point which lie on the plane defined by the basic n-simplex. After the division of the n-simplex, all the points are normalized to lie on an unit radius hypersphere. The patches after normalization defines new planes which are inscribed within the unit radius hypersphere. These planes now defines linear constraint equations which permit us to approximately represent the spherical constraint with a set of linear constraints. To illustrate this process, consider the plane defined by the points $\mathscr{P}_1$, $\mathscr{P}_2$, and $\mathscr{P}_3$ in three-dimensional space. All possible combinations ${}^3C_2$ of the points $\mathscr{P}_1$, $\mathscr{P}_2$, and $\mathscr{P}_3$ are determined which are $\{\mathscr{P}_1\,\mathscr{P}_2\}$, $\{\mathscr{P}_2\,\mathscr{P}_3\}$, and $\{\mathscr{P}_1\,\mathscr{P}_3\}$. The points which correspond to the barycentric coordinates $\begin{bmatrix} 0.5 & 0.5 \end{bmatrix}$ of the pairs of points are determined and are shown in Figure 7.32(i). Normalizing the points $\mathscr{P}_4$, $\mathscr{P}_5$, and $\mathscr{P}_6$ results in points which lie on the sphere as shown in Figure 7.32(ii). The six points define four planes which are defined by the equations:

$$0.5774x_1 + 0.5774x_2 + 0.5774x_3 = 0.8165 \tag{7.141}$$

$$0.8629x_1 + 0.3574x_2 + 0.3574x_3 = 0.8629 \tag{7.142}$$

$$0.3574x_1 + 0.8629x_2 + 0.3574x_3 = 0.8629 \tag{7.143}$$

$$0.3574x_1 + 0.3574x_2 + 0.8629x_3 = 0.8629 \tag{7.144}$$

The number of hyperplanes that approximate the n-dimensional hypersphere is given by the equation:

$$1 + n + {}^nC_2 + {}^nC_3 + \ldots + {}^nC_{n-2} \tag{7.145}$$

for each of the ℓ_1 plane and 2^n number of the ℓ_1 plane create a polytope which approximate the hypersphere. For example in three-dimension ($n = 3$), the number

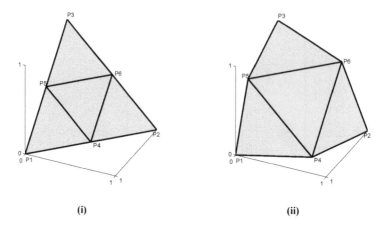

(i)   (ii)

Figure 7.32: Division of Two-Simplex

of planes to approximate the ℓ_1 plane is 4 and there are a total of $2^3 = 8$ number of these ℓ_1 planes as shown in Figure 7.19(ii). The first level (L = 1) of recursion generates a total of 24 planes which approximate the sphere and the second level (L = 2) of recursion results in $2^n(4^L) = 128$ planes.

To determine the largest error in using hyperplanes to approximate hypersphere, we need to calculate the shortest distance of every hyperplane from the origin and the difference of this distance from unity is the error.

The equation of a hyperplane in n-dimension is given by the equation:

$$\sum_{i=1}^{n} a_i x_i = b \qquad (7.146)$$

and the normal to the plane is given by the vector

$$\vec{a} = \begin{bmatrix} a_1 \\ a_2 \\ \vdots \\ a_{n-1} \\ a_n. \end{bmatrix} \qquad (7.147)$$

The equation of the normal to the plane passing through the origin is given by the equation

$$\alpha \vec{a}^T = 0 \qquad (7.148)$$

where α is a scalar. To determine the point of intersection of the normal (Equation 7.148) with the hyperplane (Equation 7.146), substitute Equation (7.148) into Equation (7.146) resulting in the equation

$$\sum_{i=1}^{n} a_i x_i = b \tag{7.149}$$

$$\alpha \sum_{i=1}^{n} a_i^2 = b \Rightarrow \alpha = \frac{b}{\sum_{i=1}^{n} a_i^2}. \tag{7.150}$$

Having determined the coefficients of the hyperplanes to approximate the hypersphere of n-dimension, the maximum error in approximating the hypersphere with a set of planes is studied as a function of the number of hyperplanes. The maximum error is given as the difference from unity (the radius of the unit hypersphere) of the shortest Euclidean distance of the plane from the origin. The point of intersection of a line that is normal to a plane passing through the origin, and the plane can be shown to be:

$$\frac{b}{\sum_{i=1}^{n} a_i^2} \begin{bmatrix} a_1 \\ a_2 \\ \vdots \\ a_{n-1} \\ a_n \end{bmatrix} \tag{7.151}$$

The shortest distance of the plane from the origin in now given by the equation

$$\frac{b}{\sum_{i=1}^{n} a_i^2} \sqrt{\sum_{i=1}^{n} a_i^2} = \frac{b}{\sqrt{\sum_{i=1}^{n} a_i^2}} \tag{7.152}$$

When approximating a circle with lines, Figure 7.33 illustrates the reduction in the error as a function of number of lines (planes) used to approximate the circle and sphere, respectively.

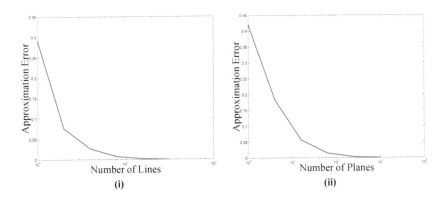

Figure 7.33: Error in Approximating a Hypersphere with Hyperplanes

Example 7.7:   The second-order spring-mass-dashpot system with an uncertain spring stiffness is considered to illustrate the LP-based approach for the determination of a control profile which minimizes the maximum ℓ_2 norm of the residual states. The cost function is defined as

$$\min \left(f = \max_i \ k^i (x^i(t_f) - 1)^2 + (\dot{x}^i(t_f))^2 \right) \tag{7.153}$$

which corresponds to the residual energy for a system whose boundary conditions correspond to a final displacement of unity and a final velocity of zero. The first approximate solution corresponds to the minimization of the ℓ_1 norm of the residual states of the set of uncertain models. The resulting solution has been illustrated in Example 7.5. The next iteration corresponds to doubling the number of lines used to approximate the ℓ_2 norm of the residual states. The resulting linear programming problem is

$$\text{minimize} \quad f \tag{7.154a}$$

$$\text{subject to}$$

$$-0.9239f < +0.9239\dot{x} + 0.3827\sqrt{k}^i x < 0.9239f \tag{7.154b}$$

$$-0.9239f < +0.3827\dot{x} + 0.9239\sqrt{k}^i x < 0.9239f \tag{7.154c}$$

$$-0.9239f < -0.9239\dot{x} + 0.3827\sqrt{k}^i x < 0.9239f \tag{7.154d}$$

$$-0.9239f < -0.3827\dot{x} + 0.9239\sqrt{k}^i x < 0.9239f \tag{7.154e}$$

$$0 \le u(m) \le 1 \ \forall m = 1, 2, \ldots, 501 \tag{7.154f}$$

$$u(m) - u(m+1) \le 0 \ \forall m = 1, 2, \ldots, 500 \tag{7.154g}$$

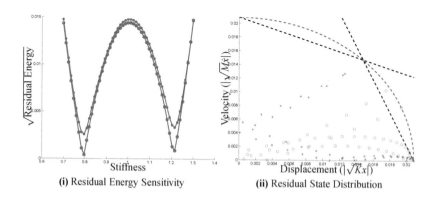

(i) Residual Energy Sensitivity   (ii) Residual State Distribution

Figure 7.34: Energy and Norm Plots

Figure 7.34(i) illustrates the variations of the square root of the residual energy as a function of the spring stiffness. The graph with the ∗ symbols corresponds to the solution determined by the linear programming problem where eight lines are used to approximate the circle which contains the residual states, and the graph with the ∘ symbols corresponds to the solution derived form a nonlinear programming problem (NLP). It can be seen that the difference between the solutions is minimal. Figure 7.34(ii) illustrates the absolute magnitude of residual displacement plotted versus the absolute magnitude of the velocity. The ∗ and the ∘ symbols corresponds to the residual states for 51 models which span the range of uncertain stiffness for the LP and NLP solutions. The dashed lines represent the lines that are used to approximate the circle. It can be seen that all the ∗ symbols lie within the polygon bounded by the dashed lines and the abscissa and the ordinate, and the ∘ symbols lie in the region bounded by the arc of a circle and the x and y axes.

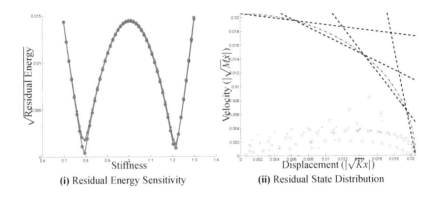

(i) Residual Energy Sensitivity   (ii) Residual State Distribution

Figure 7.35: Energy and Norm Plots

Figure 7.35 illustrates the results of the sensitivity curves and the norm of the residual states when two levels of recursion are used to generate 16 lines to approximate a circle. It can be seen that compared to Figure 7.34 the solutions are closer to that of the NLP solution, but the incremental improvement might not warrant solving a larger size LP problem.

Figure 7.36 illustrates the minimax control profile generated using 8 and 16 lines to approximate a circle, respectively. It can be noted that the solutions with an increasing number of lines to approximate the circle, tends toward the two-switch minimax solution generated by the NLP solver which is given by the equation:

$$u(t) = 0.3452 + 0.4730\mathscr{H}(t - 3.1703) + 0.1818\mathscr{H}(t - 6.3405). \qquad (7.155)$$

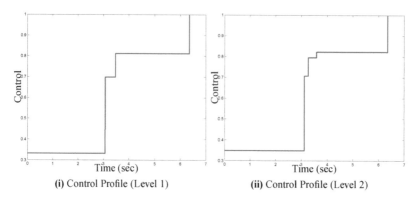

(i) Control Profile (Level 1)   (ii) Control Profile (Level 2)

Figure 7.36: Input-Shaped Control Profile

Example 7.8: The LP-based approach to solving a minimax control problem for a system with a rigid-body mode is illustrated here. The two-mass spring system with an uncertain spring stiffness is considered. The system is a described by a fourth-order model and hence a fourth-dimensional hypersphere needs to be approximated by hyperplanes. As was described earlier, planes that correspond to the ℓ_1 norm of unity are determined, which are subsequently divided to generate a series of planes which are circumscribed by the hypersphere. For illustrative purposes consider the plane defined by the four point

$$\mathscr{P}_1 = \begin{bmatrix} 1 \\ 0 \\ 0 \\ 0 \end{bmatrix}, \mathscr{P}_2 = \begin{bmatrix} 0 \\ 1 \\ 0 \\ 0 \end{bmatrix}, \mathscr{P}_3 = \begin{bmatrix} 0 \\ 0 \\ 1 \\ 0 \end{bmatrix}, \mathscr{P}_4 = \begin{bmatrix} 0 \\ 0 \\ 0 \\ 1 \end{bmatrix}, \tag{7.156}$$

which is defined by the equation

$$x_1 + x_2 + x_3 + x_4 = 1. \tag{7.157}$$

The four points $\{\mathscr{P}_1 \, \mathscr{P}_2 \, \mathscr{P}_3 \, \mathscr{P}_4\}$ can be used to represent the barycentric coordinates of a tetrahedron. The point that corresponds to the barycentric coordinate of $\begin{bmatrix} \frac{1}{3} & \frac{1}{3} & \frac{1}{3} \end{bmatrix}$ is determined for all $^4C_3$ combinations of the points $\{\mathscr{P}_1 \, \mathscr{P}_2 \, \mathscr{P}_3 \, \mathscr{P}_4\}$ which result in the points:

$$\mathscr{P}_5 = \frac{1}{3}\mathscr{P}_1 + \frac{1}{3}\mathscr{P}_2 + \frac{1}{3}\mathscr{P}_3 = \begin{bmatrix} \frac{1}{3} & \frac{1}{3} & \frac{1}{3} & 0 \end{bmatrix}^T$$

$$\mathscr{P}_6 = \frac{1}{3}\mathscr{P}_1 + \frac{1}{3}\mathscr{P}_2 + \frac{1}{3}\mathscr{P}_4 = \begin{bmatrix} \frac{1}{3} & \frac{1}{3} & 0 & \frac{1}{3} \end{bmatrix}^T$$

$$\mathscr{P}_7 = \frac{1}{3}\mathscr{P}_1 + \frac{1}{3}\mathscr{P}_3 + \frac{1}{3}\mathscr{P}_4 = \begin{bmatrix} \frac{1}{3} & 0 & \frac{1}{3} & \frac{1}{3} \end{bmatrix}^T$$

$$\mathscr{P}_8 = \frac{1}{3}\mathscr{P}_2 + \frac{1}{3}\mathscr{P}_3 + \frac{1}{3}\mathscr{P}_4 = \begin{bmatrix} 0 & \frac{1}{3} & \frac{1}{3} & \frac{1}{3} \end{bmatrix}^T$$

The tetrahedron, which is defined by the points $\{\mathscr{P}_1\ \mathscr{P}_2\ \mathscr{P}_3\ \mathscr{P}_4\}$ can be subdivided into 11 tetrahedra which are given by the following combinations:

$$\{\mathscr{P}_5\ \mathscr{P}_6\ \mathscr{P}_7\ \mathscr{P}_8\},\ \{\mathscr{P}_1\ \mathscr{P}_5\ \mathscr{P}_6\ \mathscr{P}_7\},\ \{\mathscr{P}_2\ \mathscr{P}_5\ \mathscr{P}_6\ \mathscr{P}_8\},$$
$$\{\mathscr{P}_3\ \mathscr{P}_5\ \mathscr{P}_7\ \mathscr{P}_8\},\ \{\mathscr{P}_4\ \mathscr{P}_6\ \mathscr{P}_7\ \mathscr{P}_8\},\ \{\mathscr{P}_1\ \mathscr{P}_2\ \mathscr{P}_5\ \mathscr{P}_6\},$$
$$\{\mathscr{P}_1\ \mathscr{P}_3\ \mathscr{P}_5\ \mathscr{P}_7\},\ \{\mathscr{P}_1\ \mathscr{P}_4\ \mathscr{P}_6\ \mathscr{P}_7\},\ \{\mathscr{P}_2\ \mathscr{P}_3\ \mathscr{P}_5\ \mathscr{P}_8\},$$
$$\{\mathscr{P}_2\ \mathscr{P}_4\ \mathscr{P}_6\ \mathscr{P}_8\},\ \{\mathscr{P}_3\ \mathscr{P}_4\ \mathscr{P}_7\ \mathscr{P}_8\}$$

A tetrahedon and its division into 11 smaller tetrahedra is shown in Figure 7.37 each of the vertices of the tetrahedra are normalized so as to lie on the unit hyper-

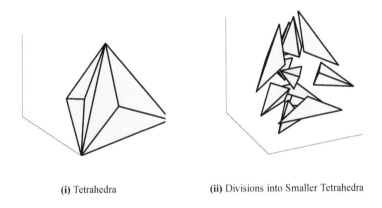

(i) Tetrahedra   (ii) Divisions into Smaller Tetrahedra

Figure 7.37: Recursive Division

sphere. These 11 normalized tetrahedra represent 11 planes which approximate part of the hypersphere. The equations for these 11 planes are:

$$0.5000x_1 + 0.5000x_2 + 0.5000x_3 + 0.5000x_4 = 0.8660$$
$$0.6830x_1 + 0.6830x_2 - 0.1830x_3 - 0.1830x_4 = 0.6830$$
$$0.6830x_1 - 0.1830x_2 + 0.6830x_3 - 0.1830x_4 = 0.6830$$
$$0.6830x_1 - 0.1830x_2 - 0.1830x_3 + 0.6830x_4 = 0.6830$$
$$-0.1830x_1 + 0.6830x_2 - 0.1830x_3 + 0.6830x_4 = 0.6830$$
$$-0.1830x_1 + 0.6830x_2 + 0.6830x_3 - 0.1830x_4 = 0.6830$$
$$-0.1830x_1 - 0.1830x_2 + 0.6830x_3 + 0.6830x_4 = 0.6830$$
$$0.8446x_1 + 0.3091x_2 + 0.3091x_3 + 0.3091x_4 = 0.8446$$

$$0.3091x_1 + 0.8446x_2 + 0.3091x_3 + 0.3091x_4 = 0.8446$$
$$0.3091x_1 + 0.3091x_2 + 0.8446x_3 + 0.3091x_4 = 0.8446$$
$$0.3091x_1 + 0.3091x_2 + 0.3091x_3 + 0.8446x_4 = 0.8446$$

The 16 $\ell_1 = 1$ planes are divided into 11 planes resulting in a total of 176 planes to approximate the four-dimensional hypersphere. Each of these planes can be recursively divided into 11 planes each for better approximation of the hypersphere. In this example, we will stay with the 176 planes to approximate the hypersphere. Using 21 points to discretize the uncertain spring stiffness space, results in an LP problem with a total of 3696 inequality constraints. Dividing the time interval into 301 time instants, results in a total of 302 variables to be solved which correspond to the magnitude of the control at 301 time instants and the variable which approximates the radius of the hypersphere which bounds the residual energy of the 21 uncertain models.

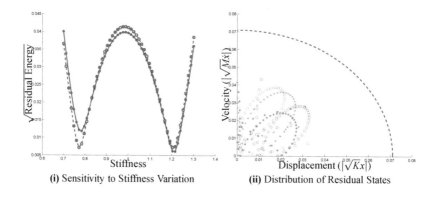

(i) Sensitivity to Stiffness Variation   (ii) Distribution of Residual States

Figure 7.38: ℓ_2 Minimax Control

Figure 7.38(i) illustrates the variations of the residual energy of the system as a function of varying spring stiffness for solutions generated by the LP problem (*) and the NLP (o). It can be seen that the sensitivity plots are nearly identical. Figure 7.38(ii) plots the absolute magnitude of the displacement error versus the velocity error of each of the modes. The asterisk and circles correspond to the errors generated by the control profile generated by the LP and NLP problems, respectively. Figure 7.39 illustrates the bang-bang ℓ_2 minimax control profiles generated by the NLP and the LP problems, respectively.

In Section 7.2.4 the minimax problem formulation was used to design control profiles which are robust to modeling uncertainties. Posing the problem in a linear programming framework required approximating the residual energy, which is a quadratic, with multiple hyperplanes. This permitted exploiting powerful linear pro-

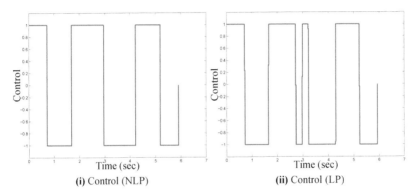

Figure 7.39: ℓ_2 Minimax Control

gramming algorithms that can solve large-order systems. Recently, there has been a tremendous growth in semidefinite programming which permit solving problems that include linear matrix inequalities (LMI). Constraints that require the feasible solution to lie within a hypersphere, which is defined by a quadratic inequality equation, can be represented as an LMI. This permits posing the minimax problem in a convex programming framework which guarantees convergence to a global optimal.

7.3 Linear Matrix Inequality

A linear matrix inequality can be described by the equation:

$$F(x) = F_0 + x_1 F_1 + \ldots + x_n F_n \prec 0 \tag{7.158}$$

where the symbol $\prec$ means negative definite. F_i are real symmetric matrices and $x = \begin{bmatrix} x_1 & x_2 & \ldots & x_n \end{bmatrix}$ is a vector of real variables. LMIs are convex constraints, i.e., if $F(x) \prec 0$ and $F(y) \prec 0$ then:

$$F(\lambda x + (1 - \lambda)y) = \lambda F(x) + (1 - \lambda)F(y) \prec 0, \ \forall \lambda \in [0, 1]. \tag{7.159}$$

LMIs are important since numerous engineering problems can be posed as LMI problems. To illustrate this, consider the linear dynamic system:

$$\dot{z} = Az \tag{7.160}$$

where z is the state vector and A is the system matrix. The system represented by Equation (7.160) is asymptotically stable if:

$$A^T P + PA \prec 0 \tag{7.161}$$

$$P \succ 0 \tag{7.162}$$

which implies that the system is stable if a positive definite P can be determined which satisfies Equation (7.161). Equations (7.161) and (7.162) constitute matrix-valued LMIs which can be represented as a combined LMI:

$$F(P) = \begin{bmatrix} A^T P + PA & 0 \\ 0 & -P \end{bmatrix} \prec 0. \tag{7.163}$$

Semidefinite program (SDP) is an optimization problem with a linear cost function and constraints that are linear equality and LMIs:

$$\text{minimize} \quad c^T x \tag{7.164a}$$

$$\text{subject to} \quad A_{eq}x = b_{eq} \tag{7.164b}$$

$$F_0 + x_1 F_1 + \ldots + x_n F_n \succeq 0 \tag{7.164c}$$

where F_i are symmetric matrices and x is the vector of parameters to be solved.

One of the most important results that is exploited by control engineers is the Schur complement for symmetric matrices. The quadratic requirement:

$$P - QS^{-1}Q^T \succ 0, \text{ where } S \succ 0 \tag{7.165}$$

can be rewritten as:

$$\begin{bmatrix} P & Q \\ Q^T & S \end{bmatrix} \succ 0. \tag{7.166}$$

To illustrate the use of the Schur complement, consider the discrete-time system:

$$x(k+1) = Gx(k) + Hu(k). \tag{7.167}$$

Assume a controller of the form $u(k) = -Kx(k)$, the closed-loop system is given by the equation

$$x(k+1) = (G - HK)x(k). \tag{7.168}$$

Consider the discrete-time Lyapunov candidate function

$$V(k) = x(k)^T Px(k), \text{ where } P \succ 0. \tag{7.169}$$

For stability, one requires

$$V(k+1) - V(k) = \qquad x(k+1)^T Px(k+1) - x(k)^T Px(k) \prec 0 \tag{7.170}$$

$$= \qquad x(k)^T \left((G - HK)^T P(G - HK) - P \right) x(k) \prec 0 \tag{7.171}$$

$$\Rightarrow P - (G - HK)^T P(G - HK) \succ 0. \tag{7.172}$$

The feedback gain K has to be selected so as to satisfy Equation (7.172) which is a quadratic inequality in the unknowns K and P. Equation (7.172) can be rewritten using Schur complement as:

$$[\mathscr{I} \ -G^T] \begin{bmatrix} P & (G-HK)^T P \\ P(G-HK) & P \end{bmatrix} \begin{bmatrix} \mathscr{I} \\ -G \end{bmatrix} \succ 0. \tag{7.173}$$

where $\mathscr{I}$ is the identity matrix. Pre- and post-multiply Equation (7.173) by P^{-1} results in the equation:

$$[\mathscr{I} \ -G^T] \begin{bmatrix} P^{-1} & P^{-1}(G-HK)^T \\ (G-HK)P^{-1} & P^{-1} \end{bmatrix} \begin{bmatrix} \mathscr{I} \\ -G \end{bmatrix} \succ 0. \tag{7.174}$$

Defining $P^{-1} = M$ and $KP^{-1} = L$, we have:

$$\begin{bmatrix} M & MG^T - L^T H^T \\ GM - HL & M \end{bmatrix} \succ 0 \tag{7.175}$$

which is an LMI and the unknowns M and L can be easily solved.

7.3.1 Time-Delay Filter

The LMI formulation can be exploited to design input-shaper profiles by parameterizing the reference profile in discrete time. The control profile in frequency domain is represented as:

$$u(s) = \frac{1}{s} \left(A_0 + A_1 e^{-sT} + A_2 e^{-2sT} + A_3 e^{-2sT} + \ldots + A_N e^{-NsT} \right) \tag{7.176}$$

where T is the sampling time. The rest-to-rest maneuver problem is stated as the selection of A_i which minimizes the cost function:

$$\underset{A_i}{\text{Min}} \ J = (x((N+1)T) - x_f)^T Q(x((N+1)T) - x_f) \tag{7.177}$$

where x_f is the desired final state, $x(TN)$ is the actual final state for a given A_i and Q is a positive definite weighting matrix. The magnitude of A_i is generally subject to constraints such as

$$0 \le A_i \le 1 \tag{7.178}$$

$$\sum_i A_i = 1. \tag{7.179}$$

The system states at time $k+1$, in terms of the parameter A_i are given by the equation:

$$\underline{x}(k+1) = G\underline{x}(k) + HA_k \text{ where } k = 0, 2, \ldots, N \tag{7.180}$$

where $\underline{x}$ is the state vector and u is the control input. The state response for the control input A_k is

$$\underline{x}(k+1) = G^k\underline{x}(1) + \sum_{i=0}^{k} G^{k-i}HA_i \tag{7.181}$$

where $\underline{x}(1)$ represents the initial state of the system. To solve control problems with specified initial and final states, in addition to the final time (T_f), the final state constraint can be represented as

$$\underline{x}(N+1) = G^N\underline{x}(1) + \sum_{i=0}^{N} G^{N-i}HA_i \tag{7.182}$$

where the maneuver time $T_f = (N+1)T_s$, i.e., the maneuver time is discretized into N intervals. Equation (7.182) can be rewritten in the standard equality constraint form:

$$\underline{x}(N+1) = G^N\underline{x}(1) + \begin{bmatrix} G^N H & G^{N-1}H & \cdots & GH & H \end{bmatrix} \begin{bmatrix} A_0 \\ A_1 \\ A_2 \\ \cdot \\ \cdot \\ \cdot \\ A_N \end{bmatrix} \tag{7.183}$$

Since the cost function is quadratic, the optimization problem is restated as:

$$\min J = \gamma \tag{7.184a}$$

subject to

$$\gamma - (x(N+1) - x_f)^T Q(x(N+1) - x_f) > 0$$

$$\underline{x}(N+1) = G^N\underline{x}(1) + \begin{bmatrix} G^N H & G^{N-1}H & \cdots & GH & H \end{bmatrix} \begin{bmatrix} A_0 \\ A_1 \\ A_2 \\ \cdot \\ \cdot \\ \cdot \\ A_N \end{bmatrix}$$

$$\sum_{i=0}^{N} A_i = 1$$

$$0 \leq A_i \leq 1 \; \forall i$$

The Schur complement can be used to rewrite the optimization problem as:

$$\min \; J = \gamma \tag{7.185a}$$

subject to

$$\begin{bmatrix} \gamma & (x(N+1)-x_f)^T \\ (x(N+1)-x_f) & Q^{-1} \end{bmatrix} \succ 0$$

$$\underline{x}(N+1) = G^N \underline{x}(1) + \begin{bmatrix} G^N H & G^{N-1}H & \cdots & GH & H \end{bmatrix} \begin{bmatrix} A_0 \\ A_1 \\ A_2 \\ \cdot \\ \cdot \\ \cdot \\ A_N \end{bmatrix}$$

$$\sum_{i=0}^{N} A_i = 1$$

$$0 \le A_i \le 1 \; \forall i$$

Efficient algorithms have been developed for SDP which result in globally optimal solutions. YALMIP [6] is a MATLAB-based toolbox to solve a variety of optimization problems including semidefinite programming. This software is used in conjunction with solvers such as SeDuMi [7] and SDPT3 [8]. YALMIP is used to solve the illustrative examples in this section.

Example 7.9:

Design a minimum time control profile for the system

$$\ddot{x} + \dot{x} + x = u \tag{7.186}$$

to satisfy the boundary conditions

$$x(0) = \dot{x} = 0, \text{ and } x(t_f) = 1, \dot{x}(t_f) = 0, \tag{7.187}$$

where u has to satisfy the constraint

$$1 \ge u \ge 0. \tag{7.188}$$

The discrete time state space representation of the simple harmonic oscillator (Equation 7.186) is

$$\begin{Bmatrix} x(k+1) \\ \dot{x}(k+1) \end{Bmatrix} = \begin{bmatrix} e^{-0.5T}\left(\mathscr{C}T + \frac{1}{\sqrt{3}}\mathscr{S}T \right) & \frac{2}{\sqrt{3}}e^{-0.5T}\mathscr{S}T \\ -\frac{2}{\sqrt{3}}e^{-0.5T}\mathscr{S}T & -e^{-0.5T}\left(\frac{1}{\sqrt{3}}\mathscr{S}T - \mathscr{C}T \right) \end{bmatrix} \begin{Bmatrix} x(k) \\ \dot{x}(k) \end{Bmatrix}$$
$$+ \begin{bmatrix} -e^{-0.5T}\left(\frac{1}{\sqrt{3}}\mathscr{S}T + \mathscr{C}T \right) + 1 \\ \frac{2}{\sqrt{3}}e^{-0.5T}\mathscr{S}T \end{bmatrix} u(k) \tag{7.189}$$

where T is the sampling interval and $\cos\left(\frac{\sqrt{3}}{2}T\right) = \mathscr{C}T$ and $\sin\left(\frac{\sqrt{3}}{2}T\right) = \mathscr{S}T$. The boundary conditions are

$$\underline{x}_1 = \left\{\begin{matrix} 0 \\ 0 \end{matrix}\right\} \text{ and } \underline{x}_{N+1} = \left\{\begin{matrix} 1 \\ 0 \end{matrix}\right\}. \tag{7.190}$$

Since, the desired maneuver is a rest-to-rest maneuver, we require

$$\underline{x}_{N+1+i} = \underline{x}_{N+1} \quad \forall i = 1,2,3,...,\infty \tag{7.191}$$

Substituting Equation (7.71), with $i = 1$, in Equation (7.69), we have

$$\left\{\begin{matrix} 1 \\ 0 \end{matrix}\right\} = \begin{bmatrix} \cos(T) & \sin(T) \\ -\sin(T) & \cos(T) \end{bmatrix} \left\{\begin{matrix} 1 \\ 0 \end{matrix}\right\} + \begin{bmatrix} -\cos(T)+1 \\ \sin(T) \end{bmatrix} u \tag{7.192}$$

or

$$\begin{bmatrix} 1-\cos(T) & -\sin(T) \\ \sin(T) & 1-\cos(T) \end{bmatrix} \left\{\begin{matrix} 1 \\ 0 \end{matrix}\right\} = \begin{bmatrix} -\cos(T)+1 \\ \sin(T) \end{bmatrix} = \begin{bmatrix} -\cos(T)+1 \\ \sin(T) \end{bmatrix} u \tag{7.193}$$

which implies that the steady state value of u is 1. This terminal equality constraint has to be included in the linear programming problem.

Since the formulation does not include a hard constraint which requires satisfaction of the boundary conditions, the numerical solution corresponds to one which minimizes γ. Figure 7.40 illustrates the variation of γ as a function of the maneuver time t_f.

Thus, one can use the bisection algorithm to determine the maneuver time which corresponds to the smallest time for which γ is less than some prespecified tolerance. Figure 7.41 illustrates the solution determined using the LMI formulation and a bisection algorithm to determine the minimum time for a rest-to-rest maneuver for a single spring-mass-dashpot system.

MATLAB Program 7.13:

```
function exitflag = example1(tf);
%
% Program to illustrate the use of LMI to solve the
% input shaping problem
%
clear all

% Single Spring-Mass-Dashpot System
A = [0 1;-1 -1];
B = [0;1];
C = [1 0];
D = [0];
Q = [1 0;0 1];
```

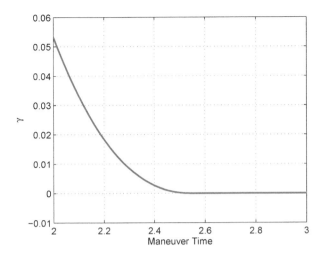

Figure 7.40: Variation of γ vs. Maneuver Time

```
xf = [1;0];
x0 = [0;0];

sys = ss(A,B,C,D);

N = 101;
Ts = tf/(N-1);


sysd = c2d(sys,Ts);
[Ad,Bd,Cd,Dd] = ssdata(sysd);

onevec = ones(N,1);
zerovec = zeros(N,1);

AAmat = eye(N);
Aineq = [];
Bineq = [];
% Constraint on the control input 0 < A_i < 1
Aineq = [Aineq;AAmat];
Bineq = [Bineq;onevec];

Aineq = [Aineq;-AAmat];
Bineq = [Bineq;zerovec];

% Final state error determination
Trmat = [];
```

```
for ind = 1:N,
    Trmat = [Ad^(ind -1)*Bd Trmat];
end;

% Declare the variables
Avec = sdpvar(N,1);
gamma = sdpvar(1);

tempmat = [gamma (Trmat*Avec-xf)';(Trmat*Avec-xf) inv(Q)];

F = set(tempmat > 0) + set(Aineq*Avec < Bineq);

options = sdpsettings('solver','sedumi');
%options = sdpsettings('solver','sdpt3');

diagnostic = solvesdp(F,gamma, options);

exitflag = 0;
if diagnostic.problem == 0,
    exitflag = 1;    % Problem converged
end;
gammaopt = double(gamma)
Aopt = double(Avec);
Aopt = [Aopt(:);1]

tvec = 0:Ts:(N)*Ts;
[YY,TT,XX] = lsim(sysd ,Aopt,tvec,x0,'zoh');
figure(1)
subplot(211), plot(TT,XX); grid
subplot(212), stairs(tvec,Aopt);
```

7.3.2   Minimax Time-Delay Filters

The LMI formulation for the design of time-delay filters to minimize residual vibrations of a rest-to-rest maneuver can be extended to design robust time-delay filters. Given a domain of uncertainty of the model parameters, a set of models ($\mathcal{M}$) are derived by sampling the uncertain space. The errors in terminal states for each of these models is determined and the maximum of the norm of the terminal state errors is minimized. The minimax problem can be stated as:

$$\min \; J = \gamma \tag{7.194a}$$

subject to

$$\begin{bmatrix} \gamma & (x_m(N+1) - x_f)^T \\ (x_m(N+1) - x_f) & Q_m^{-1} \end{bmatrix} \succ 0 \; \text{ for } m = 1,2, \ldots, \mathcal{M}$$

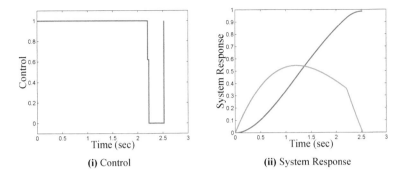

Figure 7.41: LMI Solution

(i) Control   **(ii)** System Response

$$x_m(N+1) = G_{m}^N \underline{x}(1) + \begin{bmatrix} G_m^N H_m & G_m^{N-1} H_m & \cdots & G_m H_m & H_m \end{bmatrix} \begin{bmatrix} A_0 \\ A_1 \\ A_2 \\ \cdot \\ \cdot \\ \cdot \\ A_N \end{bmatrix}$$

$$\sum_{i=0}^{N} A_i = 1$$

$$0 \le A_i \le 1 \ \forall i$$

Example 7.10: Design a minimax controller for the rest-to-rest maneuver of the benchmark problem:

$$\ddot{x}_1 + kx_1 - kx_2 = u \qquad (7.195a)$$
$$\ddot{x}_2 - kx_1 + kx_2 = 0 \qquad (7.195b)$$

where the boundary conditions are

$$x_1(0) = x_2(0) = \dot{x}_1(0) = \dot{x}_2(0) = 0 \qquad (7.196a)$$
$$x_1(t_f) = x_2(t_f) = 1, \dot{x}_1(t_f) = \dot{x}_2(t_f) = 0 \qquad (7.196b)$$

and the control input satisfies the constraint:

$$-1 \le u \le 1. \qquad (7.197)$$

The spring stiffness is uncertain and lies in the range

$$0.7 \le k \le 1.3. \qquad (7.198)$$

Parameterizing the optimal control profile as

$$u(t) = \sum_{i=0}^{N} A_i \mathcal{H}(t - iT)$$

(7.199)

where the variable A_i satisfies the constraint

$$-1 \leq A_i \leq 1, \ \forall i = 1, 2, \ldots, N.$$

(7.200)

The optimization problem is to determine A_i so the minimax problem can be solved:

$$\underset{A_i}{\text{MinMax}} \ (x_m(N+1) - x_f)^T \underbrace{\begin{bmatrix} k + K_p & -k & 0 & 0 \\ -k & k & 0 & 0 \\ 0 & 0 & 1 & 0 \\ 0 & 0 & 0 & 1 \end{bmatrix}}_{Q} (x_m(N+1) - x_f)$$

(7.201)

where K_p is a pseudo-spring constant to ensure that the Q matrix is positive definite. The subscript m refers to the mth plant in the domain of uncertainty. Sampling on an uniform grid, the uncertain space of k 21 times, and using the 21 models to minimize the maximum magnitude of the residual energy results in the control profile illustrated in Figure 7.42(i) and the corresponding sensitivity (residual energy variation) curve in Figure 7.42(ii). The maneuver time is selected to be 6 seconds and the shape of the sensitivity control profile resembles the ones derived using the minimax optimization algorithms.

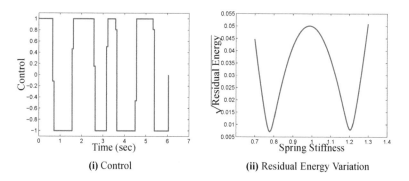

(i) Control (ii) Residual Energy Variation

Figure 7.42: LMI Minimax Time-Delay Filter

MATLAB Program 7.14:

```
%
% Program to illustrate the use of LMI to solve the minimax
% input shaping problem. The spring is considered to be
% uncertain and lies in the range 0.7 < k < 1.3
%
%
tf = 6; % Final time
% 2 Mass-spring benchmark problem
N = 101;
Ts = tf/(N-1);
onevec = ones(N,1);
zerovec = zeros(N,1);
% Declare the variables
Avec = sdpvar(N,1);
gamma = sdpvar(1);

AAmat = eye(N);
Aineq = [];
Bineq = [];
% Constraint on the control input -1 < A_i < 1
Aineq = [Aineq;AAmat];
Bineq = [Bineq;onevec];

Aineq = [Aineq;-AAmat];
Bineq = [Bineq;onevec];
F = set(Aineq*Avec < Bineq);

kvec = linspace(0.7,1.3,21);     % Sample the uncertain parameter

for k = kvec,
    M = eye(2);          % Mass Matrix
    K = [k -k;-k k];     % Stiffness Matrix
    A = [zeros(2) eye(2); -inv(M)*K zeros(2)];
    B = [0;0;inv(M)*[1;0]];
    C = [0 1 0 0];
    D = [0];

    Kpseudo = K + [1 0;0 0];
    Q = [Kpseudo zeros(2); zeros(2) M];
    xf = [1;1;0;0];
    x0 = [0;0;0;0];

    sys = ss(A,B,C,D);
    sysd = c2d(sys,Ts);
    [Ad,Bd,Cd,Dd] = ssdata(sysd);
```

```
% Final state error determination
Trmat = [];
for ind = 1:N,
    Trmat = [Ad^(ind−1)*Bd Trmat];
end;

tempmat = [gamma (Trmat*Avec−xf)';(Trmat*Avec−xf) inv(Q)];

    F = F + set(tempmat > 0);
end;
%options = sdpsettings('solver','sedumi');
options = sdpsettings('solver','sdpt3');

diagnostic = solvesdp(F,gamma,options);

gammaopt = double(gamma)
Aopt = double(Avec);
Aopt = [Aopt(:);0];

M = eye(2);
K = [1 −1;−1 1];
A = [zeros(2) eye(2); −inv(M)*K zeros(2)];
B = [0;0;inv(M)*[1;0]];
C = [0 1 0 0];
D = [0];
sys = ss(A,B,C,D);
sysd = c2d(sys,Ts);
[Ad,Bd,Cd,Dd] = ssdata(sysd);

tvec = 0:Ts:(N)*Ts;
[YY,TT,XX] = lsim(sysd,Aopt,tvec,x0,'zoh');
figure(1)
plot(TT,XX); xlabel('Time (sec)'); ylabel('System Response')
figure(2)
stairs(tvec,Aopt); xlabel('Time (sec)'); ylabel('Control')
axis([0 tf+1 −1.1 1.1])
```

7.3.3   Modal Weighted Minimax Time-Delay Filters

The minimax time-delay filter design presented in the previous section minimized the maximum magnitude of the residual energy in the system, over the domain of uncertain parameters. As compared to traditional robust input shapers where the sensitivity of the residual energy at the nominal model parameters is minimized, the minimax approach resulted in a smaller value of the worst performance over the domain of uncertainty.

Is minimizing the worst residual energy of the system the best metric to be used in the design of robust time-delay filter, or can one conceive of a different metric which generate an input shaped profile which targets the output of interest? To motivate this, consider the three mass-spring system shown in Figure 7.43. The three modes of this system

$$
\begin{Bmatrix} x_1 \\ x_2 \\ x_3 \end{Bmatrix} = \begin{Bmatrix} 0.5774 \\ 0.5774 \\ 0.5774 \end{Bmatrix}, \quad \begin{Bmatrix} -0.7071 \\ 0 \\ 0.7071 \end{Bmatrix}, \quad \begin{Bmatrix} 0.4082 \\ -0.8165 \\ 0.4082 \end{Bmatrix}
\tag{7.202}
$$

are shown in the same figure. The frequencies corresponding to the three modes are: $\omega_i = \begin{bmatrix} 0 & 1 & \sqrt{3} \end{bmatrix}$. The second-order model of the system is:

$$
\begin{bmatrix} 1 & 0 & 0 \\ 0 & 1 & 0 \\ 0 & 0 & 1 \end{bmatrix} \begin{Bmatrix} \ddot{x}_1 \\ \ddot{x}_2 \\ \ddot{x}_3 \end{Bmatrix} + \begin{bmatrix} 1 & -1 & 0 \\ -1 & 2 & -1 \\ 0 & -1 & 1 \end{bmatrix} \begin{Bmatrix} x_1 \\ x_2 \\ x_3 \end{Bmatrix} = \begin{bmatrix} 1 \\ 0 \\ 0 \end{bmatrix} u.
\tag{7.203}
$$

If the output of interest is the displacement of the third mass, the output equation is:

$$
z = \begin{bmatrix} 0 & 0 & 1 \end{bmatrix} \begin{Bmatrix} x_1 \\ x_2 \\ x_3 \end{Bmatrix}
\tag{7.204}
$$

The system model and output equation can be rewritten in modal form as:

$$
\begin{bmatrix} 1 & 0 & 0 \\ 0 & 1 & 0 \\ 0 & 0 & 1 \end{bmatrix} \begin{Bmatrix} \ddot{y}_1 \\ \ddot{y}_2 \\ \ddot{y}_3 \end{Bmatrix} + \begin{bmatrix} 0 & 0 & 0 \\ 0 & 1 & 0 \\ 0 & 0 & 3 \end{bmatrix} \begin{Bmatrix} y_1 \\ y_2 \\ y_3 \end{Bmatrix} = \begin{bmatrix} -0.5774 \\ -0.7071 \\ 0.4082 \end{bmatrix} u.
\tag{7.205}
$$

and

$$
z = \begin{bmatrix} -0.5774 & 0.7071 & 0.4082 \end{bmatrix} \begin{Bmatrix} y_1 \\ y_2 \\ y_3 \end{Bmatrix}
\tag{7.206}
$$

which clearly indicates that the second mode's contribution is the largest and the third mode's is the smallest. These modal contributions can be used to relatively weigh the modal energies to formulate a cost function whose maximum can be minimized over the uncertain domain.

Consider a second-order linear mechanical system of the form

$$
M\ddot{x} + C\dot{x} + Kx = Du
\tag{7.207}
$$

where the mass matrix M is positive definite, C is a positive semidefinite damping matrix and K is a positive semidefinite stiffness matrix. For rest-to-rest maneuvers from some prescribed input state, to a final state, the residual energy of the system at the end of maneuver time can be written as

$$
V = \frac{1}{2}\dot{x}^T M\dot{x} + \frac{1}{2}(x - x_f)^T K(x - x_f)
\tag{7.208}
$$

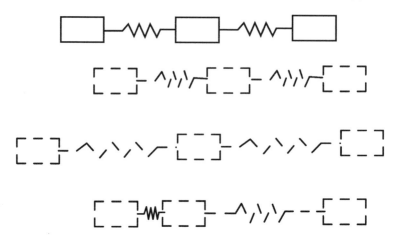

Figure 7.43: Three Mass-Spring Systems and Mode Shapes

In this approach we will transform the system to the modal form by introducing a coordinate transformation of the form

$$x(t) = \Phi y(t) \tag{7.209}$$

where Φ is the modal matrix obtained by solving the eigenvalue problem associated with the mass and the stiffness matrices. The residual energy of the system can then be expressed as

$$V = \frac{1}{2}\dot{y}^T \tilde{M}\dot{y} + \frac{1}{2}(y - y_f)^T \tilde{K}(y - y_f) \tag{7.210}$$

where $\tilde{M}$ and $\tilde{K}$ are the transformed mass and stiffness matrices given by

$$\tilde{M} = \Phi^T M \Phi = I \tag{7.211}$$

$$\tilde{K} = \Phi^T M \Phi = \begin{Bmatrix} 0 & 0 & 0 & \cdots & 0 \\ 0 & \omega_1^2 & 0 & \cdots & 0 \\ 0 & 0 & \omega_2^2 & \cdots & 0 \\ \vdots & \vdots & \vdots & \ddots & \vdots \\ 0 & 0 & 0 & \cdots & \omega_n^2 \end{Bmatrix} \tag{7.212}$$

Moving the system from a initial position of x_0 to a final position of rest x_f results in modal boundary constraint of:

$$y_0 = \Phi^T x_0 \tag{7.213}$$

$$y_f = \Phi^T x_f \tag{7.214}$$

For controlling the output with high precision, e.g., if the displacement of the final mass of a series of n mass-spring systems is the output, the output equation for n_{th} mass in modal space:

$$y = \underbrace{\begin{bmatrix} 0 & 0 & \cdots & 1 \end{bmatrix}}_{C} \Phi \begin{Bmatrix} y_1 \\ y_2 \\ \vdots \\ y_n \end{Bmatrix} \tag{7.215}$$

If one is interested in displacement precision, the residual energy of each of the mode is first weighted so as to correspond to the same final displacement. For instance, if the displacement of the first two modes are the same, we have the constraint:

$$y_1 = y_2 \tag{7.216}$$

which can be rewritten in terms of potential energies as:

$$\frac{1}{2}\omega_1^2 y_1^2 = \frac{1}{2}\omega_2^2 y_2^2 \left(\frac{\omega_1^2}{\omega_2^2}\right) \tag{7.217}$$

which states that the potential energy of the second mode has to be scaled by $\frac{\omega_1^2}{\omega_2^2}$ to satisfy the displacement precision constraint. Furthermore, since the output is a linear combination of the modal displacements, each of the modes has to be rescaled to reflect the modal contribution of every mode. For a system with an output equation in modal form:

$$y = \sum_{i=1}^{n} c_i y_i \tag{7.218}$$

the cost function for optimization can be written as:

$$V = \sum_{i=1}^{n} \frac{1}{2}\left(\dot{y}_i^2 + \omega_i^2(y_i - y_i^f)^2\right)\frac{c_i^2\omega_i^2}{\omega_i^2} \tag{7.219}$$

where y_i^f correspond to the desired final state of the the ith mode. An LMI problem can be posed to minimize the maximum magnitude of the weighted residual energy given by Equation (7.219).

Example 7.11: Design a minimax controller for the point-to-point maneuver of a double pendulum which represents the dynamics of a crane, in the presence of variations in the load being carried is presented here. Tanaka and Kouno [9] developed a linearized model of a test crane which is:

$$m_{11}\ddot{x} + m_{12}\ddot{\theta}_1 + m_{13}\ddot{\theta}_2 + c\dot{x} = u \tag{7.220}$$

$$m_{21}\ddot{x} + m_{22}\ddot{\theta}_1 + m_{23}\ddot{\theta}_2 + k_{12}\theta_1 = 0 \tag{7.221}$$

$$m_{31}\ddot{x} + m_{32}\ddot{\theta}_1 + m_{33}\ddot{\theta}_2 + k_{34}\theta_2 = 0 \tag{7.222}$$

where

$$m_{11} = m + m_1 + m_2 \qquad m_{12} = m_1 l_1 + m_2(l_1 + l_1^\circ) \qquad m_{13} = m_2 l_2$$
$$m_{21} = m_1 l_1 + m_2(l_1 + l_1^\circ) \qquad m_{22} = m_1 l_1^2 + m_2(l_1 + l_1^\circ)^2 + I_1 \qquad m_{23} = m_2 l_2(l_1 + l_1^\circ)$$
$$m_{31} = m_2 l_2 \qquad m_{32} = m_2 l_2(l_1 + l_1^\circ) \qquad m_{33} = m_2 l_2^2 + I_2$$
$$k_{33} = m_2 g l_2 \qquad k_{22} = m_1 g l_1 + m_2 g(l_1 + l_1^\circ)$$

The parameters used by Tanaka and Kouno [9] are: $m_1 = 0.5kg$, $m_2 = 6kg$, $l_1 = 1.5m$, $l_1^\circ = 0.1m$, $l_2 = 0.3m$, $I_1 = 0.0008 kgm^2$, $I_2 = 0.06 kgm^2$, $m = 8.0 kg$, and the coefficient of friction is $c = 1.0$ N/(m/s). To control the position of the trolley (Figure 7.44), a collocated PD controller with proportional gain of 10 and derivative gain of 1 was used. Robust pre-filters for shaping the position reference when there is uncertainty in the mass of payload, m_2 were designed using the aforementioned methodology. The system needs to be repositioned from the initial state of $(x_1, \theta_1, \theta_2) = (0,0,0)$ to $(x_1, \theta_1, \theta_2) = (1,0,0)$. In this scenario, we are interested in reducing the sway of the payload at the end of the trolley-positioning maneuver. So the emphasis is on the precise control of the payload displacement $(l_1 + l_1^\circ)\theta_1 + l_2\theta_2$ so as to limit the payload oscillations.

After implementing a proportional derivative controller, the system model is

$$u = -10(x - x_r) - \dot{x} \tag{7.223}$$

where x_r is the reference input, is:

$$\begin{bmatrix} 14.5 & 10.35 & 1.8 \\ 10.35 & 16.48 & 2.88 \\ 1.8 & 2.88 & 0.546 \end{bmatrix} \begin{Bmatrix} \ddot{x} \\ \ddot{\theta}_1 \\ \ddot{\theta}_2 \end{Bmatrix} + \begin{bmatrix} 2 & 0 & 0 \\ 0 & 0 & 0 \\ 0 & 0 & 0 \end{bmatrix} \begin{Bmatrix} \dot{x} \\ \dot{\theta}_1 \\ \dot{\theta}_2 \end{Bmatrix} + \begin{bmatrix} 10 & 0 & 0 \\ 0 & 101.53 & 0 \\ 0 & 0 & 17.66 \end{bmatrix} \begin{Bmatrix} x \\ \theta_1 \\ \theta_2 \end{Bmatrix} = \begin{bmatrix} 10 \\ 0 \\ 0 \end{bmatrix} x_r \tag{7.224}$$

Two cost functions are used in the design of the minimax filter. The first, is the residual energy of the system which is represented as:

$$J_1 = \frac{1}{2} \begin{Bmatrix} \dot{x} \\ \dot{\theta}_1 \\ \dot{\theta}_2 \end{Bmatrix}^T \begin{bmatrix} 14.5 & 10.35 & 1.8 \\ 10.35 & 16.486 & 2.88 \\ 1.8 & 2.88 & 0.546 \end{bmatrix} \begin{Bmatrix} \dot{x} \\ \dot{\theta}_1 \\ \dot{\theta}_2 \end{Bmatrix}$$
$$+ \begin{Bmatrix} x-1 \\ \theta_1 \\ \theta_2 \end{Bmatrix}^T \begin{bmatrix} 10 & 0 & 0 \\ 0 & 101.53 & 0 \\ 0 & 0 & 17.66 \end{bmatrix} \begin{Bmatrix} x-1 \\ \theta_1 \\ \theta_2 \end{Bmatrix}. \tag{7.225}$$

Define a transformation

$$\begin{Bmatrix} x \\ \theta_1 \\ \theta_2 \end{Bmatrix} = \begin{bmatrix} 0.2465 & -0.2528 & 0.0191 \\ 0.0186 & 0.2795 & -0.8652 \\ 0.0186 & 0.2890 & 4.8212 \end{bmatrix} \begin{Bmatrix} q_1 \\ q_2 \\ q_3 \end{Bmatrix} \tag{7.226}$$

which decouples the mass and stiffness matrices resulting in the transformed equa-

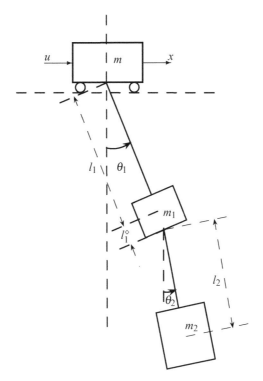

Figure 7.44: Double-Pendulum Crane

tions:

$$\begin{Bmatrix} \ddot{q}_1 \\ \ddot{q}_2 \\ \ddot{q}_3 \end{Bmatrix} + \begin{bmatrix} 0.1215 & -0.1246 & 0.0094 \\ -0.1246 & 0.1278 & -0.0097 \\ 0.0094 & -0.0097 & 0.0007 \end{bmatrix} \begin{Bmatrix} \dot{q}_1 \\ \dot{q}_2 \\ \dot{q}_3 \end{Bmatrix}$$
$$+ \begin{bmatrix} 0.65 & 0 & 0 \\ 0 & 10.05 & 0 \\ 0 & 0 & 486.44 \end{bmatrix} \begin{Bmatrix} q_1 \\ q_2 \\ q_3 \end{Bmatrix} = \begin{bmatrix} 2.465 \\ -2.528 \\ 0.191 \end{bmatrix} x_r \qquad (7.227)$$

The transformed output matrix when the position of the mass m_2 relative to trolley mass m is of interest is:

$$y = \begin{bmatrix} 0 & (l_1 + l_1^\diamond) & l_2 \end{bmatrix} \begin{Bmatrix} x \\ \theta_1 \\ \theta_2 \end{Bmatrix} = \begin{bmatrix} 0.0353 & 0.5340 & 0.0621 \end{bmatrix} \begin{Bmatrix} q_1 \\ q_2 \\ q_3 \end{Bmatrix} \qquad (7.228)$$

The modal penalties can now be calculated to be:

$$\frac{c_i^2 \omega_1^2}{\omega_i^2} = \begin{bmatrix} 1.25 \times 10^{-3} & 1.84 \times 10^{-2} & 5.13 \times 10^{-6} \end{bmatrix} \qquad (7.229)$$

which is used to define the modal weighted cost function:

$$J_2 = \frac{1}{2} \left\{ \begin{matrix} \dot{x} \\ \dot{\theta}_1 \\ \dot{\theta}_2 \end{matrix} \right\}^T \begin{bmatrix} 0.0192 & 0.0007 & 0.0001 \\ 0.0007 & 0.1574 & 0.0283 \\ 0.0001 & 0.0283 & 0.0051 \end{bmatrix} \left\{ \begin{matrix} \dot{x} \\ \dot{\theta}_1 \\ \dot{\theta}_2 \end{matrix} \right\}$$

$$+ \left\{ \begin{matrix} x-1 \\ \theta_1 \\ \theta_2 \end{matrix} \right\}^T \begin{bmatrix} 0.0234 & -0.1224 & -0.0221 \\ -0.1224 & 1.4826 & 0.2664 \\ -0.0221 & 0.2664 & 0.0480 \end{bmatrix} \left\{ \begin{matrix} x-1 \\ \theta_1 \\ \theta_2 \end{matrix} \right\}. \qquad (7.230)$$

The solution of minimizing J_1 and J_2 over the domain of uncertainty:

$$3 \leq m_2 \leq 9 \qquad (7.231)$$

results in control profiles illustrated by the dashed line and the solid line, respectively, in Figure 7.45.

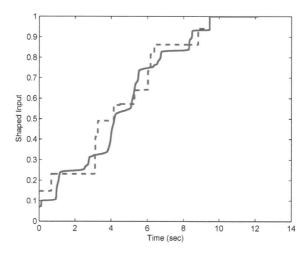

Figure 7.45: LMI Minimax Control Profiles

The final time for both the solutions is 9.46 seconds. To permit a fair comparison of both the solutions, the true residual energy of the double pendulum at 9.46 seconds is determined and plotted as a function of the uncertain mass m_2. Figure 7.46(i) illustrates that the maximum magnitude of the residual energy of the system controlled by an input profile designed by minimizing the cost function J_1 (total residual energy) is smaller than that designed based on the modal weighted residual energy, which is not a surprising result. However, if one were to plot the final mass residual energy metric defined as:

$$M = \frac{1}{2} \left(\dot{y}^2(t_f) + \ddot{y}^2(t_f) \right) \qquad (7.232)$$

as shown in Figure 7.46(ii), the control profile based on minimizing the cost function J_2 clearly outperforms the one based on minimizing the cost function J_1. In all these graphs, the dashed line corresponds to the solution when minimizing J_1 and the solid line when minimizing J_2.

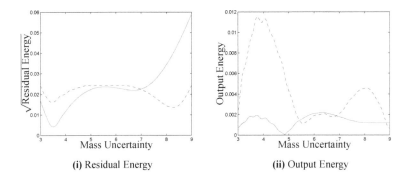

(i) Residual Energy   (ii) Output Energy

Figure 7.46: Comparison Metrics

The improved performance of using the cost function J_2 to design a shaped profile can be gauged from Figures 7.47 and 7.48, which illustrate the motion of the mass m_2 relative to the trolley after the final time $t_f = 9.46$ seconds, for various values of the uncertain mass m_2. The mesh plot in Figure 7.47 has a larger variation about the zeros position compared to the variation illustrated in Figure 7.48.

One has to note that the modal weighted approach can encounter a pathologic problem when one or more modes might have no contribution to the output of interest. For instance, in the three-mass-spring system illustrated earlier, if the output of interest is the second mass, then the second mode does not contribute to the output. In such situations, the noncontributing modes are not included in the cost function and the resulting shaped reference profile does not regulate the noncontributing modes. One can remedy this situation by including the noncontributing modes with a small weight in the cost function.

7.4 Summary

This focus of this chapter was on the use of numerical techniques for the determination of optimal controllers. The input shaper/time-delay filters, saturating controllers, and so forth, can be posed as optimization problems. If simple parame-

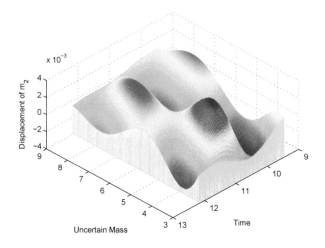

Figure 7.47: System Response after t_f vs. Uncertain Mass (Cost J_1)

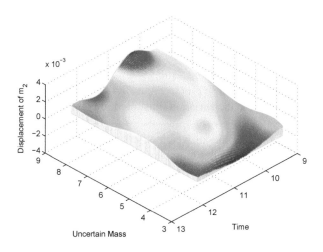

Figure 7.48: System Response after t_f vs. Uncertain Mass (Cost J_2)

terizations can be used to represent the optimal control profile such as using switch times to parameterize bang-bang controllers for minimum-time control, a nonlinear parameter optimization can be formulated and any gradient based algorithm can be used to arrive at the optimal solution. A caveat that should always be stated regarding the convergence of gradient based algorithms to local optima.

Algorithms for solving linear, quadratic, and semidefinite programming problems ensure convergence to global optima since they apply to convex problems. Optimal controller design can be posed in discrete-time domain which permits exploitation of convex programming approaches to arrive at the near-globally-optimal solutions. The sampling time selected to discretize the control profile influences the quality of the solution. Smaller the sampling time, closer is the solution to the true optimal solution. If the exact optimal solution is desired, the results of the convex optimization approach can be used to identify the structures of the optimal control profile. For example, the linear programming approach used to determine the time-optimal solution provides the structure of the control profile with the correct number of switches and accurate estimates for their values. The estimate of the number of switches and their values permit posing a nonlinear programming problem which will often converge to the globally optimal solution given that the initial guess is close to the optimal solution. Linear programming and linear matrix inequalities are used to illustrate design of various controller which include control and state constraints. Inclusion of sensitivity states and minimax formulation permit design of robust controllers for rest-to-rest maneuvers.

Exercises

7.1 Design a jerk-limited time-optimal reference profile for a rest-to-rest maneuver of a spring-mass system given by the equation of motion:

$$\ddot{x} + 2\zeta\omega\dot{x} + \omega^2 x = \omega^2 u$$

where

$$0 \le u \le 2$$

and

$$-0.2 \le \dot{u} \le 0.2$$

using linear programming. $\omega = \sqrt{2}$ and $\zeta = 0.1$. The initial and final states are

$$x(0) = \dot{x}(0) = 0, \quad x(t_f) = 2, \dot{(x)}(t_f) = 0$$

Parameterize the jerk-constrained reference profile using a time-delay filter and solve the same problem. Plot the variation in the final time of the parameterized solution and the linear programming solution as a function of the number of samples used to discretize the reference profile.

7.2 Design a velocity-constrained deflection-constrained minimum time control profile for a gantry crane using linear programming. The system model is:

$$(m_1 + m_2)\ddot{\theta} + m_2 L\ddot{\phi} = u$$
$$m_2\ddot{\theta} + m_2 L\ddot{\phi} + m_2 g\phi = 0$$

where the model parameters are listed in Table 7.2. The velocity constraint

m_1	m_2	L	g
5000 kg	250,000 kg	10 m	9.81 m/s^2

Table 7.1: Gantry Crane Parameters

is

$$-0.5m/s \le \dot{\theta} \le 0.5m/s$$

and the deflection constraint is

$$-1m \le L\phi \le 1m$$

and the control is constrained to lie within

$$-1000N \le u \le 1000N$$

7.3   Figure 7.49 illustrates a single-axis flexible-arm robot whose equations of motion are given by the equation:

$$\begin{bmatrix} M_{11}+J & M_{12} & M_{13} \\ M_{12} & M_{22} & 0 \\ M_{13} & 0 & M_{33} \end{bmatrix} \begin{Bmatrix} \ddot{\theta} \\ \ddot{q}_1 \\ \ddot{q}_2 \end{Bmatrix} + \begin{bmatrix} 0 & 0 & 0 \\ 0 & M_{22}\omega_1^2 & 0 \\ 0 & 0 & M_{33}\omega_2^2 \end{bmatrix} \begin{Bmatrix} \theta \\ q_1 \\ q_2 \end{Bmatrix} = \begin{Bmatrix} 1 \\ 0 \\ 0 \end{Bmatrix} u \quad (7.233)$$

where the model parameters are listed in Table 7.3.

$M_{11}+J$	M_{12}	M_{13}	M_{22}	M_{33}	ω_1^2	ω_2^2
0.1115+1	0.2621	0.0418	0.6347	0.6346	965.5838	3.7922e4

Table 7.2: Flexible Robot-Arm Parameters

Design a robust controller (force the sensitivity of the states with respect to the natural frequencies) for rest-to-rest maneuver of the flexible-arm robot. The input is constrained to lie in the range:

$$-10 \le u \le 10$$

The initial and final conditions of the flexible-arm states are:

$$\begin{Bmatrix} \theta \\ q_1 \\ q_2 \end{Bmatrix}(0) = \begin{Bmatrix} 0 \\ 0 \\ 0 \end{Bmatrix}, \quad \begin{Bmatrix} \dot{\theta} \\ \dot{q}_1 \\ \dot{q}_2 \end{Bmatrix}(0) = \begin{Bmatrix} 0 \\ 0 \\ 0 \end{Bmatrix}, \quad \begin{Bmatrix} \theta \\ q_1 \\ q_2 \end{Bmatrix}(t_f) = \begin{Bmatrix} \pi \\ 0 \\ 0 \end{Bmatrix}, \quad \begin{Bmatrix} \dot{\theta} \\ \dot{q}_1 \\ \dot{q}_2 \end{Bmatrix}(t_f) = \begin{Bmatrix} 0 \\ 0 \\ 0 \end{Bmatrix}.$$

and the available fuel is limited to:

$$\int_0^{t_f} |u|\, dt \le 5;$$

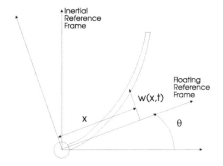

Figure 7.49: Flexible-Arm Robot

7.4 Figure 7.50 illustrates an X–Y table whose equations of motion are given by the equation:

$$\ddot{x}+0.01\dot{x}+x= u_x$$
$$\ddot{y}+0.01\dot{y}+y= u_y$$

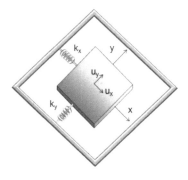

Figure 7.50: X–Y Table

Design the control profile to track a circular trajectory given by the reference model:

$$x_r = \cos(2\pi\frac{t}{t_f}), \quad y_r = \sin(2\pi\frac{t}{t_f})$$

where the final time t_f is 10 seconds. Assume that the table at initial time is at

$$x(0) = 1, \dot{x}(0) = 0, y(0) = 0, \dot{y}(0) = 0$$

and the control is constrained by the equation:

$$-1 \le u \le 1$$

Hint: Pose the problem as one where the maximum deviation from the desired trajectory is minimized.

7.5   The linearized model of a flapper-nozzle electrohydraulic servo valve is given by the transfer function [10]

$$\frac{X_s(s)}{I(s)} = \frac{-2.09e14s + 1.25e18}{s^5 + 9.90e3s^4 + 6.89e7s^3 + 3.09e11s^2 + 6.57e14s + 3.39e17}$$

where $X_s(s)$ and $I(s)$ are the Laplace transform of the servo valve spool displacement and input current, respectively. The servo value drives a hydraulic actuator whose dynamics are:

$$\frac{X_a(s)}{X_s(s)} = \frac{7.46e8}{s^3 + 1.19e3s^2 + 1.57e7s}$$

where $X_a(s)$ is the Laplace transform of the actuator displacement. Determine a current profile to track a reference actuator displacement profile shown in Figure 7.51. The current is constrained to lie in the range:

$$-10 \le i(t) \le 10$$

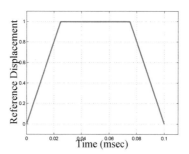

Figure 7.51: Reference Trajectory

7.6   Design a snap-limited time-optimal reference profile for a rest-to-rest maneuver of a spring-mass system given by the equation of motion:

$$\ddot{x} + 2\zeta\omega\dot{x} + \omega^2 x = \omega^2 u$$

where

$$0 \le u \le 2$$

and

$$-0.2 \le \ddot{u} \le 0.2$$

using linear programming. $\omega = \sqrt{2}$ and $\zeta = 0.1$. The initial and final states are

$$x(0) = \dot{x}(0) = 0, \quad x(t_f) = 2, (\dot{x})(t_f) = 0$$

Snap is defined as the rate of change of jerk. Study the structure snap profile as a function of permitted snap if the jerk is limited to the constraint

$$-0.5 \leq \dddot{u} \leq 0.5$$

7.7 A coordinate measuring machine is used to accurately map the topography of a surface. High speed motion of these machine excite the vibratory modes of the machine preventing a high-fidelity measurement of the surface. Seth and Rattan [11] derived a simple model to capture the dynamics of the carriage driven by a motor (Figure 7.52),

$$\frac{\Theta_c(s)}{\tau(s)} = \frac{10.10(s^2 + 3610)(s^2 + 983371)(s^2 + 25777218)}{s^2(s^2 + 60287)(s^2 + 1283877)(s^2 + 25777289)}$$

Design a controller to transition a stationary carriage to a constant velocity (8 rad/sec) in minimum time with no residual vibrations. Assume that the control input is limited to lie in the range:

$$-1 \leq \tau \leq 1$$

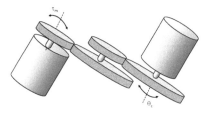

Figure 7.52: Coordinate Measuring Machine Carriage

References

[1] N. Singer and W. P. Seering. Pre-shaping command inputs to reduce system vibration. *ASME J. of Dyn. Systems, Measurements, and Cont.*, 112(1):76–82, 1990.

[2] J. C. Nash. The (Dantzig) simplex method for linear programming. *Computing in Science and Engineering*, 2(1):29–31, 2000.

[3] T. Singh. Fuel/time optimal control of the benchmark problem. *J. of Guid., Cont. and Dyn.*, 18(6):1225–1231, 1995.

[4] T. Singh. Minimax design of robust controllers for flexible systems. *J. of Guid., Cont. and Dyn.*, 25(5):868–875, 2002.

[5] J. O'Rourke. *Computational Geometry in C*. Cambridge University Press, 2000.

[6] J. Lofberg. Yalmip. http://control.ee.ethz.ch/ joloef/yalmip.php, 2006.

[7] Jos F. Strum. Sedumi. http://sedumi.mcmaster.ca/, 2006.

[8] K. C. Toh, R. H. Tütüncü, and M. J. Todd. http://www.math.cmu.edu/ reha/sdpt3.html.

[9] S. Tanaka and S. Kouno. Automatic measurement and control of the attitude of crane lifters: Lifter-attitude measurement and control. *Control Engineering Practice*, 6:1099–1107, 1998.

[10] D. H. Kim and T.-C. Tsao. A linearized electrohydraulic servovalve model for valve dynamics sensitivity analysis and control system design. *ASME J. of Dyn. Systems, Measurements, and Cont.*, 122:179–187, 2000.

[11] N. Seth and K. S. Rattan. Vibration reduction in computer controlled machines. In *IEEE International Conference on Systems Engineering*, Dayton, OH, 1991.

A

Van Loan Exponential

Van Loan[*] proposed an approach for the calculation of exponentials of block triangular matrices. Consider the block triangular matrix:

$$P = \begin{bmatrix} A_1 & B_1 \\ 0 & A_2 \end{bmatrix} \tag{A.1}$$

where A_1 is a $n \times n$, B_1 is a $n \times m$ and A_2 is a $m \times m$ matrix. For $t > 0$, the matrix exponential

$$e^{Pt} = \begin{bmatrix} F_1 & G_1 \\ 0 & F_2 \end{bmatrix} \tag{A.2}$$

where

$$F_j = e^{A_j t} \tag{A.3}$$

$$G_j = \int_0^t e^{A_j(t-s)} B_j e^{A_{j+1}s}\, ds, \quad j = 1,2 \tag{A.4}$$

If we force A_2 to be the null matrix, we have

$$e^{Pt} = \begin{bmatrix} e^{A_1 t} & \int_0^t e^{A_1(t-s)} B_1\, ds \\ 0 & I \end{bmatrix} \tag{A.5}$$

If A_1 and B_1 correspond to the system matrix and the control influence matrix, then the top left $n \times n$ submatrix of the matrix e^{Pt} corresponds to the transition matrix and the top right $n \times m$ submatrix of the matrix e^{Pt} corresponds to the response of the system with zero initial conditions to unit step inputs.

---

[*]Charles F. Van Loan. *Computing integrals involving the matrix exponential*, IEEE Trans. on Auto. Cont., (3):395404, 1978.

B

Differential Lyapunov Equation

Consider the differential Lyapunov equation

$$\dot{P} = A^T P + PA + Q \tag{B.1}$$

where P, A, and Q are all $\mathscr{R}^{n \times n}$ dimensional matrices. Writing Equation B.1 for a $\mathscr{R}^{3 \times 3}$ matrix differential Lyapunov equation, in terms of its elements, we have

$$
\begin{bmatrix} \dot{p}_{11} & \dot{p}_{12} & \dot{p}_{13} \\ \dot{p}_{21} & \dot{p}_{22} & \dot{p}_{23} \\ \dot{p}_{31} & \dot{p}_{32} & \dot{p}_{33} \end{bmatrix} = \begin{bmatrix} a_{11} & a_{21} & a_{31} \\ a_{12} & a_{22} & a_{32} \\ a_{13} & a_{23} & a_{33} \end{bmatrix} \begin{bmatrix} p_{11} & p_{12} & p_{13} \\ p_{21} & p_{22} & p_{23} \\ p_{31} & p_{32} & p_{33} \end{bmatrix} +
$$
$$
\begin{bmatrix} p_{11} & p_{12} & p_{13} \\ p_{21} & p_{22} & p_{23} \\ p_{31} & p_{32} & p_{33} \end{bmatrix} \begin{bmatrix} a_{11} & a_{12} & a_{13} \\ a_{21} & a_{22} & a_{23} \\ a_{31} & a_{32} & a_{33} \end{bmatrix} + \begin{bmatrix} q_{11} & q_{12} & q_{13} \\ q_{21} & q_{22} & q_{23} \\ q_{31} & q_{32} & q_{33} \end{bmatrix} \tag{B.2}
$$

Equation B.1 is a linear equation which can be rewritten in vector differential equation form as

$$
\begin{Bmatrix} \dot{p}_{11} \\ \dot{p}_{21} \\ \dot{p}_{31} \\ \dot{p}_{12} \\ \dot{p}_{22} \\ \dot{p}_{32} \\ \dot{p}_{13} \\ \dot{p}_{23} \\ \dot{p}_{33} \end{Bmatrix} = \begin{Bmatrix} q_{11} \\ q_{21} \\ q_{31} \\ q_{12} \\ q_{22} \\ q_{32} \\ q_{13} \\ q_{23} \\ q_{33} \end{Bmatrix} +
$$

$$
\begin{bmatrix}
2a_{11} & a_{21} & a_{31} & a_{21} & 0 & 0 & a_{31} & 0 & 0 \\
a_{12} & a_{22}+a_{11} & a_{32} & 0 & a_{21} & 0 & 0 & a_{31} & 0 \\
a_{13} & a_{23} & a_{33}+a_{11} & 0 & 0 & a_{21} & 0 & 0 & a_{31} \\
a_{12} & 0 & 0 & a_{11}+a_{22} & a_{21} & a_{31} & a_{32} & 0 & 0 \\
0 & a_{12} & 0 & a_{12} & 2a_{22} & a_{32} & 0 & a_{32} & 0 \\
0 & 0 & a_{12} & a_{13} & a_{23} & a_{33}+a_{22} & 0 & 0 & a_{32} \\
a_{13} & 0 & 0 & a_{23} & 0 & 0 & a_{11}+a_{33} & a_{21} & a_{31} \\
0 & a_{13} & 0 & 0 & a_{23} & 0 & a_{12} & a_{22}+a_{33} & a_{32} \\
0 & 0 & a_{13} & 0 & 0 & a_{23} & a_{13} & a_{23} & 2a_{33}
\end{bmatrix}
\begin{Bmatrix} p_{11} \\ p_{21} \\ p_{31} \\ p_{12} \\ p_{22} \\ p_{32} \\ p_{13} \\ p_{23} \\ p_{33} \end{Bmatrix} \tag{B.3}
$$

or for any $\mathscr{R}^{n \times n}$ matrix differential Lyapunov equation (Equation B.1), the corresponding vector differential equations is

$$\dot{\mathscr{P}} = \mathscr{A} \mathscr{P} + \mathscr{Q} \tag{B.4}$$

where

$$\mathscr{P} = \begin{Bmatrix} P(:,1) \\ P(:,2) \\ \vdots \\ P(:,n) \end{Bmatrix} \text{ and } \mathscr{Q} = \begin{Bmatrix} Q(:,1) \\ Q(:,2) \\ \vdots \\ Q(:,n) \end{Bmatrix}$$

and

$$\mathscr{A} = \begin{bmatrix} A^T + a_{11}I_{n \times n} & a_{j1}I_{n \times n} & \cdots & \cdots & a_{n1}I_{n \times n} \\ a_{12}I_{n \times n} & A^T + a_{22}I_{n \times n} & \cdots & \cdots & a_{n2}I_{n \times n} \\ \cdots & \cdots & \cdots & \cdots & \cdots \\ a_{1i}I_{n \times n} & \cdots & A^T + a_{ii}I_{n \times n} & \cdots & a_{ni}I_{n \times n} \\ \cdots & \cdots & \cdots & \cdots & \cdots \\ a_{1n}I_{n \times n} & \cdots & \cdots & \cdots & A^T + a_{nn}I_{n \times n} \end{bmatrix}$$

$\mathscr{L}(:,i)$ corresponds to the ith column of the matrix $\mathscr{L}$.

Construct the matrix Ψ which includes the matrices $\mathscr{A}$ and $\mathscr{Q}$ and calculate the matrix exponential resulting in the equation:

$$\Phi = \exp(\Psi T) = \begin{bmatrix} \exp(\mathscr{A}T) & \int_0^T \exp(\mathscr{A}(T-\tau))\mathscr{Q}\,d\tau \\ 0_{1 \times n} & 1 \end{bmatrix} \text{ where } \Psi = \begin{bmatrix} \mathscr{A} & \mathscr{Q} \\ 0_{1 \times n} & 0 \end{bmatrix} \tag{B.5}$$

The value of the $\mathscr{P}(T)$ matrix for any time T is given by the vector $\Phi(1:n, n+1)$, i.e., the first n rows of the $(n+1)^{th}$ column of the Φ matrix. This column vector can be reshaped to construct a $n \times n$ matrix which corresponds to $P(T)$. This provides an efficient technique to calculate the solution of the differential Lyapunov equation.

C

Parseval's Theorem

The inverse Fourier transform is:

$$f(t) = \frac{1}{2\pi} \int_{-\infty}^{\infty} \hat{f}(\omega)e^{i\omega t} \, d\omega \tag{C.1}$$

where $\hat{f}(\omega)$, is the Fourier transform of $f(t)$. Consider the inner product:

$$\langle f(t), f(t) \rangle = \int_{-\infty}^{\infty} f(t)f(t) \, dt = \int_{-\infty}^{\infty} f(t)f^*(t) \, dt \tag{C.2}$$

where $(.)^*$ is the complex conjugate of $(.)$. Equation C.2 can be represented as:

$$\int_{-\infty}^{\infty} f(t)f^*(t) \, dt = \int_{-\infty}^{\infty} \left[\frac{1}{2\pi} \int_{-\infty}^{\infty} \hat{f}(\omega)e^{i\omega t} \, d\omega \right] \left[\frac{1}{2\pi} \int_{-\infty}^{\infty} \hat{f}^*(\omega')e^{-i\omega' t} \, d\omega' \right] dt \tag{C.3}$$

which can be rewritten as:

$$\int_{-\infty}^{\infty} f(t)f^*(t) \, dt = \frac{1}{4\pi^2} \int_{-\infty}^{\infty} \int_{-\infty}^{\infty} \hat{f}(\omega)\hat{f}^*(\omega') \left[\int_{-\infty}^{\infty} e^{i(\omega-\omega')t} \, dt \right] d\omega \, d\omega'. \tag{C.4}$$

Equation C.4 reduces to:

$$\frac{1}{4\pi^2} \int_{-\infty}^{\infty} \int_{-\infty}^{\infty} \hat{f}(\omega)\hat{f}^*(\omega') \left[2\pi\delta(\omega-\omega') \right] d\omega \, d\omega' = \frac{1}{2\pi} \int_{-\infty}^{\infty} \hat{f}(\omega)\hat{f}^*(\omega)d\omega \tag{C.5}$$

which leads to Parseval's Theorem:

$$\int_{-\infty}^{\infty} f(t)f(t) \, dt = \frac{1}{2\pi} \int_{-\infty}^{\infty} \hat{f}(\omega)\hat{f}^*(\omega) \, d\omega \tag{C.6}$$

which can also be represented as:

$$\int_{-\infty}^{\infty} |f(t)|^2 \, dt = \frac{1}{2\pi} \int_{-\infty}^{\infty} |\hat{f}(\omega)|^2 \, d\omega \tag{C.7}$$

or

$$\langle f(t), f(t) \rangle = \frac{1}{2\pi} \langle \hat{f}(\omega), \hat{f}^*(\omega) \rangle \tag{C.8}$$

Index